Manual of British Standards in ENGINEERING METROLOGY

other books in this series

Manual of British Standards in Engineering Drawing and Design

Manual of British Standards in Building Construction and Specification

Manual of British Standards in ENGINEERING METROLOGY

Edited by KEITH BROOKER
Technical Education Officer, BSI

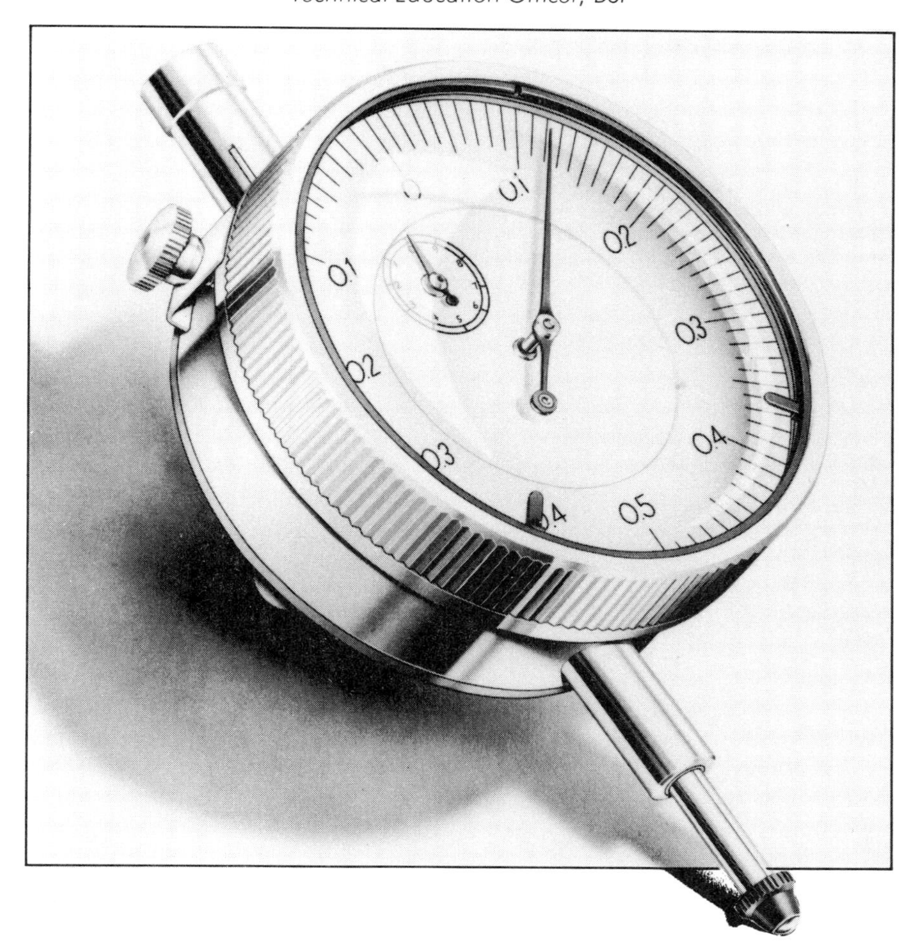

British Standards Institution
in association with Hutchinson

London Melbourne Sydney Auckland Johannesburg

Hutchinson & Co. (Publishers) Ltd

An imprint of the Hutchinson Publishing Group

17–21 Conway Street, London W1P 6JD

Hutchinson Group (Australia) Pty Ltd
30–32 Cremorne Street, Richmond South, Victoria 3121
PO Box 151, Broadway, New South Wales 2007

Hutchinson Group (NZ) Ltd
32–34 View Road, PO Box 40–086, Glenfield, Auckland 10

Hutchinson Group (SA) (Pty) Ltd
PO Box 337, Bergvlei 2012, South Africa

First published 1984

Set in Times and Univers by Preface Ltd, Salisbury, Wilts

Printed and bound in Great Britain by
Anchor Brendon Ltd,
Tiptree, Essex

British Library Cataloguing in Publication Data
British Standards Institution
Manual of British Standards in engineering metrology.
1. Mensuration
I. Title
620'.0044 T50

ISBN 0 09 151930 6 cased
0 09 151931 4 paper

Contents

Foreword 9

Preface 10

1 Standardization and the British Standards Institution 13

Introduction – The aims of standardization – BSI today – Levels of standardization

2 Measurement 16

Introduction – System of measurement – Historical note – Errors in measurement – The primary standard of length – The National Physical Laboratory – The National Engineering Laboratory – The Gauge and Tool Makers' Association – Legal metrology – Non-legal metrology – British Standards – Notes on accuracy and precision – Standards publications referred to

3 General inspection equipment 44

Introduction – Surface plates and tables – Angle plates – Engineers' squares – Engineers' parallels – Straightedges – Vee blocks – Standards publications referred to

4 Linear measurement 78

Introduction – Steel rules – Feeler gauges – Dial gauges – Dial test indicators – Vernier callipers and height gauges – Micrometers – Gauge blocks – Length bars – Comparators – Standards publications referred to

5 Angular measurement 113

Introduction – Bevel protractors – Sine bars and sine tables – Spirit levels – Standards publications referred to

6 Limits and plain limit gauges 120

Introduction to limits and fits – The ISO system of limits and fits – Introduction to plain limit gauges – Gauge blanks – Limits for limit gauges – Standards publications referred to

7 Screw threads 140

Introduction – Inspection by gauging – Inspection by measurement – Standards publications referred to

8 Gear measurement 154

Introduction – Why measure gears? – Basic gear geometry – Elemental errors – Other tests – Information to be given on the drawing – Standards publications referred to

9 Machine tool metrology 167

Introduction – British Standards – Standards publications referred to

10 Surface texture 181

Introduction – Geometry of surfaces – Measurement of surface texture – Surface roughness values produced by common production processes and materials – Process designations and guidance on suitable meter cut-offs – The indication of surface texture on engineering drawing – Summary – Roughness comparison specimens – Standards publications referred to

11 Roundness 192

Introduction – BS 3730 – Standards publications referred to

12 Air gauging 213

Introduction – Basic principles – Air gauging of rough or porous surfaces – Magnification – Air gauging system – Examples of application – Standardization – Standards publication referred to

13 Plain setting rings 230

Introduction – BS 4064 – Standards publication referred to

Index 233

I often say when you can measure what you are speaking about and express it in numbers you know something about it; but when you cannot measure it, when you cannot express it in numbers, your knowledge of it is of a meagre and unsatisfactory kind.

Lord Kelvin

Foreword

The much cited words of the eminent scientist and engineer, Lord Kelvin, leave no doubt that he carried the banner for metrology. He was also a fervent supporter of standardization and in 1906 became the first president of the International Electrotechnical Commission, the first international standards body.

Standardization is a fundamental requirement for the successful operation of modern metrology. Metrology is, in turn, an essential and inseparable part of any manufacturing process, not only as a final check on what is right or wrong, acceptable or unacceptable, but as an aid to the control of production. It is, therefore, understandable and right that so many student engineers, particularly those specializing in inspection and measurement, find their studies require a knowledge of metrology and the associated standards.

The need to provide accessible information about standards and standardization for students has long been recognized and I welcome the enterprise which brings Hutchinson and the British Standards Institution together to answer this need. Every student engineer now has ready access to the essential information selected from a whole family of metrological standards in a single volume.

I fully support the publication of this manual which I am confident will be of benefit to its readers.

D. G. Spickernell
Director General, BSI

Preface

This manual is intended for use by students and lecturers involved in technician and undergraduate courses in mechanical and production engineering with a special interest or specialism in metrology. It brings together material from over 40 British Standards and other related publications, which were current at 30 April 1983.

It should be noted that standards are subject to periodic review and may be subject to amendment, revision or withdrawal. In some cases the material selected from standards for inclusion in this manual has been heavily abridged.

In order to assist the reader two type faces have been used to differentiate between material from British Standards and additional text which introduces the standards.

Material from British Standards:

BS 5781

Measurement and calibration systems

1. Scope

This British Standard specifies requirements for a system for selecting, using, calibrating, controlling and maintaining measurement standards and measuring equipment used in the fulfilment of specified requirements.

2. References

The titles of the standards publications referred to in this standard are listed on the inside back cover.

3. Definitions

For the purposes of this British Standard, the following definition together with those given in BS 4778 and BS 5233 apply.

specified requirements. Either:

(a) requirements prescribed by the purchaser in a contract or order for material or services; or

(b) requirements prescribed by the supplier that are not subject to direct specification by the purchaser.

standards and measuring equipment necessary for the completion of this work are available and are of the accuracy, stability and range appropriate for the intended application. In particular, identification and timely action shall be taken to report, to the Purchaser's Representative, any measurement requirement that exceeds the known 'state of the art' or any new measurement capability that is needed but is not available.

4.4 Measurement limits. All measurements, whether made for purposes of calibration or product characteristic assessment, shall take into account all the errors and uncertainties in the measurement process that are attributable to the measurement standard or measuring equipment, and, as appropriate, those contributed by personnel, procedures and environment.

4.5 Documented calibration procedures. Documented calibration procedures shall be prescribed and shall be available for reference and use, as necessary, for the calibration of all measurement standards and measuring equipment used.

Additional text:

The establishment of a reputation for quality products and services depends upon the ability to demonstrate conformity to specified requirements. The control of manufacturing processes and the assessment of product quality and conformance almost invariably involves measurement of some kind. The integrity of all such measurement work is founded on the selection and correct use of suitable measuring equipment and its validation by a system of control and calibration. Any organization working to a formal series of quality system standards (such as BS 5750 'Quality systems') has to be able to demonstrate the ability to measure adequately and to certify that the results obtained can be relied upon. BS 5781 'Measurement and calibration systems' specifies requirements for a calibration system that can be used in the fulfilment of a quality system or for the supply of calibration services. It can be used for the following purposes:

(a) As a basis for evaluating the capability of a supplier's calibration system, either by a potential purchaser or by a third party, in order to provide assurance to interested parties. This may be prior to the establishment of a contract.
(b) Where invoked in a contract, to specify the calibration requirements appropriate to the particular material or services.
(c) In other documents where reference to a calibration system is appropriate.

A list of standards referred to is given at the end of each chapter.

Where standards include both metric and imperial sizes or where two standards exist, one for metric and one for imperial, the metric sizes only have been selected for inclusion in this manual. Where only imperial sizes are provided these have been included. Conversion factors are given in BS 350 'Conversion factors and tables'.*

Most equipment needs some form of protection when not in use and this is normally covered by the relevant standard specification for the equipment. For many small items, such as flats, the protection will be in the form of a case or box but, for larger items, such as surface tables, a cover is required. Protection against climatic conditions is also necessary and many of the standards for equipment refer to BS 1133 'Packaging code', Section 6 'Temporary protection of metal surfaces against corrosion (during transport and storage)'.

*BS 350 'Conversion factors and tables' consists of Part 1 'Basis of tables. Conversion factors' and Part 2 (withdrawn December 1981) 'Detailed conversion tables'. Supplement No. 1 to Part 2 'Additional tables for SI conversion' is, however, still available and was confirmed in March 1982.

No material concerning the casing, boxing, covering or protection against climatic conditions has been included in this manual and reference should be made to the complete standard for the equipment and BS 1133.

Chapter 8, *Gear measurement*, has been written by A. M. Thompson, Managing Director, Gear Technology Limited, Rotherham, whose assistance is gratefully acknowledged.

For information about BSI publications for use by students, teachers and lecturers, contact the Technical Education Officer, BSI, 2 Park Street, London W1A 2BS.

1 Standardization and the British Standards Institution

Introduction

Standardization is not new: it relates to an activity which can be seen in the natural world in every atom of oxygen or molecule of water. People have used standardization in speech, writing and music for many centuries. In AD 1215 Magna Carta stated, 'There shall be standard measures of wine, ale and corn (the London quarter), throughout the Kingdom. There shall also be a standard width of cloth. . . . Weights are to be standardized similarly'.

What is new about standardization is the approach to the subject which became necessary with the increase in industrialization. The booming markets of the Industrial Revolution brought a problem in their wake – the diversity in size and quality of similar products made in Manchester or London, Glasgow or Liverpool. The first advocate of modern standardization was probably Sir Joseph Whitworth, of screw thread fame, who pointed out in 1880 the simple domestic fact that candle butts and candlesticks came in so many sizes that they often did not match. Candles can be adapted with relative ease but this is not so with steel girders.

J. Williams Dunford, an architect, wrote in 1895 to *The Times* quoting the case of a frustrated contractor who complained that his order for iron girders had been passed from one British supplier to another and not one had been able to meet his specification. The order was eventually supplied from Belgium. A London iron merchant, John Skelton, replied:

> Rolled steel girders are imported into Britain . . . from Belgium and Germany, because we have too much individualism in this country . . . where collective action would be economically advantageous. As a result, architects and engineers generally specify such unnecessarily diverse types of sectional material for given work that anything like economical and continuous manufacture becomes impossible.

In 1900 John Skelton gave a paper on standardization to the British Iron Trade Association which so impressed a past president of the Institution of Civil Engineers, Sir John Wolfe Barry, that the Council of the Institution set up the first standards committee. The brief was to consider the advisability of standardizing iron and steel sections and the committee was extended to include representatives from the Institution of Mechanical Engineers, the Institution of Naval Architects and the Iron and Steel Institute.

It was on 26 April 1901 that the first meeting of this Engineering Standards Committee was held and modern standardization had begun.

The scope of the committee was soon enlarged to include locomotives, electrical plant and tests for engineering materials. The first results reduced the number of structural steel sections from 175 to 113 and tramway rails from 75 to 5, and produced an estimated saving of £1 million a year.

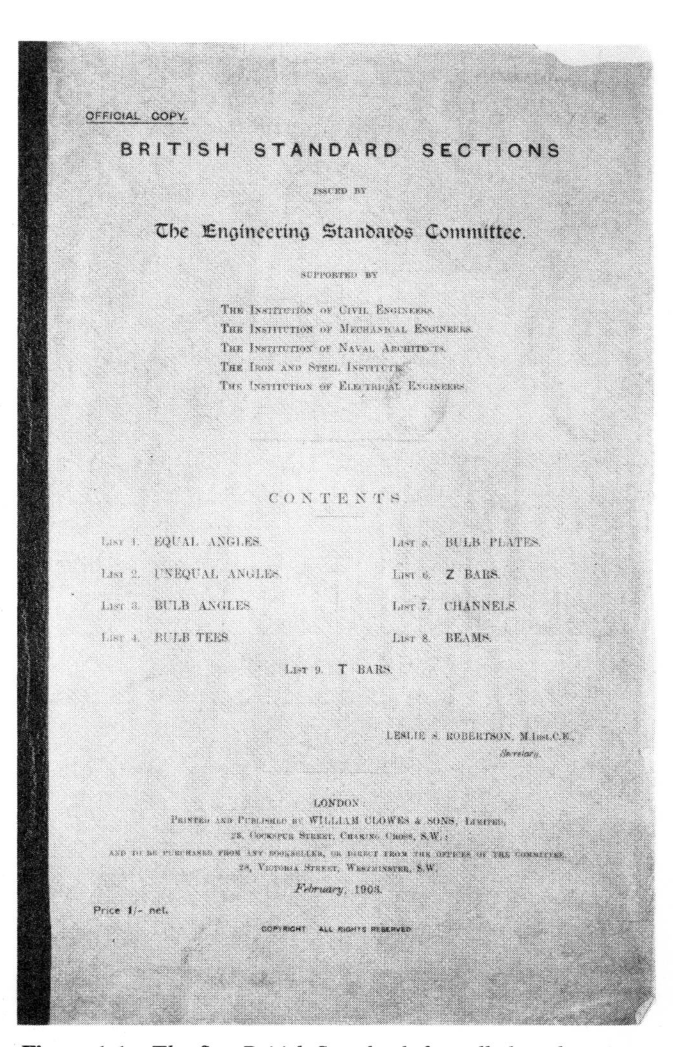

OFFICIAL COPY.

BRITISH STANDARD SECTIONS

ISSUED BY

The Engineering Standards Committee.

SUPPORTED BY

The Institution of Civil Engineers.
The Institution of Mechanical Engineers.
The Institution of Naval Architects.
The Iron and Steel Institute.
The Institution of Electrical Engineers.

CONTENTS.

List 1. EQUAL ANGLES. List 5. BULB PLATES.
List 2. UNEQUAL ANGLES. List 6. Z BARS.
List 3. BULB ANGLES. List 7. CHANNELS.
List 4. BULB TEES. List 8. BEAMS.
List 9. T BARS.

LESLIE S. ROBERTSON, M.Inst.C.E.,
Secretary.

LONDON:
Printed and Published by WILLIAM CLOWES & SONS, Limited,
23, Cockspur Street, Charing Cross, S.W.;
and to be purchased from any bookseller, or direct from the offices of the Committee,
28, Victoria Street, Westminster, S.W.
February, 1903.

Price 1/- net.

COPYRIGHT ALL RIGHTS RESERVED

Figure 1.1 *The first British Standard, for rolled steel sections, published in 1903*

In 1918 the Committee's name was changed to the British Engineering Standards Association. A Royal Charter was granted in 1929 and a Supplementary Charter in 1931, when the present name, British Standards Institution, was adopted. In 1981 a consolidated Royal Charter was issued.

The aims of standardization

Standards promote consistent quality and economic production by providing technical criteria accepted by consensus. They simplify manufacture and encourage interchangeability. They rationalize processes and methods of operation, making communication and the exchange of goods and services easier. Their use gives confidence to manufacturers and users alike.

The broad aims of standardization can be summarized as:

(a) provision of communication among all interested parties;
(b) promotion of economy in human effort, materials and energy in the production and exchange of goods;
(c) protection of consumer interests through adequate and consistent quality of goods and services;
(d) promotion of the quality of life: safety, health and the protection of the environment;
(e) promotion of trade by the removal of barriers caused by differences in national practices.

Standardization is an activity which provides solutions for repetitive application, to problems essentially in the spheres of science, technology and economics. It aims to provide the optimum degree of order in a given context which means formulating, issuing and implementing standards.

Such standards are technical specifications or other documents which are produced with the cooperation and consensus or general approval of all interests affected. The production of such documents, in the form of British Standards, is the main activity of the British Standards Institution.

Another type of standard, essential to standardization and to those involved in metrology, are the measurement standards of mass, length, time, etc. The maintenance of the primary physical standards of measurement is a function of the National Physical Laboratory (see Chapter 2).

BSI today

BSI is the national standards body in the United Kingdom which, as well as producing standards, operates product certification and capability assessment schemes, provides a wide range of testing facilities and technical information and assistance to all sectors of industry engaged in exporting. (For more information about other BSI services write to the Director, BSI, Maylands Avenue, Hemel Hempstead, Herts HP2 4SQ.)

The Board of BSI is responsible for the policies of the Institution subject to the ultimate authority of the subscribing members who support BSI by contributing to its finances. These may be local authorities, institutions or individuals, but mainly industry. Under the Board are six Councils responsible for the standards programme, and the Quality Assurance Council. Some 60 Standards Committees are responsible to the Councils for the standards projects and for deciding the priorities for work in their fields.

Under the Standards Committees operate some 1000 technical committees who prepare the standards (see Figure 1.2).

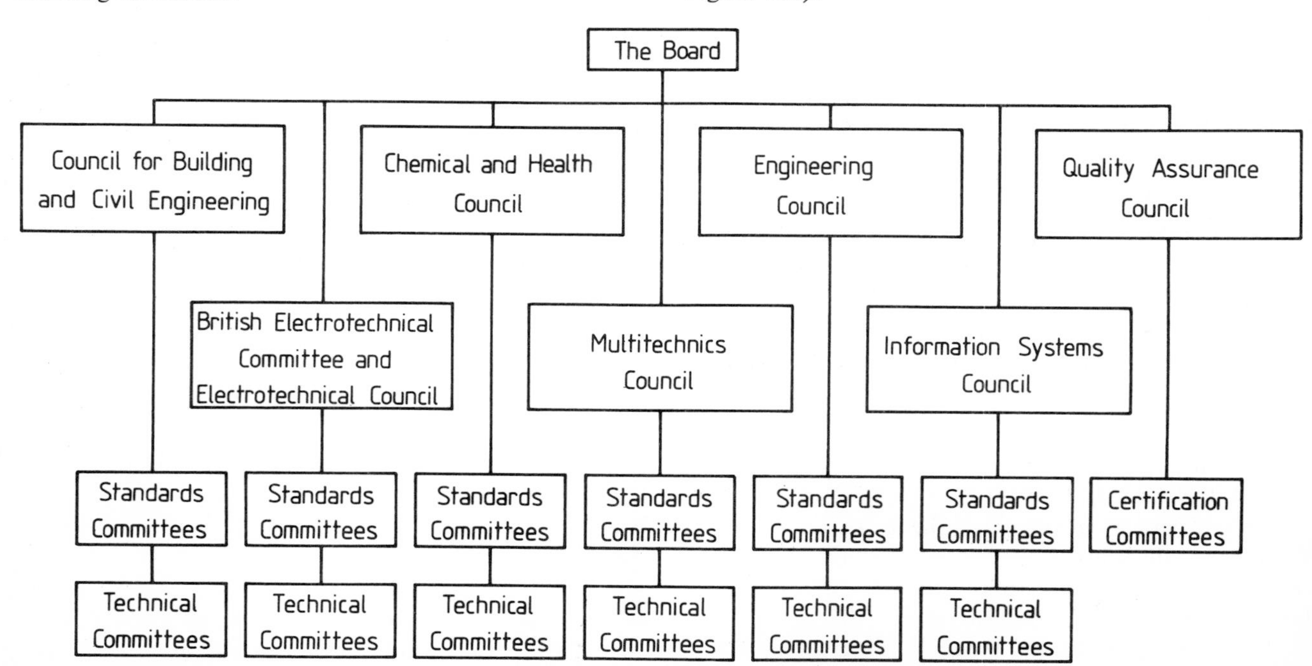

Figure 1.2 *The committee structure of BSI*

BSI publications

British Standards may be used to promulgate standardization at any of the following, normally consecutive stages:

(a) terminology, symbols;
(b) classification, designation;
(c) methods of measuring, testing, analysis, sampling, etc.; methods of declaring, specifying, etc.;
(d) specifications for materials or products: dimensions, performance, safety, etc.; specifications for processes: practices, systems, etc.;
(e) recommendations on product or process applications: codes of practice.

The best known type of British Standard is the specification, which lays down a set of requirements to be satisfied by the material, the product or process in question and embraces the relevant methods by which compliance may be determined. British Standard guides, recommendations and codes of practice are written in the form of guidance only, and are not intended to provide objective criteria by which compliance may be judged. Other types of standard commonly produced are British Standard glossaries and British Standard methods of various kinds, such as measuring and testing. These two types constitute reference documents, to be called up where appropriate in other documents. All of the above documents will have a BS number and title, for example BS 308 'Engineering drawing practice'.

Status of British Standards

British Standards are documents voluntarily agreed as a result of a process of public consultation. However, the publication of a standard does not, in itself, ensure its use. Its application depends on the voluntary action of interested parties and becomes binding only if a claim of compliance is made, if it is invoked in a contract, or if it is called up in legislation.

Levels of standardization

Standardization operates at international, national, regional and company level. Electrical engineers were among the first to realize that standardization is best undertaken at an international level and it was in 1906 that the International Electrotechnical Commission (IEC) was founded. Today the IEC is composed of over 40 national committees and covers subjects over the whole range of electrotechnical activities. The British Electrotechnical Committee, acting through BSI as its secretariat, represents the United Kingdom on the IEC.

It was not until 1926 that some of the world's leading standards bodies formed the International Federation of the National Standardizing Association (ISA), but by 1942 work had ceased as a result of World War II. In 1947 its successor, the International Organization for Standardization (ISO), was created. ISO is the international agency for standardization in all areas except those covered by the IEC. The object of ISO is to promote the development of standards in the world with a view to facilitating international exchange of goods and services, and to developing mutual cooperation in the sphere of intellectual, scientific, technological and economic activity. BSI is a leading member among some 90 national standards bodies represented on ISO.

The European Committee for Standardization (CEN) was founded in 1961, and comprises the national standards bodies of the EEC and EFTA countries, and Spain. CEN prepares European Standards that are published without variation of text as national standards in countries approving them.

The European Committee for Electrotechnical Standardization (CENELEC), which is the electrotechnical counterpart of CEN, was founded in 1973. CENELEC prepares European Standards for identical publication nationally.

The work of the international bodies is supported by the national standards bodies who are in turn supported by trade associations, companies, central government departments and local authorities.

For further information about BSI, the following publications are available from: BSI Sales, Linford Wood, Milton Keynes MK14 6LE (prices on application):

BS 0 A standard for standards
 Part 1 General principles of standardization
 Part 2 BSI and its committee procedures
 Part 3 Drafting and presentation of British Standards

BSI Yearbook – lists all the British Standards and other BSI publications, and contains information about BSI and the services it offers

The story of standards by C. Douglas Woodward, published 1972 – traces the development of standards-making in the United Kingdom

2 Measurement

Introduction

Measurement is an essential part of an engineer's work. Whether the measurements are of force or time, length or temperature, flow or angle, they will provide information upon which decisions will be made and will determine the course of action to be taken. Mechanical and production engineers are especially concerned with the measurement of length and angle, and, since it is possible to use a combination of linear measurements to obtain an angular measurement, length is of special importance.

For any measurement to be of use it must satisfy two criteria: it must use a system of measurement which is recognized and used by others in the field of application; and the degree of accuracy of the measurement must be known.

System of measurement

The standard units of measurement are those of the International System of Units (SI) and those additional units that are recognized for use with SI. The formal content of SI is determined and authorized by the General Conference of Weights and Measures (CGPM). With the exception of a very few developing countries, every country in the world has either adopted SI in legislation or is in the process of doing so. In the United Kingdom, however, successive governments have been reluctant to provide legislation to complete metrication which was started through the work of BSI.

In the earlier metric systems the decimal connection between units of different size for the same quantity aided both memory and calculation. In the SI there is the very important added advantage of coherence between the units for every quantity, thus removing all awkward factors in calculations involving quantity relations.

Historical note

The idea of decimal units of measurement was conceived by Simon Stévin (1548–1620) who also developed the even more important concept of decimal fractions. Decimal units were also considered in the early days of the French Académie des Sciences founded in 1666, but the adoption of decimal weights and measures was part of the general increase in administrative activity in Europe which followed the French Revolution. The statesman Talleyrand aimed at the establishment of a system of international decimal units of weights and measures 'à tous les temps, à tous les peuples'. On the advice of the scientists of his day, these were based on the metre as the unit of length and the gram as the unit of mass. The metre was intended to be one ten-millionth part of the distance from the North Pole to the Equator at sea level, and passing through Paris; the gram was to be the mass of one cubic centimetre of water at its maximum density, i.e. at a temperature of approximately 4 °C. A system of prefixes was developed to indicate powers of ten of the units, thus providing a flexible and convenient means of expression for a wide range of magnitudes and avoiding the need to use very large or very small numerical values. The prefixes also enabled the units of different sizes to be memorized with ease.

Although the decimal units were primarily devised as a benefit to industry and commerce, scientists soon realized their advantage and they were adopted in scientific and technical circles in Western European countries, including the United Kingdom. It was not, however, until 1960 that the CGPM formally gave the full title 'Système International d'Unités', for which the abbreviation SI is used in all languages. Since the 1960 meeting the CGPM has extended the SI base units which are now as indicated in Table 2.1.

The SI prefixes are used to form decimal multiples and sub-multiples of the SI units, as listed in Table 2.2.

The maintenance of the measurement standards and their dissemination in the United Kingdom is a function of the National Physical Laboratory (see below).

Errors in measurement

Any measurement must be made to an acceptable degree of accuracy, but it must be realized that no measurement is exact. The accuracy of determination is, therefore, as important as the measured dimension and the inherent errors in any method of measurement should be kept to a minimum. There are many types of error which need to be considered and these are given in BS 2643 'Glossary of terms relating to the performance of measuring instruments', which is extracted on pages 34–9.

Table 2.1 *The SI base units*

Quantity	*Name of unit*	*Unit symbol*
length	metre	m
mass	kilogram	kg
time	second	s
electric current	ampere	A
thermodynamic temperature	kelvin	K
amount of substance	mole	mol
luminous intensity	candela	cd

Table 2.2 *Names and symbols for the SI prefixes*

	Prefix	
Factor	*Name*	*Symbol*
10^{18}	exa	E
10^{15}	peta	P
10^{12}	tera	T
10^{9}	giga	G
10^{6}	mega	M
10^{3}	kilo	k
10^{2}	hecto	h
10^{1}	deca	da
10^{-1}	deci	d
10^{-2}	centi	c
10^{-3}	milli	m
10^{-6}	micro	μ
10^{-9}	nano	n
10^{-12}	pico	p
10^{-15}	femto	f
10^{-18}	atto	a

The primary standard of length

A standard must be reproducible with such a degree of accuracy that for all industrial and scientific applications it may be considered absolute. The metre is currently defined as the length equal to 1 650 763.73 wavelengths in vacuum of the radiation of a prescribed transition in the spectrum of the krypton-86 atom, and by the use of interferometry, the error of reproduction is of the order of 1 part in 100 million.

The National Physical Laboratory (NPL)

Within the United Kingdom, the NPL is the ultimate authority on units of measurement in the physical sciences: it interprets the rulings of the International Committee of Weights and Measures and disseminates the measurement scales implicit in SI. As one of the Department of Industry's Research Establishments, NPL not only maintains the national primary standards but undertakes research to provide better standards and methods of dissemination. Such research includes the development of lasers and interferometric methods for measuring length and wavelength.

The dissemination of the length standard to industry is via interferometric techniques and covers length standards and calibrations from gauge blocks through linear scales, metrological gratings and engineers' length bars, to geodetic tapes up to 50 m in length.

Manufacturing industry spends many millions of pounds each year in making measurements of all kinds. Such measurements are backed by a hierarchy of calibration facilities and services allowing, in principle, the calibration of every shop floor instrument to be traced to the appropriate national primary standard held at NPL.

The first link in this chain between NPL and the shop floor is often one of the laboratories approved by the British Calibration Service (BCS) which is a part of the NPL. BCS exists to accredit laboratories competent to undertake calibration work to a consistent and stated accuracy. The NPL then devolves the routine calibration to BCS laboratories whose reference standards are calibrated against the national primary standards at NPL. The laboratories are entitled to issue BCS calibration certificates (see Figure 2.1) with the backing of NPL and these are widely accepted, both in the United Kingdom and abroad, as evidence of sound measurement practice.

NPL also manages the National Testing Laboratory Accreditation Scheme (NATLAS), which accredits laboratories that perform objective testing of various kinds, but excluding calibration. The scheme is concerned with establishing the competence of testing laboratories throughout the United Kingdom, and securing their recognition by users in this country and abroad. Membership of NATLAS is open to competent laboratories of all kinds, including independent test houses, laboratories in government departments and public corporations, laboratories in research and academic institutions, as well as those of manufacturers, retailers or purchasers carrying out tests for their own organizations. The scheme is to operate initially on the chemical, electrical, mechanical and physical testing areas, but will be extended to other fields of testing as the demand arises. Further details about the work of NPL, BCS and NATLAS may be obtained from: Information Services, National Physical Laboratory, Teddington, Middlesex TW11 0LW. (See BS 5781 'Measurement and calibration systems' on pages 39–42.)

The National Engineering Laboratory (NEL)

The NEL, also a Department of Industry's Research Establishment, carries out research, development, design, consultancy and testing in most branches of mechanical engineering. The main objectives of NEL's work is to help British industry remain competitive in world markets providing firms with advanced technical expertise and resources. One of the main areas of activity is that of engineering measurement where most of the requests from industry are for trouble-shooting assistance. Further details about the work of NEL may be obtained from: National Engineering Laboratory, East Kilbride, Glasgow, G75 0QU.

HARWELL
Mechanical Standards Laboratory
Quality Assurance
Bldg. 424 A.E.R.E. Harwell, Didcot, Oxon
OX11 0RA
Telephone: Abingdon (0235) 24141 Ext. 3168 & 2615
BCS Approval No. 0020

Certificate of Calibration

This certificate is issued in accordance with the conditions of the approval granted by the British Calibration Service. It is a correct record of the measurements made. Copyright of this certificate is owned jointly by the Crown and by the issuing laboratory. The certificate may not be reproduced other than in full, except with the prior written approval of the Director, B.C.S., and of the issuing laboratory.

Date of issue 23.9.81 Certificate No. 'Audit' Page 1 of 2 pages

Submitted by British Calibration Service, National Physical Laboratory Teddington, Middlesex

Serial No. 692124

Description 10 used metric hardened steel gauge blocks, made by C.E.J.

These gauge blocks have been examined and measured in an environment of 20°C ± 1°C and the following results apply at 20°C

Condition of Measuring faces

The measuring faces were examined and the following condition noted.
100mm gauge block was found to scratched and pot marked
80mm gauge block burred on edge of measuring face
All other gauges free from adverse burrs and scratches

Measurement of Length

The errors in length of the gauges at 20°C, from their nominal lengths have been measured by comparison with laboratory standards. Three measurements were taken on each gauge, one at the centre and one towards each end. The mean measured results are given in the Table on sheet 2.

Accuracy of Measuring faces

All 10 gauge blocks were wrung to a flat steel platen. In that condition they were tested for flatness and the mean measured results were found to be as follows:
Gauges found to be better than ± 0.05 Micrometres

From the three measurements of length the parallelism of the gauges was determined and found to be as follows:

Cont/Sheet 2

Certified
Head of Laboratory
J E P Merrifield

(a)

Figure 2.1 *Certificates of calibration*
Reproduced by permission of the NPL

HARWELL Mechanical Standards Laboratory

BCS Approval No. 0020

Continuation of Certificate

Certificate No. 'Audit'

Page 2 of 2 pages

Test of a 10 used metric gauge blocks

Serial No 692124

Table

Unit of error 0.000 01 mm

Nominal Gauge Size mm	Mean Measured Error Length mm	Mean Measured Error Parallelism mm
1.001	+12	4
1.007	+1	3
1.03	-6	4
1.08	+7	4
1.3	-1	6
1.9	+2	2
3.0	+6	2
10.0	+3	6
30.0	0	1
100.0	+14	5

Uncertainty of measurement		
	Up to 10 mm	0.08 Micrometres
	Above 10 up to 25 mm	0.1 Micrometres
	Sizes 30 mm	0.12 Micrometres
	Size 100 mm	0.18 Micrometres

Head of Laboratory
J E P Merrifield

Previous Calibration Certificate
APPROVAL NO. ____
CERT. NO. ____
DATE OF ISSUE ____

Certified [signature]

Date 23. 9. 81

B.C.S. approval No. 0036

Mullard 

The Dimensional Standards Room
Mullard Mitcham
New Road, MITCHAM, Surrey. CR4 4XY
Telephone: 01-648 3471 Ext. 311

Certificate of Calibration

Date of Issue 1981 OCTOBER 26 Serial No. 02976

Date of Calibration 1981 SEPTEMBER 30 Page 1 of 1

AUDIT SAMPLE

Sent by : British Calibration Service,
National Physical Laboratory,
Teddington,
Middlesex TW11 OLW.

Description and Identification : A 12in Hilgar & Watts Standard Scale (0.1in divisions) No. 42/49.

British Calibration Service No. 3/317.

Basis of Test : British Calibration Service reference NPB/13/108/01 dated 19-3-81.

Results at 20°C :

Position in	Deviation (0.0001in)	Position in	Deviation (0.0001in)
0	0	0.8	+0.51
0.1	+0.51	0.9	+0.62
0.2	+0.47	1.0	+0.35
0.3	+0.59	2.0	+0.43
0.4	+0.39	3.0	+0.35
0.5	+0.35	4.0	+0.31
0.6	+0.43	5.0	+0.12
0.7	+0.86	6.0	-0.28

Uncertainty of measurement :-

± 0.000 06in + 0.000 003in/ in from zero.

Laboratory Note Book No. 13235A

Certified: K.H. Greenhough ....

(b)

The Gauge and Tool Makers' Association (GTMA)

The Metrology Section of the GTMA, which has been recognized as the national trade association for the gauge and tool industry for over 40 years, constitutes 64 companies. These companies have a special interest in gauging, measurement, inspection and quality control; they supply a wide range of equipment from simple bench tools to computer assisted coordinate measuring machines. For further information, contact the Gauge and Tool Makers' Association, 24 White Hart Street, High Wycombe, Bucks HP11 2HL.

Legal metrology

Legal metrology is that branch of the science of measurement used in relation to trading. It covers units of measurement, methods of measurement and measuring instruments.

The work of the Legal Metrology Branch of the Department of Trade is concerned solely with the legal metrology requirements of trading transactions (i.e. the units, methods and equipment used in trading, such as petrol dispensers, shop counter weighing machines and weighbridges).

The Weights and Measures Laboratory, which is a part of the Legal Metrology Branch, has custody of the United Kingdom secondary and tertiary standards of mass, length and capacity. It is, by statute, responsible for calibrating the standards of mass, length and capacity held by Local Authority Trading Standards Departments.

PD 6461 'Vocabulary of Legal Metrology' is an English translation of the official International Organization of Legal Metrology text, which is in French. (No part of PD 6461 is included in this manual.)

Non-legal metrology

ISO, IEC, OIML and the International Bureau of Weights and Measures (BIPM) have formed, at international level, the Joint Metrology Group which has produced a draft publication entitled *International vocabulary of basic and general terms in metrology*. At the time of writing, comments on the drafts are being considered.

British Standards

A number of British Standards have been issued which are of fundamental importance to the metrologist of which BS 5233 'Terms used in metrology' is one.

Over many years differences in usage of terms have grown up within the individual technical disciplines. These differences ranged from subtle changes of meaning of well-used terms to the extraction and misuse of everyday terms from *The Oxford English Dictionary*, by ascribing to them specific meanings within particular confined areas of use. The need for coordination of metrological terminology becomes more important as scientific and technological advances require greater and closer interrelationships.

BS 5233 provides a glossary of terms which is aimed at overcoming these problems. In the light of the outcome of the work of the Joint Metrology Group this British Standard will be reconsidered and the appropriate action taken.

BS 5233

Glossary of terms used in metrology

Section one. General terms

1. Metrology

1.1 metrology. The field of knowledge concerned with measurement.

NOTE. Metrology includes all aspects both theoretical and practical with reference to measurements, whatever their accuracy, and in whatever fields of science or technology they occur.

1.2 legal metrology. That part of metrology which treats of units of measurement, methods of measurement and of measuring instruments, in relation to mandatory technical and legal requirements.

NOTE. Attention is drawn to the 'Vocabulary of legal metrology', compiled by OIML* and published by BSI as PD 6461.

Section two. Quantities and units

2.1 quantity (measurable). An attribute of a phenomenon or a body which may be distinguished qualitatively and determined quantitatively.

NOTE. The term quantity means a quantity in the general sense (see example (a)) as well as a specific quantity (see example (b)).

Examples

(a) Quantities in a general sense: length, time, mass, temperature, hardness, electrical resistance.

(b) Quantities in a specific sense: length of a rod (4.35 mm), electrical resistance of a wire (93 Ω)

2.1.1 *quantity to be measured*

measured quantity. A quantity subjected to a process of measurement.

2.1.2 influence quantity. A quantity which is not the subject of the measurement but which influences the value of the quantity to be measured, or the indications of the measuring instrument, or the value of the material measure reproducing the quantity.

*International Organization of Legal Metrology.

NOTE. The influence quantity can arise from the ambient conditions or from the instrument itself.

Examples. Temperature, attitude of the instrument, frequency of a measured voltage, self-heating of the instrument, lapse of time.

2.2 value (of a quantity). The quantity expressed as the product of a number and the unit of measurement.

Examples. 5.3 m, 12 kg, 20 °C.

2.2.1 *true value (of a quantity).* The value which characterizes a quantity perfectly defined.

NOTE. The true value of a quantity is an ideal concept and, in general, it cannot be known.

2.2.1.1 *conventional true value (of a quantity).* A value approximating to the true value of a quantity such that, for the purpose for which that value is used, the difference between these two values can be neglected.

NOTE. The conventional true value of a quantity is generally determined by means of methods and by the use of instruments of an accuracy suitable for each particular case.

Example. When an electricity meter, the maximum permissible errors of which are ± 2 %, is verified by means of a standard meter of which the error of indication in the same range does not exceed ± 0.2 %, the standard meter will indicate the conventional true value of electrical energy for the purpose of the verification.

When, in its turn, the standard meter is itself verified by means of a standard wattmeter and chronometer, the values obtained from them will also provide the conventional true values for the purpose of this further verification.

2.2.2 *numerical value (of a quantity).* The pure number in the expression of a value of the given quantity.

Examples. The numbers 5.3, 12, 20 in the examples in **2.2**.

2.2.3 *fiducial value.* A prescribed value of a quantity to which reference is made, for example, in order to define the value of an error as a proportion of this prescribed value.

2.3 system of quantities. A group comprising a particular set of base quantities and corresponding derived quantities and covering all fields of science or possibly only one of these fields.

2.3.1 *base quantity.* Designation of the quantities which, in a system of quantities, are conventionally accepted as independent of each other, and in terms of which the quantities derived from them in this system can be expressed by equations defining them.

2.3.2 *derived quantity.* A quantity defined, in a system of quantities, as being a function of the base quantities of that system.

Examples

(a) Force F, defined in the system of quantities ℓ, m, t, by the equation:

$$F = m\,\frac{d^2\ell}{dt^2}$$

(b) The intensity of magnetic field H at a distance r from an infinitely long straight conductor carrying a current I, in the rationalized system of quantities ℓ, m, t, I, defined by the equation $H = I/2\,\pi r$.

2.3.3 *dimension (of a quantity).* An expression which represents a quantity of a system as the product of powers of the base quantities of the system with a numerical coefficient equal to 1.

Example. LMT^{-2} = dimension of force in the system of quantities ℓ, m, t.

2.3.4 *dimensionless quantity.* A quantity in the expression of which the exponents of the base quantities concerned are zero.

NOTE. The value of a dimensionless quantity differs however from a pure number by having the character of a quantity.

Examples

(a) strain

(b) angle.

2.4 reference-value scale (of a quantity). A set of values of a quantity determined in a prescribed manner and accepted by convention.

Examples

(a) The International Practical Temperature Scale based on the freezing and boiling points of an accepted group of pure substances and on the use of specified measuring instruments and interpolation formulae.

(b) The Mohs hardness scale based on the hardness of a series of specified minerals.

2.5 unit (of measurement). The value of a quantity conventionally accepted as having a numerical value equal to 1.

2.5.1 *base unit (of measurement).* A unit of one of the base quantities.

Example. The metre is the base unit of the quantity length in the International System of units (SI).

2.5.2 *derived unit (of measurement).* The unit of measurement of a derived quantity.

NOTE. For some derived units, there are special names and symbols, e.g. newton (N), joule (J) and volt (V) as the SI units of force, energy and electrical potential respectively.

It may often be convenient not to express derived units in terms of base units but in terms of other derived units.

Examples. In the International System of units (SI)

(a) $m.s^{-1}$ = unit of linear velocity.

(b) 1 Wb = 1 $m^2.kg.s^{-2}.A^{-1}$ = unit of magnetic

flux, but this unit can also be expressed by two other derived units, C.Ω for example.

2.5.3 *coherent unit (of measurement).* A unit of measurement which is expressed in terms of the base units by a formula in which the numerical coefficient is equal to 1.

NOTE 1. The expression 'coherent unit of measurement' is an abbreviation of a more precise description, 'unit of measurement in a coherent system', for with an isolated unit the concept of coherence is obviously meaningless.
NOTE 2. This concept applies only to derived units of measurement.

Example. The newton is the coherent derived unit of force in the International System of units, 1 N = 1 $m.kg.s^{-2}$.

2.5.4 *multiple (sub-multiple) unit (of measurement).* A unit of measurement which is a multiple (sub-multiple) of a unit and which is formed according to the scaling principle accepted for the unit in question.

Examples
(a) One of the decimal multiple units of the metre is the kilometre (km) and one of the decimal sub-multiple units is the millimetre (mm).
(b) One of the non-decimal multiple units of the second is the hour (h).

2.6 system of units (of measurement). A set of base and derived units corresponding to a particular group of quantities.

Examples
(a) CGS system of units;
(b) Imperial system of units;
(c) International System of units (SI).

2.6.1 *coherent system of units (of measurement).* A system of units of measurement composed of base units and coherent derived units.

Example. The following units: m, kg, s, m^2, m^3, Hz = s^{-1}, $m.s^{-1}$, $m.s^{-2}$, $m^{-3}.kg$, N = $m.kg.s^{-2}$, Pa = $m^{-1}.kg.s^{-2}$, J = $m^2.kg.s^{-2}$, W = $m^2.kg.s^{-3}$ represent a group of coherent units of measurement in mechanics belonging to the International System of units (SI).

2.6.2 *International System of units (SI).* (Système International d'Unités). The coherent system of units of measurement founded on the following seven base units:

the metre, unit of length;
the kilogram, unit of mass;
the second, unit of time;
the ampere, unit of electric current;
the kelvin, unit of thermodynamic temperature;
the candela, unit of luminous intensity;
the mole, unit of amount of substance.

adopted and recommended by the General Conference of Weights and Measures (CGPM).

Section three. Measurements

3.1 measurement. The operations having the object of determining the value of a quantity.

3.2 method of measurement. Nature of the procedure used in the measurement.

NOTE. The methods of measurement defined in **3.2.1** to **3.2.4.8** inclusive are some of the most commonly used but the list is not exhaustive. The titles are not mutually exclusive and cases of overlapping will be found between the examples of measurement quoted.

3.2.1 *direct method (of measurement).* A method of measurement by which the value of a quantity to be measured is obtained directly, i.e. without the necessity for supplementary calculations based on a functional relationship between the quantity to be measured and other quantities actually measured.

Examples
(a) Measurement of a mass on an equal-arm balance.
(b) Measurement of a length by means of a graduated rule.
(c) Measurement of a voltage on a direct-reading voltmeter.

3.2.2 *indirect method (of measurement).* A method of measurement in which the value of a quantity is obtained from measurements made by direct methods (of measurement) of other quantities linked to the quantity to be measured by a known relationship.

Examples
(a) Measurement of the density of a body on the basis of measurements of its mass and its geometrical dimensions.
(b) Measurement of the specific resistance of a conductor on the basis of measurements of its resistance, length and cross-sectional area.

3.2.3 *fundamental method (of measurement).* A method of measurement based on the measurements of the base quantities used to define the quantity.

NOTE. The method is sometimes termed 'absolute method (of measurement)'.

Examples
(a) Direct measurement of the value of a base quantity in accordance with the definition of that quantity.
(b) Indirect measurement of the acceleration due to gravity based on the distance travelled in a given time by a free falling body.
(c) Measurement of pressure with the aid of a U-tube manometer by which the pressure is deduced from measurements of density, gravitational acceleration and length, obtained in their turn from measurements of the base quantities: length, mass and time.

3.2.4 *comparison method (of measurement).* A method of measurement based on the comparison of the value of a quantity to be measured with a known value of the same quantity, or with a known value of another quantity which is a function of the quantity to be measured.

Examples
(a) The measurement of a volume of liquid by means of a material measure of capacity.
(b) The measurement of a pressure by means of a Bourdon-tube gauge.

3.2.4.1 *direct-comparison method (of measurement).* A method of measurement based on comparison in which the full value of the quantity to be measured is compared with a known value of the same quantity provided by a material measure which is used directly in the measurement.

Examples
(a) The measurement of a length by means of a graduated rule.
(b) The measurement of the volume of a liquid by means of a material measure of capacity.
(c) The measurement of a mass by means of a balance which compares the mass to be measured with the corresponding mass of a group of known weights.

3.2.4.1.1 *substitution method (of measurement).* A direct comparison method of measurement in which the value of the quantity to be measured is replaced by a known value of the same quantity chosen in such a manner that the effects on the indicating device of these two values are the same.

Examples
(a) Determination of a mass by means of a balance and known weights using a substitution method (e.g. the Borda method; see Dictionary of Applied Physics, vol. III, p. 819).
(b) Determination of temperature with a Beckmann thermometer by reproducing the same indication by means of a known temperature.

3.2.4.1.2 *transposition method (of measurement).* A method of measurement by direct comparison in which the value of the quantity to be measured is first balanced by an initial known value A of the same quantity: next the value of the quantity to be measured is put in the place of that known value, and is balanced again by a second known value B. When the balance-indicating device gives the same indication in both cases, the value of the quantity to be measured is $\sqrt{AB}$.

Examples
(a) Determination of a mass by means of a balance and known weights, using the Gauss double-weighing method.
(b) Carey-Foster resistance bridge.

3.2.4.2 *differential (comparison) method (of measurement).* A method of measurement by comparison, based on comparing the quantity to be measured with a quantity of the same kind having a known value only slightly different from that of the quantity to be measured, and measuring the difference between the values of these two quantities.

Example. Determination of a length in terms of a known length by means of a comparator.

3.2.4.2.1 *coincidence method (of measurement).* A method of differential measurement in which a very small difference between the value of the quantity to be measured and a known value of the same kind with which it is compared, is determined by observation of the coincidence of gauge or scale marks or signals.

Examples
(a) Measurement of the length of an object by means of an instrument employing a scale and vernier.
(b) Measurement of the rate of a clock by the coincidence of time signals.

3.2.4.3 *null method (of measurement).* A method of measurement in which the difference between the value of the quantity to be measured and the known value of the quantity with which it is compared, is brought to zero.

NOTE. The quantity to be measured can be replaced by a quantity of another kind which has itself been compared with a quantity of the original kind.

Examples
(a) Determination of a mass when the final position of equilibrium of the balance is the same as the initial position.
(b) Measurement of an electrical resistance by means of a Wheatstone bridge and null indicator.

3.2.4.4 *deflection method (of measurement).* A method of measurement in which the value of the quantity to be measured is indicated by the deflection of a moving element or of an indicating device.

Examples
(a) Measurement of pressure by means of a pointer-type pressure gauge.
(b) Measurement of a voltage by means of a light-spot voltmeter.
(c) Measurement of a mass by means of a self-indicating balance.

3.2.4.5 *interpolation method (of measurement).* Method of measurement by comparison in which two or more known values of the same quantity are used to determine a value of a measured quantity lying between two of these values, on the basis of the established characteristics of a measuring device.

Examples
(a) Linear interpolation between scale marks on a scale.
(b) Luminance interpolation.
(c) Establishment on the IPTS* of temperatures intermediate between the fixed points.

3.2.4.6 *extrapolation method (of measurement).* Method of measurement by comparison in which two or more known values of the same quantity are used to determine the value of a measured quantity lying outside the interval between these values, on the basis of the established characteristics of a measuring device.

Example. Establishment on the IPTS* of temperatures above the highest established fixed point.

3.2.4.7 *complementary method (of measurement).* Method of measurement by comparison in which the value of the quantity to be measured is combined with a known value of the same quantity so adjusted that the sum of these two values is equal to a predetermined comparison value.

Examples
(a) Determination of the volume of a solid by liquid displacement.
(b) Measurement of mass by means of a constant-load balance.

3.2.4.8 *resonance method (of measurement).* Method of measurement by comparison in which a known relationship between the compared values of the same quantity is established by means of the attainment of a condition of resonance.

Example. Measurement of frequencies by means of a sonometer or a vibrating-reed frequency meter.

3.3 indication (or response of a measuring instrument). The value of the quantity measured, as indicated or otherwise provided by a measuring instrument. (See also **3.4.1**.)

NOTE 1. In applying this concept to a material measure the indication is equivalent to the nominal or inscribed value of the material measure.
NOTE 2. In this context 'indication' is also applied to the case where the indication is recorded.
NOTE 3. The term 'indication' also means a presumed indication, obtained by interpolation of the position of the index of the instrument between adjacent scale marks.
NOTE 4. The value of the quantity measured can be directly indicated in units of that quantity or in other selected units. In this latter case, the direct indication must be multiplied by the instrument constant.
NOTE 5. The alternative term 'response' should not be confused with its other usages, e.g. frequency response.

3.3.1 *direct indication (of a measuring instrument).* The magnitude of the measured quantity expressed in terms of the scale marks and units displayed by the measuring instrument.

*International Practical Temperature Scale.

3.3.2 *constant (of a measuring instrument)*
instrument constant. The factor by which the direct indication must be multiplied in order to obtain the indication of the instrument.

NOTE 1. The constant of an instrument which indicates directly the value of the quantity measured, is equal to 1.
NOTE 2. When the scale of a measuring instrument does not indicate a definite unit or if it indicates units of measurement other than that related to the quantity measured, the constant is a composite number; when the scale is expressed in units of the quantity measured the constant is a pure number.
NOTE 3. Multiple-range instruments with a single scale have several constants which correspond, for example, to different positions of a selector mechanism.

3.3.3 *conventional true value of a material measure.* A value approximating to the true value of the quantity reproduced by a material measure such that, for the purpose for which that value is used, the difference between these two values can be neglected.

3.4 result of a measurement. The value of the measured quantity obtained by a measurement.

NOTE. When no confusion is possible, 'result of a measurement' can be termed 'measurement'.

3.4.1 *indicated value (of a quantity)*
uncorrected result (of a measurement). The result of a measurement before correction for systematic errors.

NOTE. In a series of measurements of the same quantity, the uncorrected result of the measurement is taken to be the arithmetic mean of the uncorrected results of the individual measurements unless otherwise qualified.

Examples
(a) A dimension of an object is measured; the indication, 14.7 mm, read on the instrument represents the uncorrected result.
(b) If in a series of 10 measurements of the same dimension the values obtained are 14.9, 14.6, 14.8, 14.6, 14.9, 14.7, 14.7, 14.8, 14.9, 14.8 mm, the uncorrected result of this series of measurements will be:

$$d = \frac{14.9 + 14.6 \ldots + 14.8}{10} = 14.77 \text{ mm}$$
$$\simeq 14.8 \text{ mm}$$

3.4.2 *measured value (of a quantity)*
corrected result (of a measurement). The result of a measurement obtained after having made the necessary corrections to the uncorrected result, in order to take account of systematic errors. (Where appropriate this result may be accompanied by an indication of the uncertainty of measurement.)

NOTE. In a series of measurements of the same quantity, the corrected result, unless otherwise qualified, is taken to be the arithmetic mean of the uncorrected results of individual measurements to which the necessary corrections have been applied.

Example. A dimension of an object is measured; the indication, 14.7 mm, read on the instrument represents the uncorrected result.

It has already been established by calibration that the correction to be applied to the results given by the instrument for this indication is −0.2 mm. In addition it is known that the random uncertainty of a single measurement is equal to ± 0.35 mm (with a probability of 99.7 %).

The corrected result of the single measurement is therefore:

$$d = (14.7 - 0.2 \pm 0.35) = (14.5 \pm 0.35) \text{ mm}$$

For the series of 10 measurements of example (b) of **3.4.1**, the corrected result of the series of measurements is therefore:

$$d = \left(14.77 - 0.2 \pm \frac{0.35}{\sqrt{10}}\right) = (14.57 \pm 0.12) \text{ mm}$$

NOTE. The random uncertainty of the result of a series of measurements is the random uncertainty of a single result divided by the square root of the number of results.

3.4.3 *weight of measurement.* A number which expresses the degree of confidence in the result of a measurement of a certain quantity in comparison with the result of another measurement of the same quantity.

3.4.4 *weighted mean.* The arithmetic mean of a series of results of measurements calculated after taking into consideration the weight of each of the individual results.

Example. The results of three measurements of the same quantity are 10.4, 10.6, 10.1 mm. If the degree of confidence attributed to the three results are expressed by the weights 1, 2 and 1 respectively,

the weighted mean

$$= \frac{10.4 \times 1 + 10.6 \times 2 + 10.1 \times 1}{4} = 10.42 \text{ mm}$$

3.5 repeatability (of measurement). A quantitative expression of the closeness of the agreement between the results of successive measurements of the same value of the same quantity carried out by the same method, by the same observer, with the same measuring instruments, at the same location at appropriately short intervals of time.

3.5.1 *reproducibility (of measurement).* A quantitative expression of the closeness of the agreement between the results of measurements of the same value of the same quantity, where the individual measurements are made under different defined conditions, e.g.:

by different methods, with different measuring instruments;

by different observers, at different locations;

after intervals of time appropriately long compared with the duration of a single measurement;

under different customary conditions of use of the instruments employed.

NOTE. The results of individual measurements are assumed to be corrected for known systematic errors.

3.6 calibration. All the operations for the purpose of determining the values of the errors of a measuring instrument (and, if necessary, to determine other metrological properties).

NOTE. The metrological usage of the term 'calibration' is often extended to include operations such as adjustment, gauging, scale graduation, etc.

Section four. Measuring instruments and their classification

4.1 measuring instrument. A device intended for the purpose of measurement which may:

reproduce one or more known values of a given quantity; or

provide an indication of the value of the measured quantity or equivalent information; or combine these functions.

Examples

(a) graduated rule;
(b) set of weights;
(c) equal-arm balance;
(d) U-tube manometer;
(e) clock;
(f) ammeter.

4.1.1 *material measure.* A name given to a measuring instrument which reproduces in a permanent fashion, during its use, one or more predetermined values of a given quantity.

Examples

(a) a weight;
(b) measure of volume, of one or several values;
(c) length standard, of one or more values, with or without a scale;
(d) standard resistor, inductor or capacitor;
(e) standard gauge block;
(f) gap gauge for checking cylinder diameters;
(g) signal generator.

4.1.2 *indicating (measuring)* instrument.* Measuring instrument which is intended to give, by means of a single unique observation, the value of a measured quantity at the time of that observation.

NOTE 1. An indicating instrument may have either continuous or discontinuous variations of indication.
NOTE 2. The term 'indicating instrument' is sometimes restricted to instruments which display the result of a measurement by the position of an index relative to a scale.
NOTE 3. Some instruments described as indicating instruments may incorporate a recording device.

*In this definition and some subsequent definitions the term 'measuring' has been placed in brackets and may be omitted if no ambiguity is thereby caused.

4.1.3 *totalizing (measuring) instrument.* Measuring instrument which determines the total value of a quantity by addition of partial values from one or more sources.

NOTE. Certain types of totalizing instruments include a zero resetting device.

Examples
(a) electrical-power-summation meter;
(b) discontinuous dispenser of liquid fuel with one or more measuring capacities.

4.1.3.1 *integrating (measuring) instrument.* Totalizing measuring instrument which determines the value of a quantity by means of integration.

NOTE 1. The summation of sufficiently small partial values is considered to be equivalent to integration.
NOTE 2. Certain types of integrating instruments include a zero resetting device.
NOTE 3. An integrating instrument enables the value of the quantity being measured to be determined by the difference between two separate indications.

Examples
(a) planimeter;
(b) continuous-belt-conveyor weighing machine.

4.1.3.1.1 *integrating meter.* Integrating measuring instrument which progressively indicates the accumulated value of the measured quantity.

NOTE. An instrument which adds together successive values of partial areas of a diagram, when the boundary of the area is followed by a tracer, is an integrating meter. By contrast, a planimeter, which gives the area of a closed curve only after having completed the trace of the curve, is not an integrating meter.

Examples
(a) positive-displacement liquid meter;
(b) gas meter;
(c) electricity meter (BS 4727*, term 104 1201);
(d) integrator attached to a flow-rate meter.

4.1.4 *recording (measuring) instrument.* Measuring instrument which is intended to provide a permanent or semi-permanent record from which the values of the measured quantity can be observed or obtained.

NOTE 1. The record can be in the form of a continuous or discontinuous graph or a sequence of numerical values.
NOTE 2. In some cases, a recording instrument may give effectively simultaneous records of the observations corresponding to the values of more than one measured quantity.
NOTE 3. The record is generally made as a function of another variable, in most cases of time. The instrument can also simply record a group or sequence of values of the measured quantity.
NOTE 4. Some instruments described as recording instruments may incorporate an indicating device.

Examples
(a) analogue chart recorder;
(b) storage oscilloscope;
(c) electrocardiograph;
(d) X-Y recorder.

4.1.5 *analogue (measuring) instrument.* Measuring instrument in which the indication is a continuous function of the corresponding value of the quantity to be measured.

Examples
(a) moving-coil voltmeter;
(b) mercury-in-glass thermometer;
(c) Bourdon-tube pressure gauge.

4.1.6 *digital (measuring) instrument.* Measuring instrument in which the quantity to be measured is accepted as, or is converted into, coded discrete signals and provides an output and/or display in digital form.

Example
Electronic digital voltmeter.

4.1.7 *moving-index (measuring) instrument.* Instrument whose indications are given by the position of a moving index in relation to a fixed scale.

4.1.7.1 *index.* A fixed or movable part of the indicating device (e.g. a pointer, a luminous spot, liquid surface, recording pen, point) whose position with reference to the scale marks enables the indicated value to be observed.

4.1.7.2 *scale.* The array of indicating marks, together with any associated figuring, in relation to which the position of an index is observed.

NOTE. The term is frequently extended to include the surface which carries the marks or figuring.

4.1.8 *moving-scale (measuring) instrument.* Instrument whose indications are given by the position of a moving scale in relation to a fixed index.

4.1.9 *measuring transducer.* A device which serves to transform, in accordance with an established relationship, the measured quantity (or a quantity already transformed therefrom) into another quantity or into another value of the same quantity, with a specified accuracy, and which may be used separately as a complete unit.

NOTE. Certain types of transducers have specific established names, e.g. microphone, thermocouple, transformer.

4.1.10 *detector.* A device or substance which responds to the presence of a particular quantity without necessarily measuring the value of that quantity.

Examples
(a) gold-leaf electroscope;
(b) indicator paper;
(c) spirit level;
(d) temperature-sensitive paint.

*BS 4727: 1971 'Glossary of electrotechnical, power, telecommunications, electronics, lighting and colour terms', Part 1 'Terms common to power, telecommunications and electronics', Group 04 'Measurement terminology'.

4.1.11 *sensor.* The part of a measuring instrument which responds directly to the measured quantity.

NOTE. The sensor is also the first transducer element, e.g. thermocouple of a thermoelectric thermometer, prime mover of a flowmeter, Bourdon tube of a pressure gauge float of a level measuring instrument.

4.2 measuring apparatus. The ensemble of technical equipment required for carrying out a particular defined measurement task, and including all the measuring instruments and auxiliary devices, assembled according to a definite scheme, necessary for the application of a given method.

Examples

(a) Apparatus for measuring the electrical resistivity of materials.
(b) Apparatus for the volumetric measurement of hydrocarbons by a continuous meter.
(c) Orifice-meter equipment in accordance with BS 1042*.

4.2.1 *measuring installation.* The ensemble of measuring apparatus required for carrying out measurements of one or several kinds of quantities.

NOTE. While a 'measuring apparatus' consists of the devices necessary for a given measurement, a 'measuring installation' is a collection of instruments or apparatus to enable various measurements to be carried out.

Examples

(a) The measuring installation of a Weights and Measures Office includes standards of length, of capacity, of mass; weighing instruments; supporting bases; water tanks; etc.
(b) Boiler-house instrumentation.

4.3 measurement standard†. A measuring instrument, or measuring apparatus, which defines, represents physically, conserves or reproduces the unit of measurement of a quantity (or a multiple or sub-multiple of that unit) in order to transmit it to other measuring instruments by comparison.

4.3.1 *base standard.* A measurement standard of one of the base quantities of a system of quantities.

4.3.2 *derived standard.* A measurement standard of one of the derived quantities of a system of quantities.

4.3.3 *fundamental standard*
absolute standard. A base or derived standard of which the values provided have been established in terms of the relevant base units without recourse to another standard of the same quantity.

NOTE 1. 'Fundamental' or 'absolute', as applied to a standard, refer to the method of establishment and not to the metrological quality of the standard.

*BS 1042 'Code for flow measurement'.
†In this and subsequent descriptions the term 'measurement' may be omitted if no ambiguity is thereby caused.

NOTE 2. The expression 'calculable standard' is sometimes used to refer to a particular case of 'fundamental' standard, e.g. a calculable capacitor.

4.3.4 *collective standard.* A group of measuring instruments of the same pattern and the same metrological characteristics brought together to fulfil, by their combined use, the role of the standard.

NOTE. The value given by a collective standard is commonly taken as the weighted arithmetic mean of the values given by the individual instruments.

Examples

(a) Collective standard of voltage consisting of a group of saturated Weston cells.
(b) Collective standard of pressure consisting of a group of pressure balances.
(c) Collective standard of luminous intensity consisting of a group of incandescent lamps.

4.3.5 *primary standard.* A standard of a particular quantity which has the highest class of metrological qualities in a given field.

NOTE. The concept of a primary standard is equally valid for base units and for derived units.

4.3.6 *international standard (of measurement*)*. A standard recognized by international agreement as the basis for fixing the values of all other standards of the given quantity.

4.3.7 *national standard (of measurement*)*. A standard recognized by a national decision as the basis for fixing the value, in a country, of all other standards of the given quantity.

NOTE. In general, the national standard in a country is also the primary standard to which other standards are traceable.

4.3.8 *secondary standard.* A standard the value of which is determined by direct or indirect comparison with a primary standard.

NOTE. 'Secondary' is used here in the sense of 'subsidiary', but the alternative meaning of 'second in a hierarchical chain' is also used.

4.3.9 *reference standard.* A standard, generally the best available at a location, from which the measurements made at the location are derived.

4.3.10 *working standard.* A measurement standard, not specifically reserved as a reference standard, which is intended to verify measuring instruments of lower accuracy.

4.3.11 *transfer standard.* A measuring device used to compare measurement standards, or to compare a measuring instrument with a measurement standard by sequential comparison.

4.3.12 *travelling standard.* A measuring device, sometimes of special construction, used for the comparison of values of a measured quantity at different locations.

*The description '(of measurement)' may be omitted if no ambiguity is thereby caused.

4.3.13 *reference material.* A material or subtance officially recognized as a standard, and characterized by a high degree of stability of one or more of its physical, chemical or metrological properties.

Examples

(a) Pure water used in triple-point cell.

(b) Pure mercury for use in barometers, manometers, etc.

(c) Normal heptane as used in BS 4696†.

4.3.13.1 *certified reference material.* A reference material the properties of which have been certified by a body having appropriate standing in the field.

4.3.14 *reference-valued standard*
reference-value method. A mode of reproducing a unit of measurement (or a multiple or sub-multiple of that unit) which enables measurements to be made in terms of:

either fixed values of certain properties or bodies;

or physical constants.

Examples. Reference-value standards (methods) for reproducing:

(a) the metre in wavelengths of light;

(b) the kelvin by one of a number of fixed points on a thermometric scale;

(c) the unit of density of a liquid.

NOTE 1. In certain cases, a reference-value standard (method) is adopted as the basis of the definition of the unit of the quantity concerned.

NOTE 2. A reference-value standard (method) is intended to enable measurements to be made without direct use of a primary or other reference standard.

4.4 traceability. The concept of establishing a valid calibration of a measuring instrument or measurement standard, by step-by-step comparison with better standards up to an accepted or specified standard.

NOTE. In general, the concept of traceability implies eventual reference to an appropriate national or international standard.

Section five. Errors in the results of measurements and errors of measuring instruments

5.1 error of measurement. The discrepancy between the result of the measurement and the true value of the quantity measured.

NOTE 1. In general, 'true value' may be replaced by 'conventional true value'.

NOTE 2. The discrepancy can be expressed as either:
the algebraic difference between these two values, i.e. (error of measurement) = (result of measurement) − (true value); or
as the quotient of that difference and the value of the quantity measured.

These two forms of expression are often identified as 'absolute error' and 'relative error' respectively.

NOTE 3. The term 'absolute value of an error of measurement' is used* to describe the value of an error without regard to sign, i.e. the modulus of the error.

Example. If the maximum permissible error of a measuring instrument is ±0.2 mm, the absolute value of this error is ±0.2 mm = 0.2 mm.

5.1.1 *systematic error.* An error which, in the course of a number of measurements of the same value of a given quantity, remains constant when measurements are made under the same conditions and remains constant or varies according to a definite law when the conditions change.

NOTE 1. The causes of systematic error may be known or unknown.

NOTE 2. Absence of systematic error is sometimes referred to as 'freedom from bias'. (See definition 9.5 of PD 6461: 1971†.)

5.1.1.1 *correction.* A value which must be added algebraically to the indicated value (uncorrected result) of a measurement to obtain the measured value (corrected result).

NOTE. An assumed error and its associated correction are numerically equal but are of opposite sign.

5.1.2 *random error.* An error which varies in an unpredictable manner, in magnitude and sign, when a large number of measurements of the same value of a quantity are made under effectively identical conditions.

NOTE. It is not possible to take account of random error by the application of a correction to the uncorrected result of the measurement; it is only possible to fix limits within which, with a stated probability, the error will lie, on completion of a series of measurements made under effectively identical conditions (using the same measuring instrument, with the same observer and under the same environmental conditions, etc.).

5.1.3 *law of combination of errors.* The law connecting the error associated with the result of an indirect measurement of a quantity, with the errors associated with the direct measurements of the component quantities.

5.1.3.1 *partial error.* The part of the error of indirect measurement of a quantity which results from the error associated with a component quantity.

5.1.4 *deviation.* The divergence of the value of a quantity from a standard or reference value.

NOTE. Particularly in statistics, the reference value is frequently the arithmetic mean of the results in a series of measurements.

5.1.5 *uncertainty of measurement.* That part of the expression of the result of a measurement which states the range of values within which the true value or, if appropriate, the conventional true value is estimated to lie.

†BS 4696 'Method for determination of asphaltenes in petroleum products (precipitation with normal heptane)'.

*The reader is warned of the danger of confusion of this term with 'absolute error' (see note 2).

†PD 6461 'Vocabulary of legal metrology. Fundamental terms'.

NOTE. In cases where there is adequate information based on a statistical distribution, the estimate may be associated with a specified probability. In other cases, an alternative form of numerical expression of the degree of confidence to be attached to the estimate may be given.

5.2 error (of indication, or of response)* of a measuring instrument. The difference $v_i - v_c$ between the value indicated by (or the response of) the measuring instrument v_i and the conventional true value of the measured quantity v_c.

Example. The indication of an ammeter is $v_i = 40$ A; if the conventional true value of the current is $v_c = 41$ A, the error of indication of the ammeter is $e_i = 40\text{ A} - 41\text{ A} = -1$ A.

5.2.1 *error (of indication) of a material measure.* The difference $v_i - v_c$ between the indicated value v_i and the conventional true value v_c reproduced by the material measure.

Example. A capacity measure contains $v_c = 1005$ ml and it is marked 1 litre, that is $v_i = 1000$ ml. The error of indication of the measure is $e_i = 1000\text{ ml} - 1005\text{ ml} = -5$ ml.

5.2.2 *instrumental error.* That part of the error of indication originating from the instrument used for the measurement.

Example. Error which arises, among other causes, from friction between the moving parts of the instrument, from the incorrect positioning of the scale marks, from an inequality between the arms of a balance.

5.2.3 *bias error (of a measuring instrument) total systematic error (of a measuring instrument).* Algebraic sum of the systematic errors affecting the indications or responses of a measuring instrument under defined conditions of use.

Example. If an electric current is measured with a microammeter and if the 10 successive indicated values in microamperes are found:

6.27, 6.38, 6.62, 6.18, 6.31, 6.43, 6.36, 6.00, 6.35 and 6.50, the arithmetic mean of these values is:

$$\bar{v}_i = \frac{6.27 + 6.38 + 6.62 + \ldots 6.50}{10}$$

$= 6.34\ \mu\text{A}$ (indicated value, see note to **3.4.1**)

and if in measuring the same current by a much more precise method,

$v_c = 6.52\ \mu\text{A}$ (conventional true value) is obtained, the bias error of the microammeter is equal to:

$$e_j = \bar{v}_i - v_c = 6.34 - 6.52 = -0.18\ \mu\text{A}.$$

5.2.4 *repeatability (of a measuring instrument).* The quality which characterizes the ability of a measuring instrument to give identical indications, or responses, for repeated applications of the same value of the measured quantity under stated conditions of use.

5.2.4.1 *scatter (of indications or responses) dispersion (of indications or responses).* The phenomenon exhibited by a measuring instrument whereby it gives different indications (responses) in a series of measurements of the same value of the quantity measured.

NOTE. This concept can be expressed quantitatively by the standard deviation or by the spread (range of dispersion) of the indications (responses).

5.2.4.2 *spread (range of dispersion) of indications (responses).* A measure of the dispersion of the indications (responses) of a measuring instrument expressed by the difference between the largest v_i max. and the smallest v_i min. of the indications which, in a series of measurements, correspond to the same value of the measured quantity:

$$W = v_i\text{ max.} - v_i\text{ min.}$$

5.2.4.3 *repeatability error (of a measuring instrument).* The error which results from the dispersion of the indications (responses) of a measuring instrument under stated conditions of use.

NOTE. This may be expressed in various ways, e.g. root-mean-square-error, average error, etc. The root-mean-square-error (standard deviation) is often adopted.

Example. In the example given under **5.2.3**, the differences between the results obtained v_i and their mean value $\bar{v}_i$ are:

−0.07, +0.04, +0.28, −0.16, −0.03, +0.09, +0.02, −0.34, +0.01, +0.16 μA, the r.m.s. repeatability error of one of the results of the measurement is:

estimated standard deviation $s =$

$$\frac{\sqrt{0.07^2 + 0.04^2 \ldots + 0.16^2}}{10 - 1} = 0.17\ \mu\text{A}$$

NOTE. This quantity is an estimate of the real standard deviation, since the average from which the deviations are calculated is based on a finite number of measurements.

5.2.4.4 *limits of repeatability error.* The limiting errors (confidence limits with a stated probability) of a single measurement of a given value of the quantity to be measured under the specified conditions of use, not taking bias errors into consideration.

NOTE. The limits of repeatability error are usually calculated as the product of the r.m.s. repeatability error and a number t (whole or fractional) giving $e_1 = +ts$ and $e_2 = -ts$, where s is the estimate of the standard deviation and the number t is determined as a function of the probability P of not exceeding the values e_1 and e_2. (See BS 2846*.)

*In this definition and in some subsequent definitions the term 'of indication, or of response' has been placed in brackets and may be omitted if no ambiguity is thereby caused.

*BS 2846 'The reduction and presentation of experimental results'.

5.2.5 *quantization error (of a measuring instrument).* The error which may result from the measurement of the value of a quantity, by a process which converts the quantity into a digital indication, or response.

NOTE. If the least significant bit of the digitally-expressed value is n, the limits of quantization error will be $\pm\frac{1}{2}n$.

Example. A digital voltmeter having a full scale of 1.00 V would, if free from other errors, respond to all values of measured quantity lying between 0.995 V and 1.005 V with an indication of 1.00 V. If the conventional true value of the measured quantity is 1.00 V the quantization error is zero, whereas for conventional true values of 0.995 V and 1.005 V the quantization errors are +0.005 V and −0.005 V respectively. These latter values are the limits of quantization error; for example, a decrease of the conventional true value of the measured quantity to, say, 0.994 V will result in a digital indication of 0.99 V.

5.2.6 *discrimination (of a measuring instrument).* The property which characterizes the ability of a measuring instrument to respond to small changes of the quantity measured.

NOTE. In some fields of measurement, the term 'resolution' is used as synonymous with 'discrimination', but attention is drawn to **6.2.1** 'sensitivity'.

5.2.6.1 *discrimination error (of a measuring instrument).* That part of instrumental error which results from imperfect discrimination of a measuring instrument, and which is usually stated as that change of the measured quantity which just fails to cause a detectable change of indication (for an average operator and under defined conditions of use of the instrument).

NOTE 1. The discrimination error may not be constant, and the most appropriate form of numerical statement may depend on the type and manner of use of the instrument.
NOTE 2. In many cases it may be appropriate to state the results as the average and range of the discrimination errors obtained in a representative series of trials.

5.2.6.2 *maximum discrimination error.* The largest discrimination error obtained in a representative series of trials.

5.2.7 *hysteresis (of a measuring instrument).* That property of a measuring instrument whereby it gives different indications, or responses, for the same value of the measured quantity, according to whether that value has been reached by a continuously increasing change or by a continuously decreasing change of that quantity.

NOTE. By convention, the concept of hysteresis as applied to a measuring instrument is taken to include 'dead band' (see **5.2.7.4**).

5.2.7.1 *hysteresis error (of a measuring instrument).* That part of the instrumental error which results from hysteresis.

NOTE. The magnitude of the hysteresis error depends on the value of the measured quantity, the previous sequence of values and their rates of increase or decrease.

5.2.7.2 *hysteresis band.* For a particular value of the measured quantity, the value of the difference between the indications, or responses, when the same value of the measured quantity is reached on the one hand by increasing, on the other hand by decreasing that quantity.

NOTE. This difference is stated in the form $(I_{dec} - I_{inc})$ where I_{inc} and I_{dec} are the values of the indication, or response, corresponding respectively to increase and decrease of the value of the measured quantity.

5.2.7.3 *maximum hysteresis band.* The greatest value of the hysteresis band within a specified range of the measuring instrument.

5.2.7.4 *dead band.* The range through which a stimulus can be varied without initiating a change in response.

NOTE 1. The term 'threshold sensitivity' is sometimes used as being synonymous with dead band.
NOTE 2. Dead band is sometimes introduced deliberately, for example, to reduce unwanted change in the response for small changes in the stimulus.

5.2.8 *response-law error.* That part of instrumental error whereby a change of the indication, or response, departs from the intended relationship to the corresponding change of the value of the measured quantity, over a defined range.

NOTE 1. This term may be qualified in accordance with the intended response law, e.g. 'square-law response error', 'logarithmic-law response error', etc., as appropriate.
NOTE 2. This term is also used to describe a departure from the intended form of the law relating the response of a device to a particular parameter of the stimulus, e.g. the departure from its intended form of the frequency-response curve of a microphone for a constant sound pressure level.

5.2.8.1 *non-linearity error.* That part of the instrumental error whereby a change of the indication, or response, departs from proportionality to the corresponding change of the value of the measured quantity over a defined range.

NOTE. Various methods are employed for the numerical expression of non-linearity error.

(a) Departure from independent linearity, i.e. the 'best fit' straight line by 'least squares' for the available data. (See figure 1.)
(b) Departure from zero-based linearity, i.e. the 'best fit' straight line by 'least squares' for the available data, but conditional upon there being no error of response for zero stimulus. (See figure 2.)
(c) Departure from terminal-based linearity, i.e. the straight line joining the responses of the instrument at the terminal points of its span. (See figure 3.)

5.2.9 *datum error.* The error (of indication) of a measuring instrument for a value of the quantity measured which is chosen as a datum point for checking the instrument.

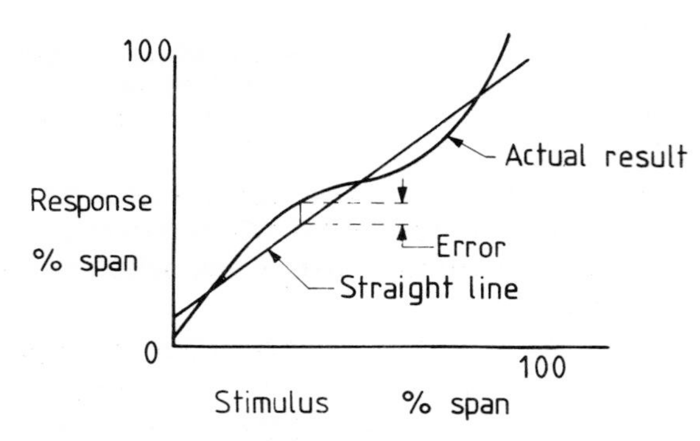

Figure 1. Independent linearity

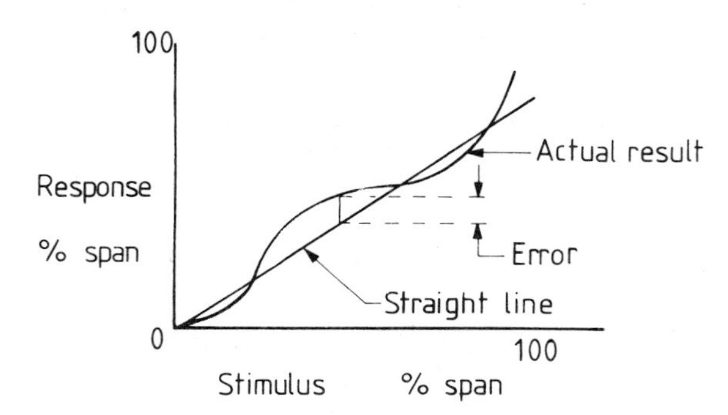

Figure 2. Zero-based linearity

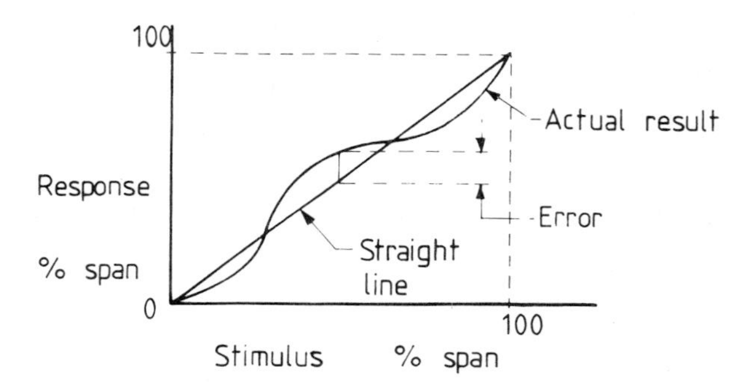

Figure 3. Terminal-based linearity

NOTE. The choice of datum point will depend upon the characteristics and intended use of the particular instrument.

Examples

(a) The error of indication associated with the temperature of 100 °C applied to a temperature gauge scaled from 100 °C to 200 °C.

(b) The error of indication associated with atmospheric pressure applied to a pressure gauge scaled from −100 kPa to +150 kPa gauge pressure.

5.2.9.1 *zero error.* The departure from zero of the indication of a measuring instrument for zero value of the quantity measured.

NOTE 1. This is a particular case of datum error where the zero scale mark is intended to correspond to zero value of the measured quantity.

NOTE 2. This definition is not intended to apply to mechanical or other offset zero conditions associated with some instruments.

Example. The error of indication associated with zero current applied to an ammeter scaled 0 to 10 A.

5.2.10 *intrinsic error.* The error of a measuring instrument when used under reference conditions.

5.2.11 *influence error (of a measuring instrument).* The error resulting from the departure of one of the influence quantities from its reference conditions.

5.2.11.1 *temperature (influence) error (of a measuring instrument).* The influence error due to a departure of the temperature of the instrument from its reference value.

5.2.11.2 *temperature-error coefficient (of a measuring instrument).* The relative variation, under steady state conditions, of the indications of a measuring instrument when its temperature has changed by the unit of temperature.

Example. If the temperature-error coefficient of a pressure gauge employing an elastic deformation element (the reference temperature of which is 20 °C), is, for example, 4×10^{-4} /°C and if the instrument is used at a temperature of 30 °C, then its indication will have an error (in this case positive) of 0.4 %.

5.2.12 *supplementary error (of a measuring instrument).* The error of a measuring instrument arising from the fact that the values of the influence quantities differ from those corresponding to the reference conditions.

5.2.13 *accuracy (of a measuring instrument).* The quality which characterizes the ability of a measuring instrument to give indications equivalent to the true value of the quantity measured.

NOTE. The quantitative expression of this concept should be in terms of uncertainty.

5.2.14 *total error (of a measuring instrument)*
overall error (of a measuring instrument). The whole error of a measuring instrument, under specified conditions of use.

NOTE 1. This includes bias error as well as repeatability error.

NOTE 2. These terms are equivalent to 'inaccuracy (of a measuring instrument)' as defined in 9.10.1 of PD 6461: 1971*.

5.2.14.1 *limits of total (overall) error (of a measuring instrument).* Set of two values obtained under specified conditions of use:

by adding to the bias error, the modulus of the value of the limit of repeatability error;

by subtracting from the bias error, the modulus of the value of the limit of repeatability error.

*PD 6461 'Vocabulary of legal metrology. Fundamental terms'.

5.3 error curve (of a measuring instrument). A curve which represents the error of a measuring instrument as a function either of the quantity measured or of any other quantity which has an influence on this error.

5.3.1 *calibration curve (of a measuring instrument).* A curve which expresses the correspondence between the values of the quantity measured and the values indicated by the instrument.

5.3.2 *correction curve (of a measuring instrument).* A curve which expresses, as a function of the indicated value of the measured quantity, or of the value of an influence quantity, the corrections to be applied to the results obtained with a measuring instrument.

5.4 maximum permissible errors (of a measuring instrument). The extreme values of the error (positive and negative) permitted by specifications, regulations etc., for a measuring instrument.

5.5 error of method. The error due to the use of a method of measurement which is inappropriate, or in which circumstances prevent the application of the best procedure.

5.6 error of observation. An error of measurement which is attributable to an observer when making an observation.

Example. In a measurement of luminous intensity by means of a contrast photometer, the error due to imperfect equalization of illumination of the two zones.

5.6.1 *reading error.* An error of observation resulting from the incorrect reading of the indication of the measuring instrument by the observer.

5.6.1.1 *parallax error.* A reading error which is produced when, the index being at a certain distance from the surface of the scale, the observation of the reading is not made in the appropriate direction.

5.6.1.2 *interpolation error.* A reading error resulting from an inexact evaluation of the position of the index with reference to the two adjoining scale marks between which the index is located.

5.7 fiducial error. Quotient of the error of indication of a measuring instrument and the appropriate fiducial value.

Section six. Conditions of use and metrological properties of measuring instruments.

6.1 conditions of use. The conditions which must be fulfilled in order to use a measuring instrument correctly, taking account of its design, construction and purpose.

NOTE. The conditions of use can refer, among other things, to the type and condition of the subject of the measurement, the value of the quantity measured, the values of the influence quantities, the conditions under which the indications are observed, etc.

6.1.1 *reference value.* A particular value of an influence quantity stated in the specification of a measuring instrument as a basis for determining its intrinsic error.

6.1.2 *reference range.* A particular range of values of an influence quantity stated in the specification of a measuring instrument as a basis for determining its intrinsic error.

6.1.3 *reference conditions.* The ensemble of reference values or reference ranges of different influence quantities affecting a measuring instrument.

NOTE. It is customary to select the reference conditions to correspond to a conveniently established environment for testing and to fall within the range of conditions expected in the use of the instrument.

6.1.4 *rated conditions of use.* The ensemble of the ranges of influence quantities and of the measured quantity within which the performance of the instrument is specified.

6.1.5 *limiting conditions of use.* The ensemble of the ranges of values of influence quantities and of the measured quantity which an instrument can withstand without consequential damage or degradation of performance when it is subsequently operated under rated conditions of use.

6.2 transfer function. The mathematical function expressing the relationship between a stimulus and the corresponding response over a defined range of conditions.

6.2.1 *sensitivity (of a measuring instrument).* Definition 1. The relationship of the change of the response to the corresponding change of the stimulus.

NOTE. The exact form of expression of the sensitivity will depend upon the type and conditions of use of the instrument. It is normally expressed as a quotient.

Definition 2. The value of the stimulus required to produce a response exceeding, by a specified amount, the response already present due to other causes, e.g. noise.

6.2.2 *gain (of a measuring instrument).* The sensitivity (according to the first definition in **6.2.1**), expressed as a quotient in the case where the stimulus and response are quantities of the same kind and expressed in the same units of measurement.

NOTE. In cases where increments of stimulus and response are involved, the term 'incremental gain' is often used.

6.2.3 *amplification.* A term sometimes used as an alternative to 'gain', particularly where the gain is not less than 1.

6.2.4 *attenuation.* The reciprocal of gain where the gain is less than 1.

6.3 drift. That property of a measuring instrument as a result of which its metrological properties change with time, under defined conditions of use.

NOTE. It should be noted that the term 'drift' is used in the general sense to describe the changes in performance of a measuring instrument due to changes in influence quantities. In such cases ambiguity may be caused unless the influence quantities concerned are stated, e.g. drift with time, temperature, etc.

6.3.1 *stability.* The property which characterizes the ability of a measuring instrument to maintain constant certain specified metrological properties.

NOTE. Although it is usual to describe stability in terms of lapse of time, lack of stability may arise from other influence quantities.

6.4 response time (of a measuring instrument). The time which elapses after a step change in the quantity measured, up to the point at which the measuring instrument gives an indication equal to the expected indication corresponding to the new value of the quantity, or not differing from this by more than a specified amount.

6.4.1 *maximum (minimum) repetition rate of measurement.* The number of measurements in unit time above (below) which the results of measurement obtained from the instrument are liable to be in error by more than a specified amount.

NOTE. This concept applies to repetitive measuring instruments with discontinuous operation, e.g. volumetric or gravimetric filling machines, weighing machines.

6.5 range (of a measuring instrument). The interval between the lower and upper range-limits.

6.5.1 *lower range-limit.* The minimum value of the quantity to be measured, for which the instrument has been constructed, adjusted or set.

6.5.2 *upper range-limit.* The maximum value of the quantity to be measured, for which the instrument has been constructed, adjusted or set.

6.6 effective range
measuring range
working range. The range of values of the measured quantity for which any single measurement, obtained under specified conditions of use of a measuring instrument, should not be in error by more than a specified amount.

NOTE. The effective range may be the whole or a specified part of the interval between the lower and upper range-limits.

6.7 suppressed-zero range. A range such that its lower limit is above the zero value of the measured quantity.

NOTE. The converse term 'elevated-zero range' may also be used. In this case the lower range-limit will be below the zero value of the measured quantity.

Examples

(a) *Suppressed zero:* a 0 to 250 voltmeter so arranged that the lower range-limit has been raised to 200 V while the upper range limit remains at 250 V.

(b) *Elevated zero:* a thermometer so arranged that the lower and upper range limits are −90 °C and −20 °C, respectively.

6.8 span. The algebraic difference between the upper and lower values specified as limiting the range of operation of a measuring instrument.

Example. A thermometer intended to measure over the range −40 °C to +60 °C has a span of 100 °C.

6.9 accuracy class. A classification of particular types of measuring instruments into groups, the members of which are required to have certain metrological properties within specified limits of error.

NOTE. The class is usually identified by one of a set of symbols adopted by convention and usually referred to as the 'class index'.

BS 2643 'Glossary of terms relating to the performance of measuring instruments' includes information on errors in instruments as well as providing a glossary of terms. In the light of the outcome of the work of the Joint Metrology Group this British Standard will be reconsidered and the appropriate action taken.

BS 2643

Glossary of terms relating to the performance of measuring instruments

---

. . . .

Introductory notes

1. Scope

The instruments which the compilers of this Glossary have had in mind are those designed to measure a range of values of some variable quantity. A scale of values is thus a typical, though not an essential, part of these instruments. An engineer's limit gauge is clearly not a measuring instrument in the sense intended here, neither is an assembly of apparatus such as an Orsat gas analyser, yet the distinction between a gauge, a measuring instrument, and an apparatus cannot be sharply drawn. This is of little consequence for the present purpose since it is hoped that the appropriateness of the definitions given will prove self-evident.

It should be understood that the definitions in this Glossary have been designed, as far as possible, to cover general application to instruments of a wide variety and that no attempt has been made to define particular types of instrument or terms associated with particular types of accuracy.

Instruments may be designed to register the magnitude of a quantity either without reference to its previous values or as the result of a process of integration, usually with respect to time. Many so-called integrating instruments are simply counters; for example, the domestic gas-meter counts units of gas volume as they are passed through the meter. A true integrating flow meter makes a continuous or periodic measurement of the rate of flow and integrates the rate with respect to time. Both the counter instrument and the true integrator are, by common usage, integrating instruments.

An example of spatial integration is afforded by those optical pyrometers which incorporate an integrating sphere.

Summation instruments are not to be confused with integrating instruments; they add together the simultaneous values of a quantity measured in separate circuits or flow systems.

Those instruments which inscribe a permanent or semi-permanent record of the values of the measured quantity are called recording, or in some cases, graphic, instruments. Those which only exhibit momentary values are indicating instruments; they are, strictly speaking, indicating *measuring* instruments, but as this Glossary is concerned exclusively with measuring instruments, it will be clear that indicators which merely show the presence or absence or a phenomenon, or are used only to indicate direction, without performing a measurement in the sense indicated in the first paragraph, are not under consideration.

Instead of measuring a quantity by means of the deflection of a calibrated instrument the measurement may be made by a direct comparison of the unknown quantity (or of another parameter directly related to it), with a known value of the same quantity. The function of the instrument is then to provide known values and to indicate, by means of a null-point indicator, when equality of values is achieved. An electrical resistance bridge is one example of this 'null-point' class of instruments; another is the chemical balance in its elementary form. Null-point instruments are also frequently used in conjunction with servo-mechanisms.

2. Errors in instruments

The choice of suitable terms to describe the performance of a measuring instrument depends on an appreciation of the various possible causes of errors of indication. The following notes are intended to give a general survey of some of the many factors that have to be taken into account.

The construction of a perfectly accurate instrument would require an exact knowledge of:

(a) the physical phenomena utilized in the principles of the instrument;
(b) the properties of the materials used in its construction;
(c) the influence of environmental conditions on (a) and (b).

In addition, the physical and mechanical design of the instrument would have to make complete use of this knowledge. Finally, the manufactured instruments would have to be in perfect agreement with the design, and the scale correctly graduated. The attainment of complete accuracy would then require proper application, operation, and observation of the instrument. These notes deal with errors introduced by the instrument itself and not with those due to the method of application or operation.

Physical principles. The physical laws covering the principle of an instrument are usually known to a degree of accuracy higher than that required of the instrument. There are, however, some instruments in which the law of response of the detecting element is empirical and is sometimes liable to unpredictable variations which mainly affect the stability of calibration.

Hysteresis is inherent in some physical phenomena such as ferromagnetism, elastic distortion, vapour absorption and change of electrical resistance in strained conductors; although choice of materials can minimize the effect of hysteresis it must often be tolerated. Since the effect of hysteresis on the indication of a particular value of the measured quantity depends on the magnitude and direction of previous changes in the measured quantity an error is introduced which although predictable in theory is in most cases unknown and so affects repeatability.*

Conditions of use. Errors due to the method of application of an instrument often greatly exceed the intrinsic errors. An appreciation of the physical principles of the instrument and, particularly, of its interaction with the system into which it is introduced is, strictly speaking, necessary to achieve accuracy in the interpretation of the instrument reading, but this is a question of errors of application rather than errors of indication.

*A change in shape of the meniscus in mercury barometers and similar instruments when the movement of the mercury changes direction causes a comparable error which is not properly described as hysteresis since it does not depend on the extent of displacement except for very small changes of pressure.

Properties of constructional materials. Dimensional changes in constructional materials occurring under load or as a result of ageing lead to distortion of the components and to variations in the law of response of the instrument; the law of response is also affected by changes in physical properties of the materials. The result of these variations is to introduce in greater or lesser degrees errors of every type, but the most prominent are likely to be zero error and change of sensitivity. Serious random errors may be introduced by the presence of electrostatic charge on instrument windows.

Design and construction. Practical considerations enforced by the conditions under which the instrument is to be used lead to a compromise in design, which consequently cannot be perfect from the point of view of accuracy. Such quantities as thermal capacity, mechanical inertia and viscosity introduce time lag in response. Mechanical defects such as bearing friction and play, and backlash in gears and linkages, cause random errors which are accentuated by the tolerance necessary in manufacture and which increase as a mechanism wears. Some instruments are designed and calibrated for use only in a particular orientation with respect to a gravitational, electric or magnetic field. In other cases, where no specific orientation is specified, or implied, errors arising from limitations in the design or construction may vary with the orientation of the instrument in these fields.

In electrical instruments thermoelectric voltages and resistance of other changes arising from the heating effect of the current passing through the instrument may not be negligible. The error depends on the duration and magnitude of the current. Instruments employing electronic devices may show considerable drift during the period of 'warming up' after being switched on.

Most instruments are pointed and ranged by comparison with an instrument of which the errors are known, though in the graduation of instruments such as the thermometers, standard physical quantities are sometimes used to give fixed scale points directly. It is impracticable to establish the position of every scale mark in this way so the intervals between the pointing marks together with any extrapolated scale are divided according to a presumed law of response, and an indication error results from any deviation of the actual response from the presumed law. If, for example, in a mercury-in-glass thermometer the displacement of the end of the mercury column is presumed to be proportional to the change of temperature, the scale marks will be equally spaced between the fixed points. Any inequalities in the size of the bore will then result in a departure from this presumed law and so an indication error will be caused. If the presumed law is followed by the instrument but the division of the scale is not carried out with sufficient accuracy the scale error so introduced will appear as an indication error when the instrument is verified.

Errors due to environment. The law of response of the instrument is in general dependent on the conditions of its environment. The significant conditions include temperature, pressure, humidity, external electrostatic and magnetic fields, and vibration. Changes in any electrical power supplies may also be included under this heading.

The effect of variations in any one of these conditions may be complex. Thus, in a moving-coil ammeter change of temperature will cause changes in linear dimensions of the coil, its mounting, and the magnetic system, in coil and shunt resistances, in magnetic field strength, and in the length and torque of the control spring.

The detecting elements of some instruments are particularly vulnerable to contamination or corrosion by agents inseparable from the conditions of use. Examples are the elements of various types of hygrometer and the orifices of flow meters. This leads to systematic errors due to changing sensitivity, while the ingress of dust also causes random errors through variable friction.

The performance of an instrument under changeable conditions of environment (constancy) must be distinguished from its performance when the test conditions are fixed (stability).

Observation errors. Instruments in which the index is not in the same plane as the scale are subject to parallax error which is minimized by various well-known devices.

The estimation of fractions of a scale division is subject to errors which, for trained observers, do not usually exceed one-tenth of a division under good conditions. The estimation error is influenced by the relative widths of index, scale marks, and distance between scale marks.

3. Glossary

3.1 Design details

measuring instrument. An apparatus for ascertaining and exhibiting in some suitable manner the magnitude of a physical quantity or condition presented to it.

NOTE 1. Essentially, every measuring instrument includes a detecting element and a measuring element, with means for indicating or correcting, and some form of coupling element interposed. These three elements may, or may not, be physically separable.

NOTE 2. The value of the measured quantity may be exhibited by the relative position of an index and scale, or by a counter mechanism: or ascertained by observing the adjustments required to bring an indicator to 'zero'; or by other means.

indicating instrument. A measuring instrument in which the value of the measured quantity is visually indicated, but is not recorded.

Examples: ammeter; pressure-gauge.

recording instrument. A measuring instrument in which the values of the measured quantity are recorded on a chart.

Examples: barograph; surface-finish recorder.

index. The pointer, light-spot, liquid surface, recording pen, stylus or other means, by the position of which, in relation to the instrument scale or chart, the value of the measured quantity, or the point of balance of a null-point instrument, is indicated.

scale. On an indicating instrument the array of marks, together with any associated figuring, with relation to which the position of the index is observed.

NOTE. The scale may indicate values either directly in units of the quantity measured, or in arbitrary units requiring appropriate conversion to give the values of the measured quantities.

chart. On a recording instrument, the prepared sheet or surface on which the record is made. On most recording instruments the chart is ruled with continuous lines corresponding to the scale marks on an indicating instrument, these lines being crossed at intervals by a succession of scale-base curves. So far as they may be applicable, references to scale marks in this glossary are to be read as including equivalent chart-rulings.

index path. The line of travel relative to the scale of that point on the index, or on the geometric prolongation of the index which is effective in indicating the value of the quantity measured.

NOTE. In some instruments the point on the index which is thus effective may vary with the position of the index.

scale base. The line, actual or implied, running from end to end of the scale, which defines, or corresponds with, the index path.

scale mark, graduation mark, graduation line, graduation (alone), *deprecated*. One of the marks constituting a scale.

scale length. The distance between the centres of the terminal scale marks, measured along the scale base.

maximum scale value. The greatest value of the measured quantity which the scale is graduated to indicate.

NOTE. In multi-range instruments this definition is to be read as applying to the particular range which the instrument is set up to measure.

minimum scale value. The smallest value of the measured quantity which the scale is graduated to indicate.

NOTE. In multi-range instruments this definition is to be read as applying to the particular range which the instrument is set up to measure.

scale division, scale sub-division, *deprecated*. A part of a scale delimited by two adjacent scale marks.

scale spacing. The distance between the centres of two adjacent scale marks, measured along the scale base.

scale interval. The increment of the measured quantity corresponding to the scale spacing.

scale range. The difference between the nominal values of the measured quantities corresponding to the terminal case marks.

NOTE. Scale range is conveniently expressed in the form 'A to B' where A is the minimum scale value and B is the maximum scale value.

instrument range. The total range of values which an instrument is capable of measuring. In a single-range instrument, the instrument range corresponds to the scale range. In a multi-range instrument, the instrument range is expressed by the difference between the maximum scale value for scale of highest values and the minimum scale value for the scale of lowest values, provided that the maximum and minimum of adjacent ranges are equal or overlap.

NOTE 1. Instrument range is conveniently expressed in the form 'A to B continuously', or 'A to B with gap C to D' for an instrument in which the whole range A to B is not covered.
NOTE 2. A multi-purpose instrument capable of measuring more than one physical quantity has a separate range for each such quantity.

effective range. That portion of the scale over which the instrument purports to comply with specified limits of error.

amplitude. In an oscillatory motion, the departure of the extreme point of the movement of the index from the zero or the rest position.

NOTE 1. The amplitude of movement of the index of a measuring instrument may be expressed in terms either of displacement along the index path, or of the equivalent measured quantity.
NOTE 2. The use of the term 'amplitude' in relation to sustained oscillations (e.g. those of a clock pendulum) to signify the *total* distance between the extreme points of the movement on opposite sides of the rest position is deprecated. It is considered that the present definition (which corresponds with the value of the constant 'a' in the formula $d = a \sin pt$, for a simple harmonic motion) is the better adapted for general use, particularly when regard is made to the extension of its application to damped oscillation. To avoid the confusion which results from the current use of the term 'amplitude' with two different meanings it is strongly recommended that the distance between the extreme points of movement on opposite sides of the rest position should be described as the 'double amplitude' or 'peak to peak amplitude'.

damping. The progressive reduction or suppression of the free oscillation of a system.

NOTE. Damping may be an inherent factor in the operation of an instrument or may be introduced deliberately as a feature in its design.

damper. A device introduced into the design of an instrument to produce damping.

damped. An instrument is said to be damped when there is a progressive diminution in the amplitude, or complete suppression, of successive oscillations of the index, after an abrupt change in the value of the measured quantity.

critically damped. An instrument is said to be critically damped when it is subject to the minimum degree of damping which will suffice to prevent oscillation of the index after an abrupt change in the value of the measured quantity.

under-damped. An instrument is said to be under-damped when the degree of damping is insufficient to prevent oscillation of the index after an abrupt change in the value of the measured quantity.

over-damped. An instrument is said to be over-damped when the degree of damping is more than sufficient to prevent oscillation of the index after an abrupt change in the value of the measured quantity.

aperiodic. An instrument is described as aperiodic when the motion of its index is critically damped or over-damped.

3.2 Calibration and graduation

calibration. The process of determining the characteristic relation between the values of the physical quantity applied to the instrument and the corresponding positions of the index.

NOTE. Calibration provides, typically, a chart giving the values of the measured quantities in terms of reading on an arbitrary scale.

graduation. The process of setting out a scale.

NOTE. The use of the word 'graduation' alone in the sense of a 'graduation mark' is deprecated.

pointing. The operation of locating certain marks on the scale, by reference to which the division of the scale is afterwards completed.

dividing. The process of completing the graduation of the scale after the pointing marks have been fixed.

setting. The process of adjusting the instrument so that it indicates correctly at a particular prescribed mark on the scale.

ranging, calibration, *deprecated*. The process of adjusting the instrument so that the index movement conforms to a pre-established scale at two or more points.

scale factor. (a) For an instrument having an arbitrary scale, the factor by which the indication has to be multiplied to obtain the nominal value of the quantity measured.

(b) For multi-range instruments having a scale (or scales) graduated in units of the quantity measured, the factor by which the indicated value has to be multiplied to obtain the nominal value of the measured quantity in the range being used.

verification. The process of testing an instrument for the purpose of assessing the indication errors. The assessment may be employed to determine whether an instrument complies with a prescribed specification.

3.3 Sensitivity and accuracy

sensitivity. The sensitivity of an instrument at any indicated value is the relation between the movement of the index and the change in the measured quantity that produces it. It may be expressed as a numerical ratio if the units of measurement of the index path and units of the measured quantity are stated.

discrimination. The sensitiveness of an instrument. The smallest change in the quantity measured which produces a perceptible movement of the index.

NOTE. It is appreciated that the condition of viewing may affect the movement of the index which is assessed as perceptible. The discrimination of an instrument should, therefore, be related to the conditions under which the index is observed.

settling time. The time taken for the index of an instrument to attain, and remain within, a specified deviation from its final steady value, after an abrupt change in the measured quantity.

repeatability. The reproducibility of the readings of an instrument when a series of tests is carried out in a short interval of time under fixed conditions of use.

stability. The reproducibility of the mean readings of an instrument when tested under defined conditions of use, repeated on different occasions separated by intervals of time which are long compared to the time of taking a reading.

constancy. The reproducibility of the (uncorrected) reading of an instrument over a period of time, when the quantity to be measured is presented continuously, and the conditions of test are allowed to vary within specified limits.

NOTE 1. Degrees of reproducibility, whether repeatability, stability, or constancy, are expressed by reference to their opposites, i.e. in terms of the limits of variation in the readings under the respective conditions stated. The smaller the variation the better the repeatability, etc., and vice versa.

NOTE 2. In assessing stability the readings are assumed to be corrected, if necessary, for any departure of the actual test conditions from the defined conditions. 'Constancy' is an all-embracing term, intended to

include the effects due to variations (within the specified limits) in the conditions of use, as well as those associated with repeatability and stability. In assessing constancy, therefore, the readings are not corrected.
NOTE 3. Apart from any over-all limits which may be assigned for constancy, specifications for individual types of instrument may prescribe limits for the variation in reading corresponding to any particular variation in the conditions of use.

accuracy. A general term describing the degree of closeness with which the indications of an instrument approach the true values of the quantities measured.

accuracy grade. The category into which an instrument falls by virtue of the specified limits of error with which it purports to comply.

observation error, reading error, *deprecated*. The error committed by the observer when reading the indication of an instrument.

NOTE. Observation errors may be due, for example, to parallax, to faulty estimation of the fractional part of a scale interval, or to simple misreading of the indication.

scale error. The difference between the position of a scale mark and its theoretical position on a scale correctly graduated in accordance with the assumed law of operation of the instrument.

indicating error, instrument error, reading error, *deprecated*. The difference obtained by subtracting the true value of the quantity measured from the indicated value, due regard being paid to the sign of each.

NOTE. For the purposes of this definition, instruments such as centre-zero ammeters marked 'Charge' and 'Discharge', or surveying aneroids marked in height and depth above and below sea-level, are to be regarded as embodying two separate instruments reading in opposite directions. In any statement or chart of errors resulting from the verification of an instrument in which plus and minus signs are applied to the indications, these signs have to be taken into account where they apply.

datum error. The indication error when the instrument is in the specified conditions of use and a physical quantity, of a magnitude prescribed for the purpose of verifying the setting, is presented to it.

zero error. The indication when the instrument is in the specified conditions of use and the magnitude of the physical quantity presented to it is zero.

correction. The amount which should be added to the indicated value, due regard being paid to sign, to obtain the true value of the quantity measured.

NOTE. The correction to the indication of an instrument is equal in magnitude to the indication error, but is of opposite sign.

interval error. The difference obtained by subtracting the indication error at one of two scale marks on the scale from that at the other, due regard being paid to the signs of each.

NOTE. In certain cases – e.g. surveying aneroids marked in height and depth above and below sea level – where the parts of the scale on either side of zero are treated as independent, it may be necessary to *add* the indication errors, instead of subtracting them, in order to determine the interval error. This is of importance only in the application of limits to the permissible interval errors. In use the individual readings will normally be adjusted separately by the application of the appropriate corrections and the true value of the interval ascertained by adding or subtracting the corrected readings according to self-evident conditions.

limits of error. The positive and/or negative values of the errors which must not be exceeded under test.

(a) For indication errors the limits may be expressed either
- (i) directly in units of the quantity measured, or
- (ii) as percentages of the scale range, maximum scale reading, or indicated value.

(b) For interval errors the limits apply to the differences between the indication errors at two scale marks which may be either
- (i) any two specified marks
- (ii) any two marks separated by a specified interval, or
- (iii) any two marks in the scale range.

NOTE 1. In the application of limits of error it is assumed throughout that the instrument is first adjusted to the prescribed conditions of use.
NOTE 2. In certain cases the sums of the indication errors may have to be taken, instead of their differences, in ascertaining interval errors.
NOTE 3. The use of the term 'tolerance' in describing the limits of error of an instrument is deprecated.

The establishment of a reputation for quality products and services depends upon the ability to demonstrate conformity to specified requirements. The control of manufacturing processes and the assessment of product quality and conformance almost invariably involves measurement of some kind. The integrity of all such measurement work is founded on the selection and correct use of suitable measuring equipment and its validation by a system of control and calibration. Any organization working to a formal series of quality system standards (such as BS 5750 'Quality systems') has to be able to demonstrate the ability to measure adequately and to certify that the results obtained can be relied upon. BS 5781 'Measurement and calibration systems' specifies requirements for a calibration system that can be used in the fulfilment of a quality system or for the supply of calibration services. It can be used for the following purposes:

(a) As a basis for evaluating the capability of a supplier's calibration system, either by a potential purchaser or by a third party, in order to provide assurance to interested parties. This may be prior to the establishment of a contract.

(b) Where invoked in a contract, to specify the calibration requirements appropriate to the particular material or services.
(c) In other documents where reference to a calibration system is appropriate.

BS 5781

Measurement and calibration systems

---

1. Scope

This British Standard specifies requirements for a system for selecting, using, calibrating, controlling and maintaining measurement standards and measuring equipment used in the fulfilment of specified requirements.

. . . .

3. Definitions

For the purposes of this British Standard, the following definition together with those given in BS 4778 and BS 5233 apply.

specified requirements. Either:
(a) requirements prescribed by the purchaser in a contract or order for material or services; or
(b) requirements prescribed by the supplier that are not subject to direct specification by the purchaser.

4. Requirements

4.1 Calibration system. The supplier shall establish and maintain an effective system for the control and calibration of measurement standards and measuring equipment used in the fulfilment of specified requirements. The system shall be designed to ensure that all measurement resources have the capability of making measurements within limits designated as appropriate to these requirements. It shall provide for the prevention of inaccuracy by prompt detection of deficiencies and timely action for their correction.

NOTE. Complete in-plant measurement or calibration capability is not essential if these services are obtained from sources that comply with the requirements of this standard.

Objective evidence that the system is effective shall readily be available to the purchaser or his authorized representative, hereinafter referred to as the 'Purchaser's Representative'.

4.2 Periodic review of the calibration system. The system established in accordance with the requirements of this standard shall be periodically and systematically reviewed by the supplier to ensure its continued effectiveness and the results shall be recorded.

4.3 Planning. The supplier shall review the technical requirements specified at the earliest practicable stage before commencing new work on material or services, and shall establish a programme to ensure that measurement standards and measuring equipment necessary for the completion of this work are available and are of the accuracy, stability and range appropriate for the intended application. In particular, identification and timely action shall be taken to report, to the Purchaser's Representative, any measurement requirement that exceeds the known 'state of the art' or any new measurement capability that is needed but is not available.

4.4 Measurement limits. All measurements, whether made for purposes of calibration or product characteristic assessment, shall take into account all the errors and uncertainties in the measurement process that are attributable to the measurement standard or measuring equipment, and, as appropriate, those contributed by personnel, procedures and environment.

4.5 Documented calibration procedures. Documented calibration procedures shall be prescribed and shall be available for reference and use, as necessary, for the calibration of all measurement standards and measuring equipment used.

NOTE. Procedures may be, but are not necessarily, limited to the compilation of published standard measurement practices and purchaser's or instrument manufacturer's written instructions. Exceptions may be allowed for measuring equipment when it is uneconomical and technically unnecessary to require a detailed procedure.

The exceptions given in the above note shall be recorded. Procedures shall contain sufficient instruction to provide data adequate to ensure valid measurement when the equipment is used for the measurement of product or process characteristics or calibration.

4.6 Records. The supplier shall maintain records of all measurement standards and measuring equipment used to establish conformance to specified requirements. These records shall demonstrate that each measurement standard and item of measuring equipment is capable of performing measurements within the designated limits. When measurement standards or measuring equipment are found to be outside these limits, the extent of the errors shall be recorded and appropriate action taken.

NOTE. The amount of data to be recorded is dependent on the nature of the measuring equipment, the measurement standards in use and on their application.

Functionally simple and rugged measuring equipment shall be the subject of records (individual or collective) or other means of showing that they are within their calibration interval and designated limits.

All other measuring equipment and measurement standards shall be the subject of records which shall always include the following information:

(a) the description of equipment and unique identification;
(b) the date on which each calibration was performed;
(c) the results obtained from calibration;
(d) the planned calibration interval;

and, when appropriate, the following additional information:

(e) the designated permissible limits of error;
(f) the reference to calibration procedures;
(g) the source of calibration used to establish traceability;
(h) the environmental condition for calibration and the measurement data as measured and as corrected to reference conditions;
(j) a statement of the cumulative effect of uncertainties on the data obtained in the calibration;
(k) details of any maintenance (servicing, adjustment, repairs) or modifications that could affect the calibration status;
(l) any limitations in use.

4.7 Calibration labelling. All measurement standards and measuring equipment shall be labelled, coded or otherwise identified to indicate their calibration status. Any limitation of calibration or restriction of use shall be clearly indicated on the equipment. When neither labelling nor coding is practicable, or is not considered essential for control purposes, other procedures shall be established to ensure conformance to these requirements.

4.8 Sealing for integrity. Access to adjustable devices on measurement standards and measuring equipment, which are fixed at the time of calibration, shall be sealed or otherwise safeguarded to prevent tampering by unauthorized personnel. Seals shall be so designed that tampering shall destroy them.

NOTE. This requirement does not apply to adjustable devices that are intended to be set by the user without needing external references, e.g. zero adjusters.

4.9 Intervals of calibration. Measurement standards and measuring equipment shall be calibrated at periodic intervals established by the supplier on the basis of stability, purpose and usage. Intervals shall be established so that re-calibration occurs prior to any probable change in accuracy that is of significance to the use of the equipment. Depending upon the results of preceding calibrations, intervals of calibration shall be shortened, if necessary, to ensure continued accuracy.

NOTE. The intervals may be lengthened if the results of previous calibrations provide definite indications that such action will not adversely affect confidence in the accuracy of the equipment.

4.10 Invalidation of calibration. The supplier shall provide for the immediate removal, or conspicuous identification to prevent use, of any measurement standard or measuring equipment that is:

(a) outside its designated calibration period; or
(b) has failed in operation; or
(c) is suspected of being or is known to be outside its designated limits; or
(d) shows evidence of physical damage that may affect its accuracy.

The Purchaser's Representative shall be notified immediately if suspected or actual equipment failures affect the acceptability of the product.

Details of such invalidation of calibration requiring corrective action shall be recorded.

4.11 Sub-contractors. The supplier shall ensure that sub-contractors employ a system which complies with the requirements of this standard.

4.12 Storage and handling. The supplier shall establish and maintain a system for handling, transporting and storing all measurement standards and measuring equipment to prevent abuse, misuse, damage or change in dimensional or functional characteristics.

4.13 Traceability. All measurement standards and measuring equipment shall be calibrated using measurement standards that are traceable to national or international measurement standards, except where they have been derived from acceptable values of natural physical constants or by the ratio type of self-calibration techniques.

NOTE. Measurement standards in use need not be referred directly to national measurement standards, but may be calibrated against intermediate standards provided that the requirement for traceability is satisfied.

All measurement standards used in the calibration system shall be supported by certificates, reports or data sheets attesting to the date, accuracy, and conditions under which the results were obtained and are valid. All such documents shall be signed by a responsible person.

4.14 Cumulative effect of errors. The cumulative effect of the errors in each successive stage of a calibration chain shall be taken into account for each measurement standard or item of measuring equipment calibrated. Action shall be taken when the total uncertainty is such that it significantly compromises the ability to make measurements within the required limits. The details of the components of the total uncertainty shall be recorded. The method of combining these shall also be recorded.

4.15 Environmental control. Measurement standards and measuring equipment shall be calibrated and used in an environment controlled to the extent necessary to ensure valid measurements. Due consideration shall be given to temperature, rate of change of temperature, humidity, lighting, vibration, dust control, cleanliness and other factors affecting measurement. When pertinent, these factors shall be continuously monitored and recorded, and, when necessary, compensating corrections shall be applied to measurement data. Records shall contain both the original and the corrected data.

4.16 Evaluation of calibration system. Where specified in a contract, the calibration system shall be open to evaluation by the Purchaser's Representative and reasonable access and facilities shall be made available for this purpose.

4.17 Training. All personnel performing calibration functions shall have appropriate experience or training.

. . . .

Tests performed on presumably identical materials in presumably identical circumstances do not, in general, yield identical results. This is attributed to unavoidable random errors inherent in every test procedure; the factors that may influence the outcome of a test cannot all be completely controlled.

In the practical interpretation of test data this variability has to be taken into account. For instance, the difference between a test result and a value specified by contract may be within the scope of unavoidable random errors, in which case a true deviation from specification has not been established. Similarly, comparing test results from two batches of material will not indicate a fundamental quality difference if the difference between them can be attributed to inherent variation in the test procedure.

Many different factors (apart from sampling error) may contribute to the variability of a test procedure, for example:

(a) the operator;
(b) the instruments and equipment used;
(c) the calibration of the equipment;
(d) the environment (temperature, humidity, air pollution, etc.).

The variability will be greater when the tests to be compared are performed by different operators and/or with different instruments from those when they were carried out by a single operator using the same instrument(s). Hence many different measures of variability are conceivable according to the circumstances under which the tests were performed.

Two extreme measures of variability, termed repeatability and reproducibility, have been found sufficient to deal with most practical cases. Repeatability refers to tests performed at short intervals in one laboratory by one operator using the same equipment, while reproducibility refers to tests performed in different laboratories, which implies different operators and different equipment. Under repeatability conditions, factors (a) to (d) listed above are considered constants and do not contribute to the variability, while under reproducibility conditions they vary and contribute to the variability of the test results.

BS 5497 'Precision of test methods', Part 1 'Guide for the determination of repeatability and reproducibility for a standard test method' is concerned exclusively with test methods whose results are expressed quantitatively, and is primarily intended to be applied to test methods that have previously been standardized and are used in different laboratories. No part of BS 5497 is reproduced in this manual.

Notes on accuracy and precision

BS 5532 'Statistics – vocabulary and symbols' gives the following definition for 'true value':

> The value which characterizes a quantity perfectly defined in the conditions which exist at the moment when that quantity is observed (or the subject of a determination).
> It is an ideal value which could be arrived at only if all causes of measurement errors were eliminated and the population was infinite.
> This definition is illustrated in Figure 2.2.

BS 4778 'Glossary of terms used in Quality Assurance (including reliability and maintainability terms)' gives the following definition for 'accuracy':

> The closeness of an observed quantity to the defined or true value.
> *Note:* Accuracy is usually expressed in terms of error or uncertainty.
> This definition is illustrated in Figure 2.3.

BS 5532 gives the following definition for 'accuracy of the mean':

> The closeness of agreement between the true value and the mean result which would be obtained by applying the experimental procedure a very large number of times.

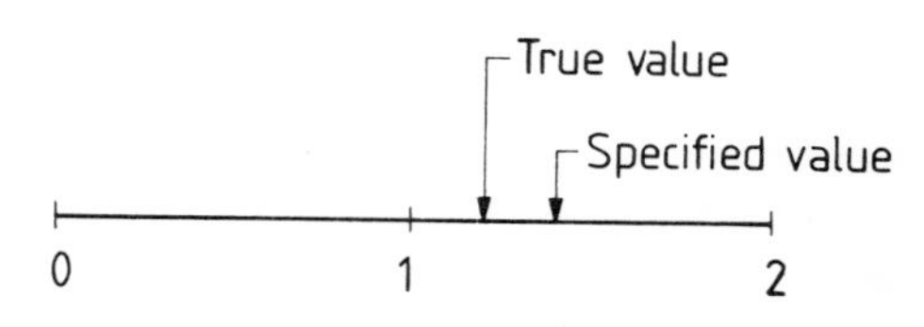

Figure 2.2 *This is a single real value and may only be measured with ideal equipment in ideal conditions. It may or may not be equal to the specified value*

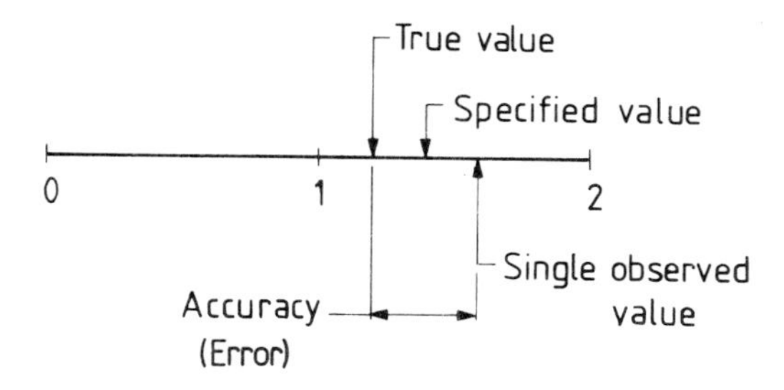

Figure 2.3 *This is a single observed value, and is related to the true value. It contains both systematic and random errors*

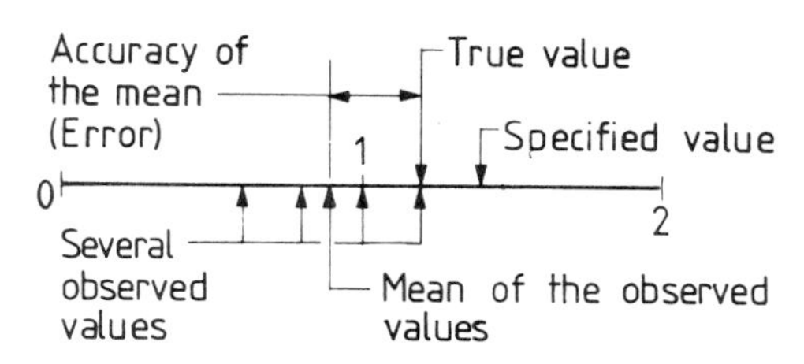

Figure 2.4 *This is the mean of several observed values, and is related to the true value. As the number of observations tends to infinity, so the random errors tend to zero, so it is a measure of systematic errors*

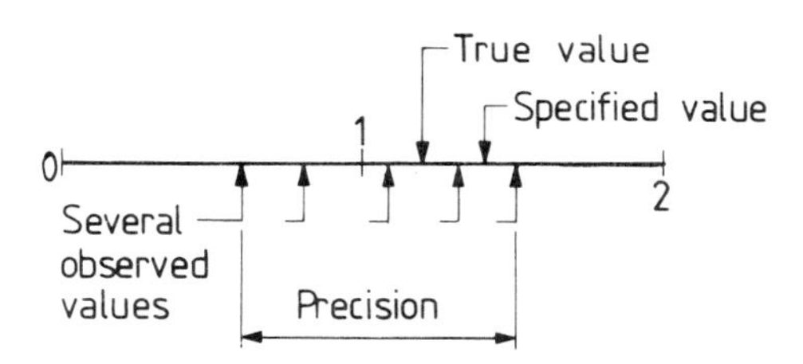

Figure 2.5 *This is a measure of the scatter of results of several observations and is NOT related to the true value. It is a comparative measure of the test results and is a measure of only the random errors*

The smaller the systematic part of the experimental errors which affect the results, the more accurate is the procedure.
This definition is illustrated in Figure 2.4.

BS 5532 gives the following definition for 'precision':

The closeness of agreement between the results obtained by applying the experimental procedure several times under prescribed conditions.
The smaller the random part of the experimental errors which affect the results, the more precise is the procedure.
This definition is illustrated in Figure 2.5.

Standards publications referred to in this chapter

BS 2643	Glossary of terms relating to the performance of measuring instruments
BS 4778	Glossary of terms used in Quality Assurance (including reliability and maintainability terms)
BS 5233	Terms used in metrology
BS 5497	Precision of test methods
	Part 1 Guide for the determination of repeatability and reproducibility for a standard test method
BS 5532	Statistics – Vocabulary and symbols
BS 5750	Quality systems (in 3 parts)
BS 5781	Measurement and calibration systems
PD 6461	Vocabulary of legal metrology
Handbook 22	Quality assurance – a compendium of seven quality assurance standards

Note: The above list does not include all the standards referred to in the extracts.

3 General inspection equipment

Introduction

Although some areas of metrology involve the use of 'special' equipment, the majority of inspection procedures require, in some way, the use of general inspection equipment. That this equipment should be suitable for the task in hand is obvious and it is to satisfy the demands of manufacturers and users that a number of British Standards have been published over the years. While much of the contents of these standards is of particular interest to the manufacturer, they also contain information which should be borne in mind when both purchasing and using such equipment. The material from standards selected for inclusion in this chapter has, therefore, been made to cover such areas as grades, accuracy, sizes, marking and testing, which will be of interest to the user.

For information on protective boxes and covers and protection against climatic conditions see the Preface to this manual.

Surface plates

BS 817 'Surface plates' specifies requirements for rectangular or square, cast iron or granite surface plates ranging from 160 mm × 100 mm to 2500 mm × 1600 mm in four grades of accuracy (grades 0, 1, 2 and 3). It applies to new surface plates, plates in use and those which, after wear, have been resurfaced.

The standard specifies the calculation of permitted values for grades and size of deviation from flatness overall of the working surface and of variation in local flatness of the working surface.

The standard does not specify the size of surface plates, but gives preferred sizes and specifies a tolerance on the nominal lengths of sides. It does not specify the tolerance for straightness, mutual parallelism or squareness of the edges of the surface plates.

BS 817

Surface plates

. . . .

2. Nomenclature and definitions

For the purposes of this British Standard the nomenclature shown in figure 1 and the following definitions apply.

2.1 surface plate. An iron casting or block of granite with a worked surface which is used as a plane and a datum.

NOTE. Surface plates of larger sizes which are mounted on a stand are usually known as surface tables but for the purposes of simplicity in this standard the term 'surface plate' has been used throughout.

2.2 deviation from flatness overall of the working surface. The minimum distance separating the two parallel planes between which the whole surface can just be contained.

2.3 variation in local flatness of the working surface. The variation in the indicator reading when the working surface is explored with the variation gauge shown in figure 2.

2.4 variation gauge. A freely moving base with three co-planar pads for bearing on the surface being tested, and carrying an indicator recording variations in deviation from flatness from adjacent areas of that surface through a fourth co-planar pad, which is itself mounted on an intermediate sprung block (see figure 2).

3. Classification

Surface plates are classified according to their accuracy as grade 0, 1, 2 or 3. The grades represent the deviation from flatness based on diagonal length; each grade has twice the permitted deviation from flatness of the preceding grade, grade 0 being the most accurate. The calculation of permitted deviation from flatness overall is given in appendix B, and table 1 gives the permitted deviations from flatness overall for grades 0 to 3 for preferred sizes.

. . . .

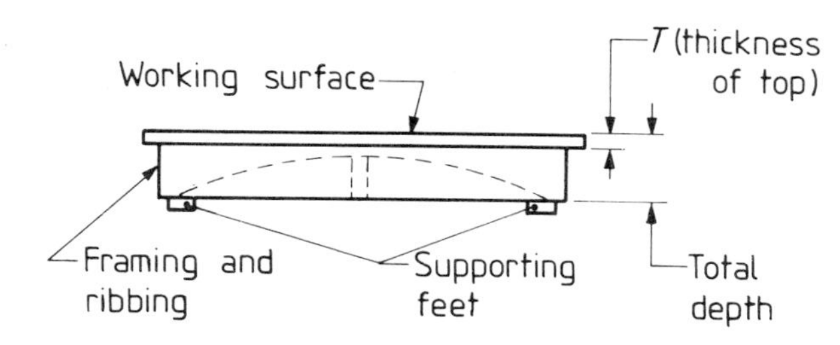

(a) Typical cast iron surface plate

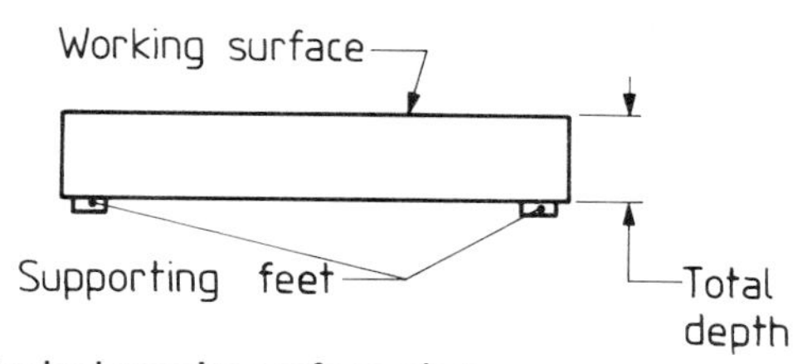

(b) Typical granite surface plate

Figure 1. Nomenclature

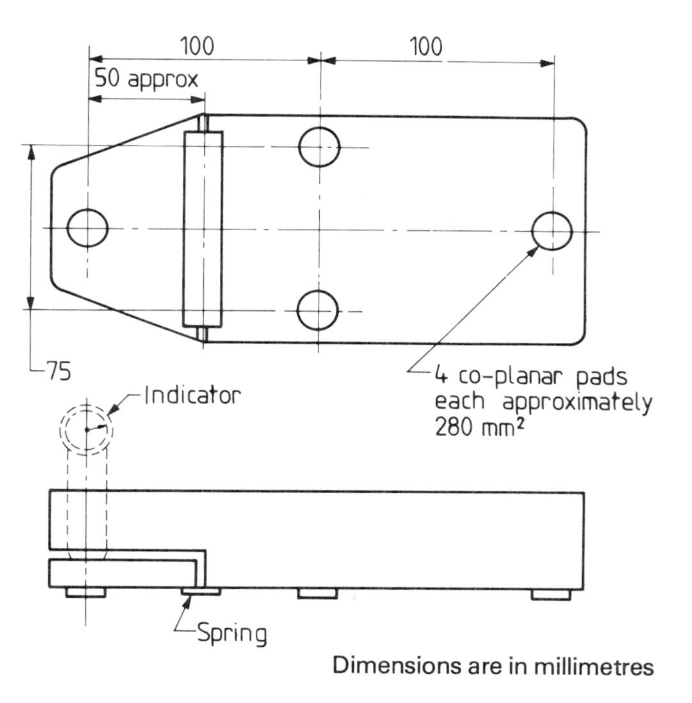

Dimensions are in millimetres

Figure 2. Variation gauge

7. Finish

7.1 Cast iron surface plates. The working surface of grades 0 and 1 surface plates shall be finished either by scraping or by other process that results in a surface similar to that obtained by scraping. The working surface of grades 2 and 3 surface plates shall be finished either by one of the same processes or by machining.

The proportion of bearing area, when tested in accordance with appendix D or by any other proven method, shall be not less than 20 % for grades 0 and 1, and not less than 10 % for grades 2 and 3 surface plates. High spots on scraped surfaces shall be uniformly distributed and the percentage of bearing area shall not be as high as to cause wringing.

Unmachined surfaces of cast iron surface plates shall be finished by painting.

7.2 Granite surface plates. The working surface of grades 0 and 2 surface plates shall be finished by lapping. The working surface of grades 2 and 3 surface plates shall be either finished by lapping or left as ground.

. . . .

9. Accuracy

9.1 Permitted deviation from flatness overall of the working surface. The deviation from flatness overall of the working surface of the surface plate, except for a border zone, when tested in accordance with one of the methods described in appendix E or by any other proven method shall not exceed the appropriate value for grade and size as calculated in accordance with appendix B. The width of the border zone shall not exceed 2 % of the length of the shorter side, or 10 mm, whichever is smaller, and no point on the border zone shall project above the rest of the working surface.

NOTE. Permitted deviations from flatness overall for the preferred sizes are given in table 1.

9.2 Variation in local flatness of the working surface

NOTE 1. It is important to regulate variations in local flatness as abrupt local changes in contour, convexities and concavities in the working surface could cause undesirable variations in the measurements gained using different positions on the working surface. The apparatus used for testing for variation in local flatness is also important as the readings obtained using different forms of apparatus may vary according to the dispositions of their contacts with the working surface. The response of the variation gauge specified below corresponds with the influence of variations in local flatness in practice. It should be used in workshop or inspection room to survey the whole working surface of a surface plate for variations in local flatness, methodically and expeditiously.

Variations in local flatness of the working surface of surface plates of 400 mm × 250 mm and above, shall be determined by variation gauge (see figure 2) of which the readings shall not exceed the values given in table 2.

NOTE 2. Corresponding variations for surface plates smaller than 400 mm × 250 mm have not been specified as the deviations from flatness overall for these sizes do not permit significant variation in local flatness.

9.3 Lengths of sides. The lengths of side of surface plates shall be within ±5 % of nominal size.

Table 1. Permitted deviations from flatness overall of the working surface for preferred sizes

Preferred size	Diagonal length	Width of border zone	Permitted deviations from flatness overall			
			Grade 0	Grade 1	Grade 2	Grade 3
mm	mm	mm	mm	mm	mm	mm
160 × 100	188	2	0.0030	0.006	0.012	0.024
250 × 160	296	3	0.0035	0.007	0.014	0.028
400 × 250	471	5	0.0040	0.008	0.016	0.032
630 × 400	745	8	0.0045	0.009	0.018	0.036
1000 × 630	1180	10	0.0055	0.011	0.022	0.044
1600 × 1000*	1880	10	0.0075	0.015	0.030	0.060
†2000 × 1000*	2236	10	0.0085	0.017	0.034	0.068
2500 × 1600*	2960	10	0.0100	0.020	0.040	0.080
250 × 250	354	5	0.0035	0.007	0.014	0.028
400 × 400	566	8	0.0040	0.008	0.016	0.032
630 × 630	891	10	0.0050	0.010	0.020	0.040
1000 × 1000*	1414	10	0.0065	0.013	0.026	0.052
Deviation expressed to			0.0005	0.001	0.001	0.001

NOTE. The nomal sizes given in column 1 are preferred. It is recommended that other sizes are ordered only when it is not practicable to adopt one of these.
* These surface plates are provided with more than three supporting feet.
†It will be seen that, with this exception, the nominal lengths of side are taken from the R5 series of preferred numbers (see BS 2045). The 2000 mm × 1000 mm surface plate is included because it is an established, widely used size.

Table 2. Permitted variations of readings of variation gauge for grades 0 to 3

Grade	Permitted variations of readings of variation gauge
	mm
0	0.004
1	0.004
2	0.016
3	0.032

NOTE. Variations are expressed to 0.002.

. . . .

11. Marking

Each surface plate shall be either legibly and permanently marked, or shall bear a designation plate attached to one side, with the following inscription in characters not less than 3 mm high:

(a) the manufacturer's name or trademark;
(b) the number of this standard, i.e. BS 817;
(c) the grade of accuracy;
(d) the year of manufacture;
(e) an identification number.

Example.

X and Co.
BS 817
Grade 0
1982
No. 1234

Appendix A

The care and use of surface plates

A.1 The surface plate is a datum and should be protected against damage. The top should be frequently wiped clean from dust and other particles. When measurements are being made, a wiping cloth should be spread on the surface plate to receive small tools and gauge blocks. When the surface plate is not in use the top should always be kept covered.

A.2 A surface plate should be located in a stirred atmosphere under constant temperature control. Accordingly, it should be kept clear of direct sunlight or sudden draughts: in particular it is important that these should not cause a vertical gradient of temperature such that the top and underside of the surface plate are at different temperatures.

A.3 The surface plate should be supported firmly and levelled. Stands should be located on a stable foundation.

A.4 Care should be taken that, wherever possible, the load on a surface plate is distributed over the working surface.

A.5 Point engagement should not be made with scraped or machined surface plates because of the high proportion of low, unrepresentative areas of the working surface.

To avoid direct point engagement with the surface plate, contact should be made through an intermediate precision gauge block, preferably not more than 10 mm tall, or a distance piece of similar precision and approximately the same surface area.

A.6 Use should be made of the full available area of the working surface and not concentrated on any one area.

The variations in local flatness of the working surface should be checked occasionally by using a variation gauge.

A.7 A common form of damage to cast iron surface plates is burrs on the working surface: the excess metal may be stoned away by attention confined to the burr.

This operation should be followed by thorough cleaning from abrasive dust.

If the surface plate is not required for some days, the surface should be coated with a corrosion preventive such as petroleum jelly.

Rusting is a sign of neglect and misuse. Its effects can be reduced by frequently wiping the top when in use and by occasionally gently rubbing with another surface plate using a paste of paraffin and a little jewellers' rouge as a lubricant.

A.8 A common form of damage to granite surface plates is cuts in the working surface: these can be minimized only by care in the use of the plates.

A.9 Surface plates wear as a result of use. The user can detect evidence of wear by rubbing the plate with a superior grade surface plate and studying the rubbed appearance, by testing for deviation from straightness along lines on the surface plate (see appendix D), and/or by using a variation gauge.

A.10 Users are advised to take advantage of the specialized services of surface plate manufacturers to have surface plates reconditioned.

Appendix B

Calculation of permitted deviation from flatness overall of the working surface

The permitted deviation, in micrometres, from flatness overall of the working surface t, for grade 0 surface plates, is calculated from the equation:

$$t = 2.5\left(1 + \frac{d}{1000}\right)$$

where

d is the nominal length of the diagonal of the surface plate (in mm) rounded up to the nearest 100 mm;

The result should be rounded to the nearest 0.5 μm.

Each succeeding grade has double the permitted deviation of the preceding grade.

Appendix C

Test for rigidity of surface plate

C.1 Apparatus (see figure 4)

C.1.1 *Beam comparator* incorporating a sensitive

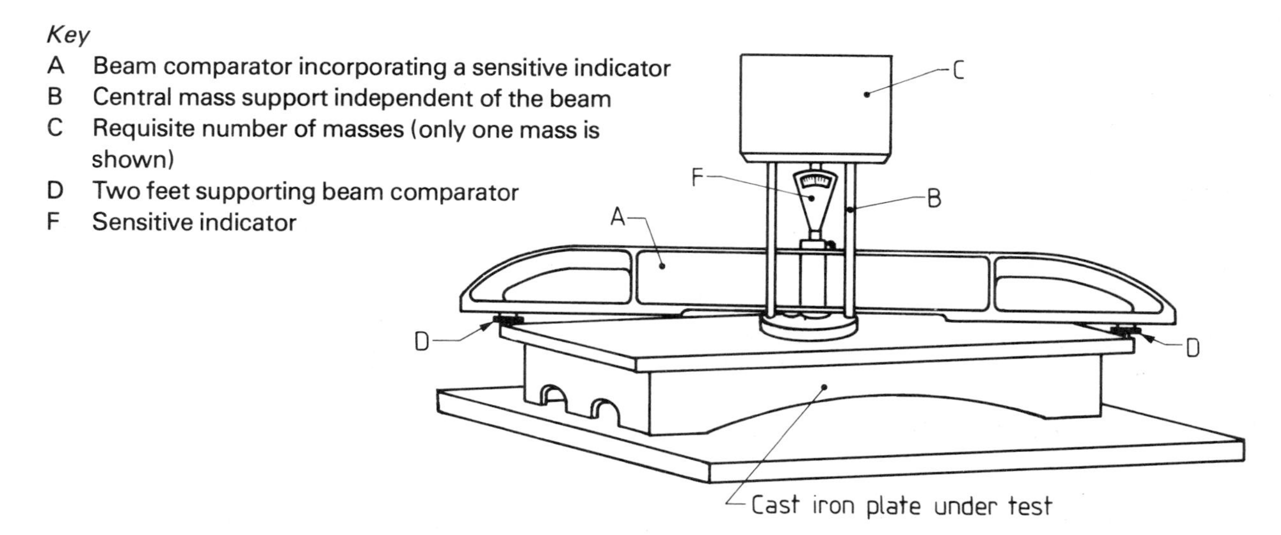

Figure 4. Cast iron surface plate being subjected to rigidity test

Key
A Beam comparator incorporating a sensitive indicator
B Central mass support independent of the beam
E Central third foot of beam comparator
F Sensitive indicator

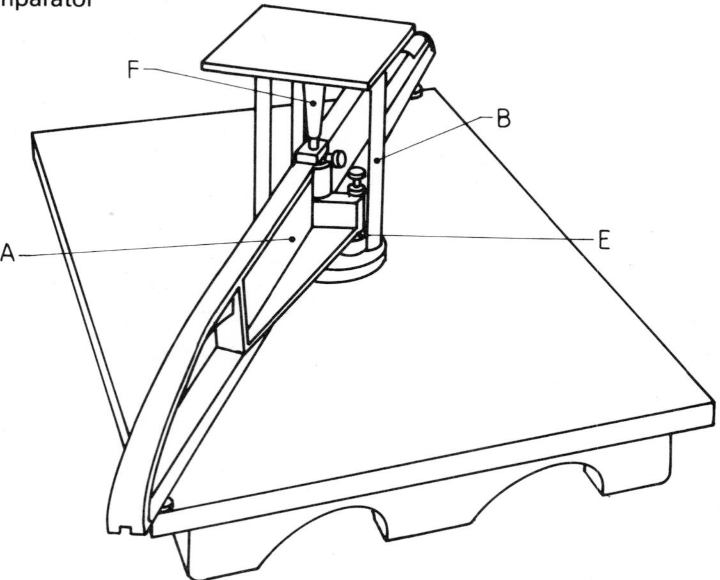

Figure 5. End-on view of rigidity test apparatus

indicator. The beam comparator is a rigid structure which is supported on two feet, the positions of which can be adjusted along the length of the beam. A third foot, which is positioned centrally along the beam and slightly offset, is provided to keep the beam stable (see figure 5). The sensitive indicator, with its contact tip pressed against the surface plate, is rigidly clamped to the centre of the beam. (If the indicator is very slightly offset from the centre line of the two supporting feet, the offset central foot can, in addition to its principal function as a stabilizer, also serve as a useful fine adjustment for setting the zero of the instrument. This offset has to be very small relative to the offset of the central foot if inaccuracies in measurement are to be minimized.)

C.1.2 *Central mass support* which is independent of the beam and can be moved, within limits, on the surface plate relative to the beam.

C.1.3 *Requisite number of masses* to form the applied load.

NOTE. Figure 5 illustrates an end-on view of the apparatus with the central support unloaded.

C.2 Procedure

C.2.1 Apply a load reasonably proportionate to the size of the surface plate and not so large as to deflect the surface plate by more than half the tolerance for deviation from flatness overall. Apply the load over a central area ranging from about a 120 mm diameter circle for the smaller surface plates to a 300 mm diameter circle for the larger surface plates.

C.2.2 Adjust the supporting feet of the beam comparator lengthwise to span the diagonal length of the surface plate under test. When the beam is in position, set the indicator to read on the surface plate and note its reading. Then load the central mass support and again note the indicator reading. Finally, repeat the initial unloaded reading. Calculate the difference between the indicator readings for the loaded and unloaded states to obtain the deflection of the surface plate under the load applied.

Appendix D

Assessment of bearing area of cast iron surface plates

D.1 Apparatus

D.1.1 *A second surface plate.*

D.1.2 *Engineer's blue.*

D.1.3 *Small glass plate* on which an area 50 mm × 50 mm has been ruled into 400 squares 2.5 mm × 2.5 mm. Such ruled glass plates can readily be produced like lantern slides by transferring on to the glass a photographic image of a chart drawn on paper.

D.2 Procedure. First blue the surface of the scraped surface plate and rub with the second surface plate so that the small bearing areas are brought clearly into view.

Then place the small glass plate upon the working surface. Observe each small square in turn and note the estimated fraction of its area (in tenths) which is occupied by a high spot of the working surface underneath.

Divide the sum of all these fractions by four to obtain the percentage of the bearing area of the working surface over the region tested.

NOTE. After a few surface plates have been tested by this method, a comparison of the few results obtained with the general appearance of the surface plates in terms of bearing area will enable a fairly close estimate of the proportion of bearing area of a surface plate to be made from its general appearance.

Appendix E

Tests for permitted deviation from flatness overall of the working surface

NOTE. Various methods are available and details can be found in technical publications. Selected methods are given in this appendix.

E.1 By comparator (grades 1, 2 and 3).

E.1.1 *Apparatus*

E.1.1.1 *Robust comparator stand* from which a measuring arm having a light operating pressure is carried on a rigid extension arm.

E.1.1.2 *Precision gauge block* or similar distance piece.

E.1.1.3 *Second surface plate* of larger area than and superior grade to the surface plate under test.

E.1.2 *Procedure*. Place the comparator stand on the second surface plate and the precision gauge block on the surface plate under test. Bring the measured head into contact with the gauge block.

Move the comparator stand on a region of the second surface plate selected for minimum deviation from flatness, whilst moving the gauge block on the surface plate under test, and record the deviation from flatness.

E.2 By straightedge (grades 1, 2 and 3)

E.2.1 *Apparatus*

E.2.1.1 *Straightedge*.

E.2.2 *Procedure*. Measure deviations from straightness along various lines parallel to the sides, and along the diagonals, by comparison with the reference straightedge. Then integrate the results into deviations from flatness by relating the results at the centre point of the surface plate where the diagonal surveys intersect, and at other points where lines of test intersect.

E.3 By straightness test using the inclination method (all grades)

E.3.1 *Apparatus*

E.3.1.1 *Exploring bridge block* of an appropriate size for exploring the surface plate in some detail.

E.3.1.2 *Spirit or electronic level, or autocollimator* giving a sensitivity of reading of at least one second of arc of tilt for grade 0 surface plates.

NOTE. Less sensitive instruments may be used on surface plates of lower grades.

E.3.2 *Procedure*. Measure the deviation from flatness by moving the exploring bridge block along a line, step by step, and recording the tilts of the block by one of the following methods.

(a) By means of a spirit level or electronic level, where the exploring bridge block is large enough to carry the instrument, the surface plate substantial enough not to be deflected by the load of the level and exploring bridge block, and the foundation sufficiently stable to remain untilted by the movements of the observer and apparatus.

(b) By autocollimator where the surface plate has to be sufficiently substantial so that it is not distorted by the exploring bridge block and where the autocollimator has to be mounted rigidly on the surface plate without distorting it.

NOTE. The results may now be integrated into permitted deviations from flatness.

The accuracy of steel toolmakers' flats and small high precision cast iron or granite surface plates is covered in BS 869 'Toolmakers' flats and high precision surface plates'. The standard includes only such requirements as are essential to ensure that the flats up to 200 mm diameter and surface plates up to 400 mm diameter will be suitable for high precision work and be sufficiently robust to retain their original accuracy.

BS 869

Toolmakers' flats and high precision surface plates

. . . .

3. Material

. . . .

3.4 Defects. The repair of defects in the working surfaces of plates is not permitted.

4. Features of design and recommended sizes

4.1 Toolmakers' flats. Toolmakers' flats shall be circular and of solid steel or granite and of an overall thickness not less than that given in table 1, column 3. Flats may have one or both surfaces finished as working surfaces. Any non-working surface may be recessed, as shown in figure 1, to approximately the dimensions given in table 1, columns 4 and 5. The base and the working face, or the two working faces, shall be parallel to within 0.0025 mm.

All sharp edges shall be removed.

It is recommended that a shallow groove be provided around the periphery of the larger flats to facilitate handling.

4.2 High precision cast iron surface plates. Cast iron plates shall be circular and of robust design, with adequate framing and ribbing underneath, so that distortion when in use is reduced to a minimum.

Each plate shall be supported on three feet which shall be smoothly machined. The plane of the feet shall be parallel to the working face to within 0.012 mm unless means of adjustment of the feet is incorporated.

The top of each plate shall project slightly, to at least 20 mm depth, beyond the framing and shall be machined round the outside.

The thickness of the top of each plate after machining and finishing shall be not less than the appropriate amount given in table 1, column 6.

The total depth of the top and framing of each plate shall be not less than the appropriate amount given in table 1, column 7.

4.3 High precision granite surface plates. These plates shall be solid and machined all over. The top of each plate shall project slightly for convenience in lifting.

The total depth of each plate shall be not less than the appropriate amount given in table 1, column 7.

4.4 Recommended sizes. The recommended sizes of flats and plates are listed in columns 1 and 2 of table 1.

5. Finish

The working surface (or surfaces) of the flats and plates shall be finished by high grade lapping* free from noticeable scratches.

Working surfaces shall be free from embedded abrasives.

All unmachined parts of cast iron plates shall be painted.

6. Accuracy

Each working surface (exclusive of the margin specified in table 2) shall everywhere lie between two parallel planes whose distance apart does not exceed the amount given in column 3 of table 2.

7. Cases for flats

Each flat shall be supplied in a case which shall provide adequate protection for the faces and edges of the flat.

*A non-wringing surface (or surfaces) may be specified by the purchaser.

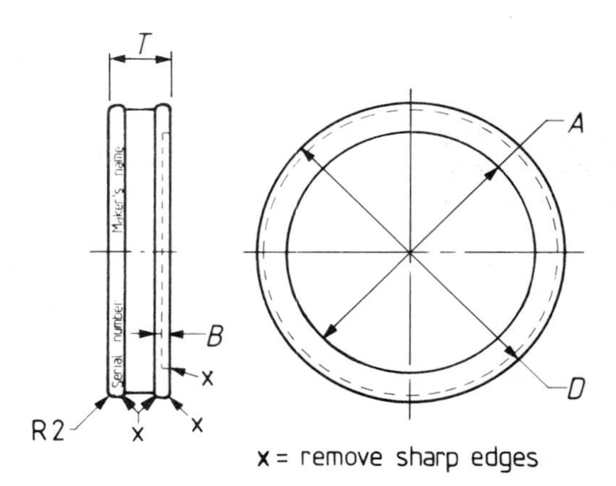

Figure 1. Toolmakers' flat showing recess in base and groove round periphery

Table 1. Dimensions of flats and plates

Dimensions in millimetres.

1	2	3	4	5	6	7
Recommended size and diameter of flat or plate *D*		**Minimum overall thickness of flat *T***	**Recess**		**Minimum thickness of top of plate**	**Minimum total depth of plate**
			Dia. *A*	**Depth *B***		
Flats	63	16	40	2	—	—
	100	20	80	3	—	—
	160	32	120	6	—	—
	200	40	160	6	—	—
Plates	250	—	—	—	20	70
	400	—	—	—	30	100

Table 2. Tolerances

1	2	3
Diameter of flat plate	**Marginal width which may be disregarded**	**Separation of limiting planes**
	mm	µm
Flats up to 200 mm	2	0.5
250 mm plates	6	0.8
400 mm plates	10	1.0

8. Covers for plates

Each plate shall be supplied with a protective cover of a suitable material which shall be so constructed as to protect both the surface and the edges of the plate.

. . . .

10. Marking

Each flat and plate shall have legibly and permanently marked on the side in characters not less than 3 mm high, the manufacturer's name or trademark and serial number.

Marking shall not be of such a nature as to impair the surface of the plate, e.g. stamping.

NOTE. As an alternative to the marking of the manufacturer's name or trademark on the machined edge, the name or mark may, in cases of cast iron plates, be legibly cast on the framing.

....

Angle plates

Angle plates, because they offer one or more right angles for obtaining squareness, are supplementary to machine tool work tables and to surface plates. They are accordingly used extensively in various manufacturing methods and processes and in inspection rooms. BS 5535 'Right angle and box angle plates' provides for three grades of angle plates (viz AA, A and B) ranging in size from 60 mm × 50 mm to 600 mm × 500 mm made from steel, cast iron or granite.

BS 5535
Right angle and box angle plates

....

3. Nomenclature and definitions

For the purposes of this standard, the term 'angle plates' shall include both right angle plates and box angle plates. The nomenclature given in figures 1, 2 and 3 has been adopted and the following definitions apply.

3.1 working faces. The term 'working faces' applies to the exterior faces, end faces and longitudinal faces, and to the interior faces when they are required to be finished accurately.

NOTE 1. See figures 1, 2 and 3.
NOTE 2. The working face of a scraped surface is defined as the bearing surface, known as high spots.

3.2 flatness tolerance. The maximum permissible distance between two parallel planes which just enclose the surface under consideration. Any departure from flatness shall be such that the angle plate shall not rock when any of its surfaces are supported on a plane.

3.3 squareness tolerance. The maximum permissible distance separating two parallel planes, which just enclose one of the surfaces under consideration and which are perpendicular to a plane in contact with the other surface. Squareness tolerance is expressed over the total length of the long surfaces.

NOTE. See figure 4.

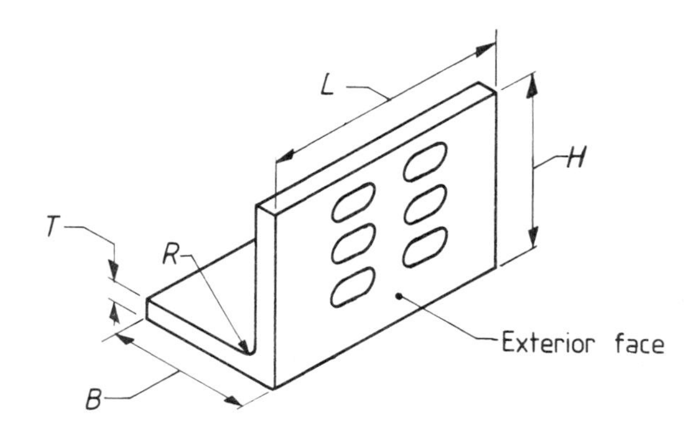

Figure 1. Right angle plate: unwebbed

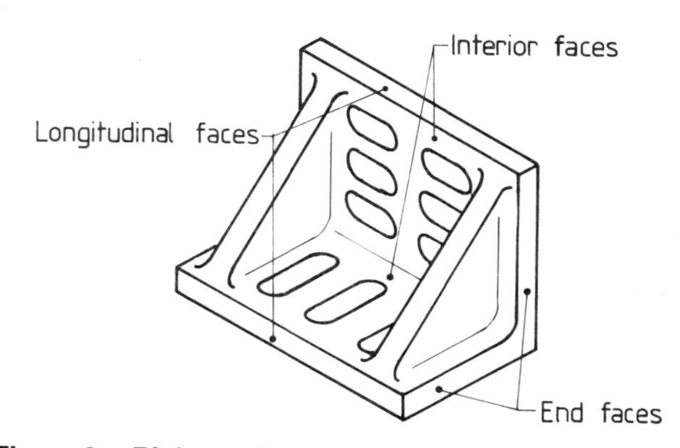

Figure 2. Right angle plate: webbed

NOTE. Designation of faces and dimensions apply to both figures 1 and 2.

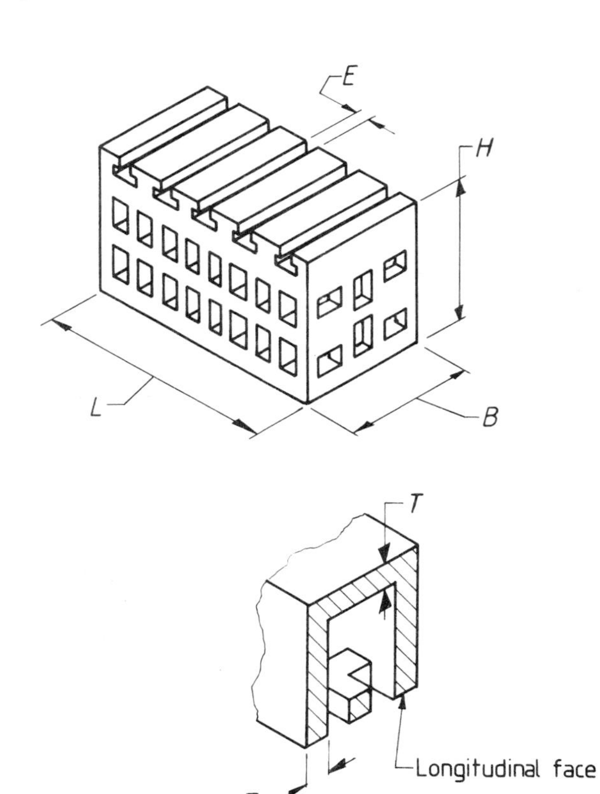

Figure 3. Box angle plate

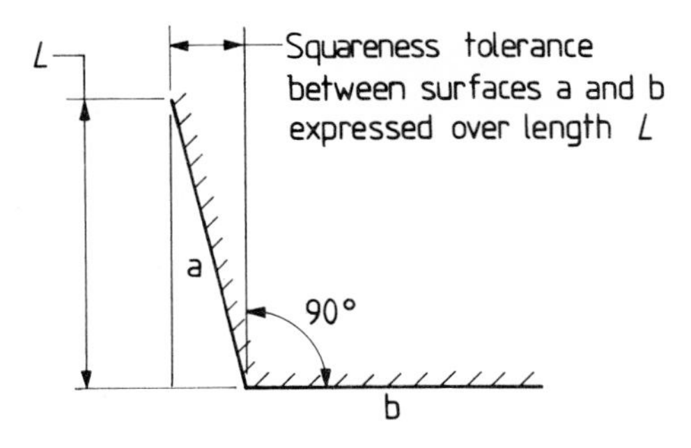

Figure 4. Exaggerated illustration of squareness tolerance

3.4 parallelism tolerance. The maximum permissible distance separating two planes which just enclose one of the surfaces and which are parallel to a plane in contact with the other surface.

. . . .

5. Finish

5.1 Grade AA and grade A plates shall be finished by lapping, fine grinding or scraping.

Cast iron plates, when finished by grinding or scraping, shall have working surfaces consisting of uniformly distributed high spots (bearing area) which give an area of not less than 20% of the total surface area.

5.2 Grade B plates shall be finished by coarse grinding or smooth machining.

5.3 All sharp edges shall be removed from steel and cast iron plates, and edges of granite plates shall be well chamfered to prevent chipping.

5.4 All non-working faces and slots of steel and cast iron plates shall be smooth and painted.

5.5 When the interior faces of unwebbed plates are required to be finished to high accuracies, the accuracies controlling the exterior faces shall apply.

. . . .

7. Recommended sizes

The first and second choices of recommended sizes are shown in column 1 of table 1. They are also the designation sizes and generally correspond with the R10 series of preferred numbers (see BS 2045). Should it be necessary to extend the range of sizes in either direction, it is recommended that this be done by continuing with the relevant number from the series. If sizes other

Table 1. General dimensions of right angle plates and box angle plates

Dimensions in millimetres

1	2	3	4	5	6	7	8	9	10
				T min.					
				Steel or cast iron right angle plates		Box angle plates	Granite angle plates		
Designation size	*H*	*B*	*L*	Webbed	Unwebbed			Radius *R*	Width *E* (for tee slots) See BS 2485
60 × 50	60	50	80	12	16	10	30	8	6
(80 × 60)	80	60	100	15	18	10	40	8	8
100 × 80	100	80	125	18	20	13	50	8	10
(125 × 100)	125	100	160	20	22	13	60	8	14
160 × 125	160	125	200	22	25	16	75	10	14
(200 × 160)	200	160	250	25	30	20	100	10	18
250 × 200	250	200	315	30	35	20	100	10	18
(300 × 250)	300	250	400	35	40	20	100	13	22
400 × 300	400	300	500	35	—	25	125	—	22
(500 × 400)	500	400	630	35	—	30	125	—	22
600 × 500	600	500	800	40	—	35	150	—	28

NOTE. Sizes in brackets are second choice.

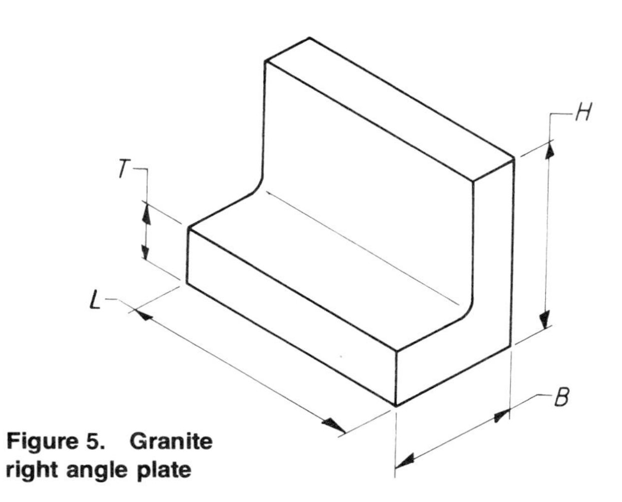

Figure 5. Granite right angle plate

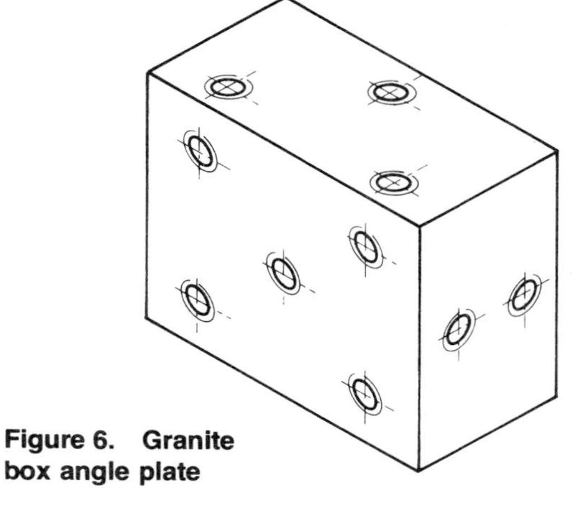

Figure 6. Granite box angle plate

than those shown in table 1 are required or supplied, the tolerances for the next standard size lower are to be applied.

NOTE. Threaded inserts are optional, see **6.2.2,** not included in this manual.

8. Accuracy

The tolerances for flatness, squareness and parallelism shall be as shown in table 2.

9. Matched pairs of box angle plates

Matched pairs of box angle plates shall comply with tolerances equivalent to those shown in column 8, 9 or 10 of table 2 for equality of corresponding dimensions *L*, *B* and *H*.

The form, dimensions and positions of slots shall be the same on each plate of a matched pair.

. . . .

Table 2. Tolerances for right angle plates and box angle plates

Dimensions in millimetres

1	2	3	4	5	6	7	8	9	10	11	12	13
	Flatness of working faces			**Mutual squareness of working faces as measured over dimension *H***			**Parallelism of working longitudinal faces and opposite end faces over their total length**			**Squareness of working faces to end faces as measured over dimension *L***		
Designation size	**AA**	**A**	**B**	**AA**	**A**	**B**	**AA**	**A***	**B***	**AA**	**A**	**B**
60 × 50	0.0015	0.003	0.015	0.004	0.008	0.040	0.0013 per 25 mm	0.008	0.046	0.004	0.008	0.045
(80 × 60)	0.002	0.004	0.020	0.004	0.008	0.040		0.010	0.054	0.005	0.010	0.050
100 × 80	0.0025	0.005	0.025	0.005	0.010	0.050		0.013	0.065	0.006	0.013	0.060
(125 × 100)	0.0025	0.005	0.025	0.005	0.010	0.050		0.013	0.065	0.006	0.013	0.065
160 × 125	0.003	0.006	0.030	0.006	0.012	0.060		0.015	0.075	0.007	0.015	0.075
(200 × 160)	0.004	0.008	0.035	0.007	0.015	0.075		0.017	0.085	0.008	0.017	0.085
250 × 200	0.004	0.008	0.035	0.007	0.015	0.075		0.017	0.085	0.008	0.017	0.085
(300 × 250)	0.005	0.010	0.050	0.008	0.017	0.085		0.020	0.100	0.010	0.020	0.100
400 × 300	0.005	0.010	0.050	0.009	0.017	0.085		0.020	0.100	0.010	0.020	0.100
(500 × 400)	0.005	0.010	0.050	0.009	0.017	0.085		0.020	0.100	0.010	0.020	0.100
600 × 500	0.006	0.012	0.060	0.010	0.020	0.100		0.022	0.110	0.011	0.022	0.110

NOTE. Sizes in brackets are second choice.

*These tolerances represent longitudinal parallelism. Transversely the tolerance given shall be a quarter of the value shown for the total length over dimension *T*.

11. Marking

Each right angle plate or box angle plate shall be legibly and permanently marked with the particulars given below.

The marking shall be applied in such a manner that it does not affect the accuracy of the plate.

(a) The manufacturer's name or trademark.
(b) The number of this British Standard (BS 5535).
(c) Grade of plate. (Grade AA, A or B.)
(d) Matched pairs of box angle plates shall bear a common serial number. In addition, all angle plates and box angle plates which are certified by a testing authority are required to bear a serial number.

. . . .

Engineers' squares

BS 939 'Engineers' squares (including cylindrical and block squares)' specifies requirements for engineers' try squares (excluding adjustable squares), cylindrical squares and block squares of solid or open form. The standard is in five sections as follows:

Section one	General requirements applicable to all types of squares
Section two	Engineers' try squares in three grades of accuracy (viz AA, A and B) with inner blade lengths from 50 mm to 1000 mm
Section three	Cylindrical squares in accuracy grade AA in sizes from 75 mm to 750 mm
Section four	Solid form block squares in accuracy grade AA and A in sizes from 50 mm × 40 mm to 1000 mm × 1000 mm
Section five	Open form block squares in accuracy grades A and B in sizes from 150 mm × 100 mm to 600 mm × 400 mm

---

BS 939

Engineers' squares (including cylindrical and block squares)

Section one. General

. . . .

3. Nomenclature and definitions

For the purposes of this British Standard the nomenclature given in figures 2, 6, 7 and 8 for the various types of square has been adopted, and the following definitions apply.

3.1 deviation from straightness and flatness. The minimum distance between two parallel planes that just enclose the surface under consideration.

3.2 deviation from parallelism. The difference between the maximum and minimum distances separating the surfaces or edges under consideration.

3.3 squareness of two surfaces

3.3.1 *deviation from squareness.* The minimum distance between two parallel planes that will just enclose one surface (A), including any straightness error of surface (A), and that are perpendicular to a datum plane in contact with the other surface (B) (see figure 1).

3.3.2 *squareness tolerance.* The maximum permissible deviation from squareness (see figure 1).

3.4 tolerances. The maximum permissible deviations from straightness, flatness, parallelism and squareness.

. . . .

Section two. Engineers' try squares

5. Designating size

The designating size of an engineers' try square is the length of the blade from its tip to the inner working face of the stock (see figure 2).

6. Recommended sizes

Recommended sizes of engineers' try squares are:

50, 75, 100, 150, 200, 300, 450, 600, 800 and 1000 mm.

Squares of other sizes should be ordered only when it is not practicable to adopt one of the recommended sizes. If squares of intermediate sizes are supplied they should be made to the same tolerances as those specified for the next smaller recommended size.

. . . .

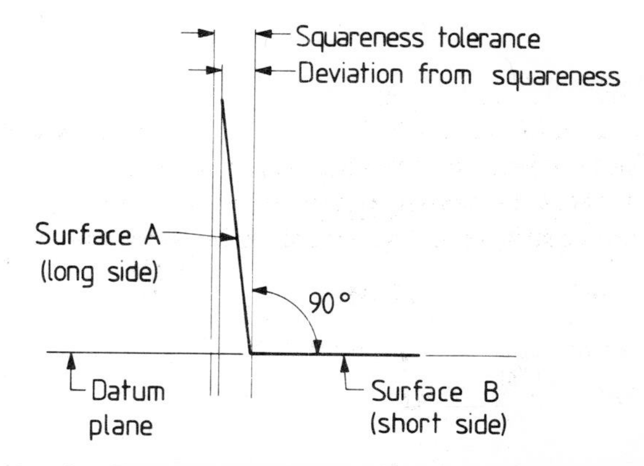

Figure 1. Exaggerated illustration of deviation from squareness and squareness tolerance

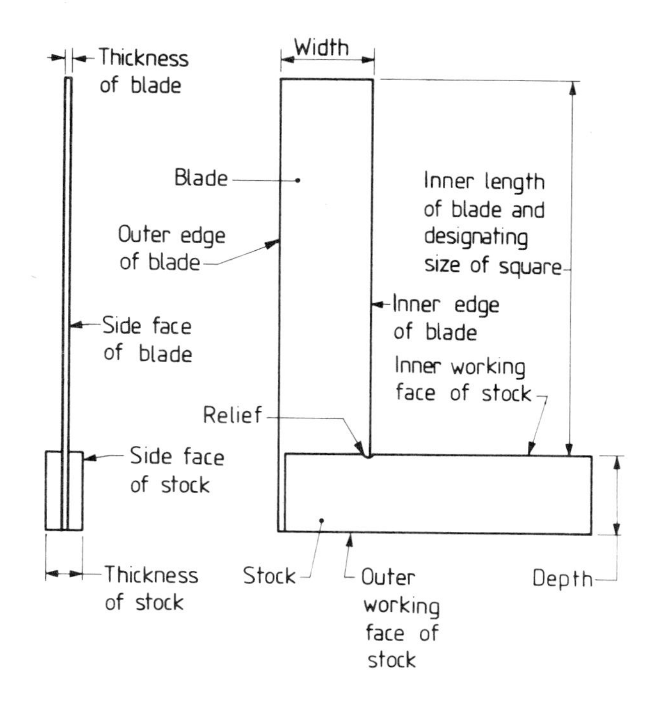

Figure 2. Nomenclature for engineers' try squares, also showing alternative form of relief

9. Finish

All working surfaces of the blade and stock of the three grades of engineers' try square shall have a lapped, finely ground or polished finish, having a surface roughness value of 1.0 μm R_a maximum, and all sharp edges shall be removed (see BS 1134).

10. Accuracy

10.1 Straightness of inner and outer edges of blade. Each edge of the blade shall be straight to within the tolerance specified in table 1 when the outer working face or side face of the stock is resting on a horizontal surface.

10.2 Parallelism of blade edges. The blade edges shall be parallel to each other to within the tolerances specified in table 2.

10.3 Flatness of working faces of stock. Each working face of the stock shall be flat to within the tolerance specified in table 3. Deviations from flatness shall be of concave nature to prevent rock.

10.4 Parallelism of working faces of stock. The working faces of the stock shall be parallel to within the tolerances specified in table 4.

Table 1. Tolerances on straightness of blade edges

1	2	3	4
	Tolerance		
Size of square	**Grade AA**	**Grade A**	**Grade B**
mm	μm	μm	μm
50 and 75	2	4	8
100 and 150	2	4	8
200	2	4	8
300	3	6	12
450	4	8	16
600	6	12	24
800	8	16	32
1000	10	20	40

Table 2. Tolerances on parallelism of blade edges

1	2	3	4
	Tolerance		
Size of square	**Grade AA**	**Grade A**	**Grade B**
mm	μm	μm	μm
Up to 150	3	5	8
200 and 300	3	8	12
400 and 600	7	12	18
800 and 1000	12	18	24

Table 3. Tolerances on flatness of working faces of stock

1	2	3	4
	Tolerance		
Size of square	**Grade AA**	**Grade A**	**Grade B**
mm	μm	μm	μm
50 and 75	1	2	4
100 and 150	1.5	3	6
200	2	4	8
300	2	4	8
450	2.5	5	10
600	4	8	16
800	5	10	20
1000	6	12	24

Table 4. Tolerances on parallelism of working faces of stock

1	2	3	4
	Tolerance		
Size of square	**Grade AA**	**Grade A**	**Grade B**
mm	μm	μm	μm
Up to 150	2	3.5	5
200 and 300	3	5	8
450 and 600	5	8	12
800 and 1000	8	12	16

10.5 Parallelism of side faces of stock. The side faces of the stock shall be parallel to within the tolerance specified in table 5.

10.6 Squareness of side faces of stock to working faces of stock. The side faces of the stock shall be square to the working faces as follows.

Grade AA to within 1 μm/mm
Grade A to within 2 μm/mm
Grade B to within 3 μm/mm

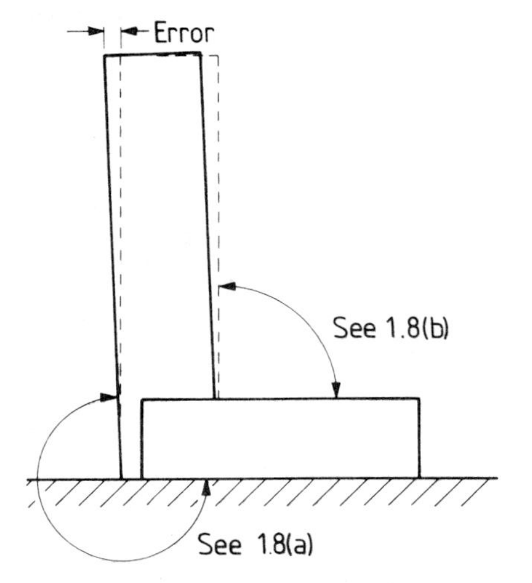

Figure 3. Squareness of blade edges to working faces of stock (see **10.8**)

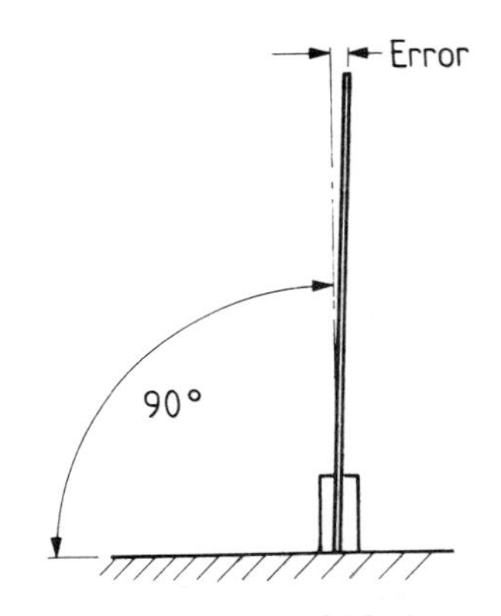

Figure 4. Lateral squareness of blade (see **10.9**)

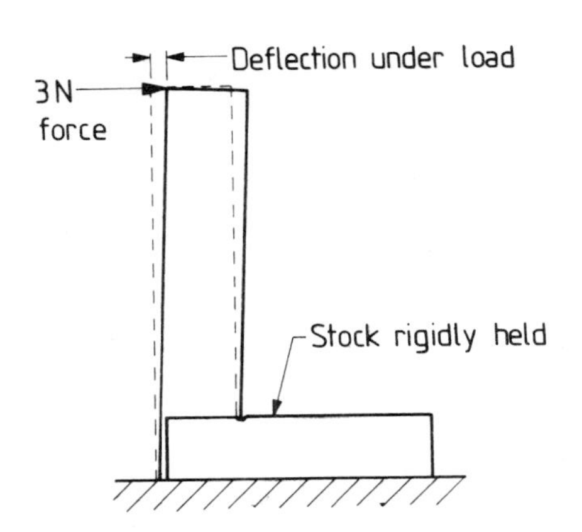

Figure 5. Rigidity

10.7 Squareness of edges of blade to side faces of stock. The inner and outer edges of the blade shall be square to the side faces of the stock to within 15 μm over the thickness of the edges.

10.8 Squareness of edges of blade to working faces of stock. When the square is placed upright on a horizontal surface the maximum departure from squareness of

(a) the outer edge of the blade to the outer

Table 5. Tolerances on parallelism of side faces of stock

Size of square	Tolerance
	All grades
mm	μm
Up to and including 150	25
200 and 300	50
450 and 600	100
800 and 1000	150

Table 6. Tolerances on squareness of blade edges to working faces of stock

1	2	3	4
	Tolerance measured at tip of blade		
Size of square	Grade AA	Grade A	Grade B
mm	μm	μm	μm
50 and 75	4	8	16
100 and 150	4	8	16
200	4	8	16
300	6	12	24
450	8	16	32
600	12	24	48
800	16	32	64
1000	20	40	80

Table 7. Tolerances on lateral squareness of blade to base of stock

1	2	3	4
	Tolerance measured at tip of blade		
Size of square	Grade AA	Grade A	Grade B
mm	μm	μm	μm
50 and 75	50	75	125
100 and 150	75	125	250
200 and 300	125	200	375
450	200	300	500
600	250	300	500
800	300	375	600
1000	300	375	650

working face of the stock, and
(b) the inner edge of the blade to the inner working face of the stock

shall not exceed the tolerance specified in table 6 (see figure 3).

10.9 Lateral squareness of blade. The blade shall be laterally square to the base of the stock (see figure 4) to within the tolerances specified in table 7.

11. Rigidity

When the stock of the try square is rigidly clamped and a force of 3 N is applied to the free end of the blade in a horizontal direction parallel to the length of the stock (see figure 5) the end of the blade shall not deflect in this direction by an amount greater than the tolerance specified in table 6.

NOTE. This test is independent of any errors in the try square in its 'free' condition.

12. Marking

Each try square shall have legibly and permanently marked upon it the following particulars:

(a) the number of this British Standard, i.e. BS 939;
(b) the grade designation;
(c) an identification number for grade AA try squares;
(d) the manufacturer's name or trademark.

Section three. Cylindrical squares

13. Designating size

The designating size of a cylindrical square is its length (see figure 6).

14. Recommended sizes and dimensions

Recommended sizes and dimensions of cylindrical squares are specified in table 8. Squares of other sizes should be ordered only when it is not practicable to adopt one of the recommended sizes. If squares of intermediate sizes are supplied they should be made to the same tolerances as those specified for the next smaller recommended size.

. . . .

17. Finish

External surfaces of cylindrical squares shall have a lapped or finely ground finish.

18. Accuracy

18.1 General. Cylindrical squares shall be accurate to within the tolerances specified in table 9.

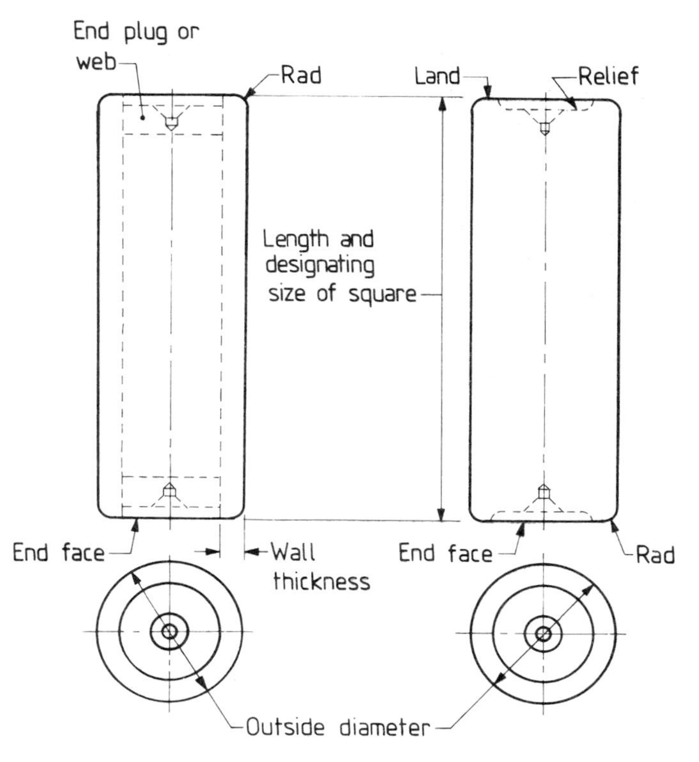

Figure 6. Nomenclature for cylindrical squares

Table 8. Recommended sizes and dimensions of cylindrical squares

1	2	3
Length	**Outside diameter**	**Wall thickness** (hollow section, see **16.1**)
mm	mm	mm
75	50	—
150	70	—
220	90	—
300	95	12
450	120	16
600	140	20
750	160	22

Table 9. Tolerances on cylindrical squares

1	2	3	4
Size of square	**Straightness of sides of cylinder**	**Flatness of end faces**	**Squareness of end faces with cylindrical surface over length of cylinder**
mm	μm	μm	μm
75	1	1	1.5
150	1.5	1.5	2
220	2	2	3
300	3	2.5	5
450	4.5	4	7
600	6	5	9
750	7.5	6	11

18.2 Straightness of sides. The sides of the cylinder shall be straight when the square is standing on its base, or when supported horizontally at the positions of minimum flexure, to within the tolerances specified in column 2 of table 9.

18.3 Flatness of ends. The ends shall be flat within the tolerances specified in column 3 of table 9. Any departure from flatness shall be of a concave configuration.

18.4 Squareness of cylindrical surface to end faces. The squareness of the cylindrical surface shall be tested by placing the cylinder on a grade AA or A surface plate, first on one end face and then on the other*. The maximum error found in the squareness of the cylindrical surface along any generator to the surface plate shall not exceed the tolerances specified in column 4 of table 9.

19. Marking

Each square shall have legibly and permanently marked upon it the following particulars, and the marking shall not protrude above the working surfaces of the square:

(a) the number of this British Standard, i.e. BS 939;
(b) the designating size and grade AA;
(c) an identification number;
(d) the manufacturer's name or trademark.

Section four. Solid form block squares

20. Designating size

The designating size of a solid form block square is its length by its width in millimetres, e.g. 150 mm x 100 mm (see figure 7).

21. Recommended sizes and dimensions

Recommended sizes and dimensions of solid form block squares are specified in table 10. Squares of other sizes should be ordered only when it is not practicable to adopt one of the recommended sizes. If squares of intermediate sizes are supplied they should be made to the same tolerances as those specified for the next smaller recommended size.

. . . .

24. Finish

24.1 All working faces of solid form steel block squares shall have a lapped finish, preferably of high reflectivity, suitable for optical applications. The front and back surfaces shall have a lapped or finely ground finish.

24.2 All working faces of solid form granite or cast iron block squares shall have a lapped or finely ground finish and shall be free from surface defects.

24.3 All sharp edges shall be removed and corners shall be rounded.

25. Accuracy

All working faces of solid form block squares shall be accurate to within the tolerances specified in tables 11 and 12.

*Granite cylindrical squares may be designated 'square to base end only', one end being required for marking purposes.

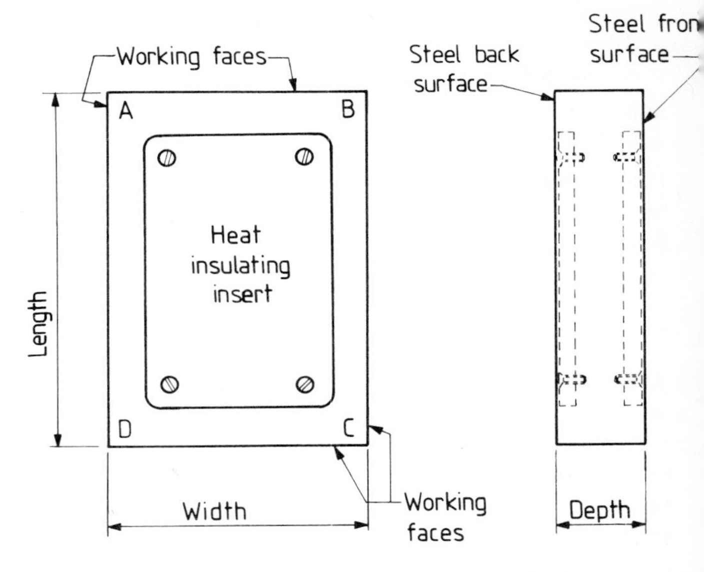

Figure 7. Nomenclature for solid form block squares

Table 10. Recommended sizes and dimensions of solid form block squares

1	2	3	4
	Length	**Width**	**Depth**
	mm	mm	mm
Steel squares	50	40	12
	75	50	14
	100	75	20
	150	100	22
Granite (diabase) or cast iron squares	150	150	50
	250	250	75
	350	350	75
	450	450	75
	600	600	100
	750	750	100
	1000	1000	100

Table 11. Tolerances on flatness, parallelism and squareness of working faces (solid iron block squares)

1	2	3	4	5	6	7
	Tolerance					
	Grade AA			Grade A		
Size of square	Flatness	Parallelism	Squareness over length	Flatness	Parallelism	Squareness over length
mm	μm	μm	μm	μm	μm	μm
50 × 40	0.5	0.8	1.5	1	1.5	3
75 × 50	0.5	0.8	1.5	1	1.5	3
100 × 75	1	1	1.5	2	2	3
150 × 100	1.5	1.5	2	3	3	4
150 × 150	1.5	1.5	2	3	3	4
250 × 250	2.5	2.5	4	5	5	8
350 × 350	3.5	3.5	5	7	7	10
450 × 450	4.5	4.5	7	9	9	14
600 × 600	6	6	9	12	12	18
750 × 750	7.5	7.5	11	15	15	22
1000 × 1000	10	10	15	20	20	30

Table 12. Tolerances on flatness and squareness of front and back surfaces

1	2	3	4	5
	Tolerances			
	Grade AA		Grade A	
Size of square	Overall flatness of front and back surfaces	Squareness over depth of working face	Overall flatness of front and back surfaces	Squareness over depth of working face
mm	μm	μm	μm	μm
Up to 150 × 100	5	3	10	6
Above 150 × 100 to 450 × 450	8	5	16	10
Above 450 × 450	12	10	24	20

26. Marking

Each solid form block square shall have legibly and permanently marked on it the following particulars. The marking may be on the front heat insulating insert of steel squares or on an insert in a recess on granite squares. The marking or insert shall not protrude above the front surface:

(a) the number of this British Standard, i.e. BS 939;
(b) the designating size;
(c) the grade designation AA or A, as appropriate;
(d) an identification number for grade AA squares;
(e) the manufacturer's name or trademark.

In addition, the four corners of the front surface shall be identified by the letters A, B, C and D respectively.

Section five. Open form block squares

27. Designating size

The designating size of an open form block square is its length by its width in millimetres (see figure 8).

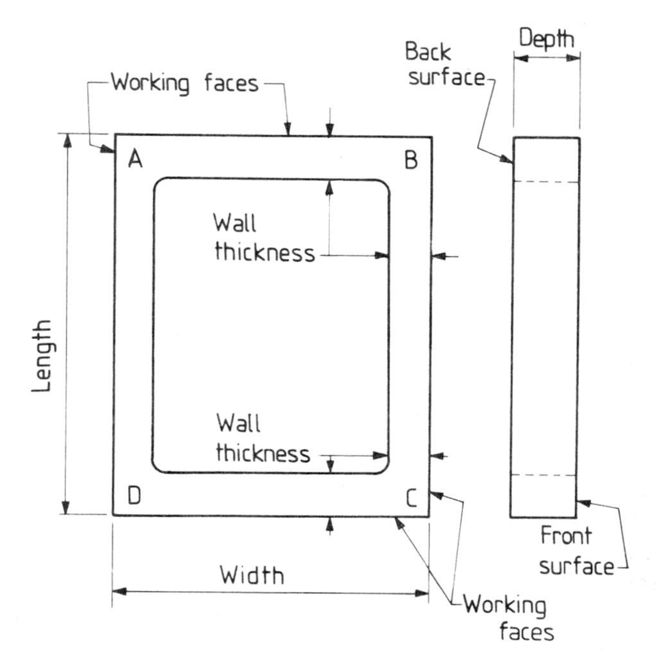

Figure 8. Nomenclature for open form block squares

28. Recommended sizes and dimensions

Recommended sizes and dimensions of open form block squares are given in table 13.

Squares of other sizes should be ordered only when it is not practicable to adopt one of the recommended sizes. If squares of intermediate sizes are supplied they should be made to the same tolerances as those specified for the next smaller recommended size.

....

31. Finish

The working faces and front and back surfaces of open form block squares shall have a lapped or finely ground finish.

All sharp edges shall be removed and corners shall be rounded.

The inside surfaces shall have the skin removed and shall be painted.

Table 13. Recommended sizes and dimensions of open form block squares

1	2	3	4
Dimensions			
Length	**Width**	**Depth**	**Wall thickness**
mm	mm	mm	mm
150	100	50	20
200	125	50	25
250	150	50	30
300	200	50	35
450	300	75	40
600	400	75	45

Table 14. Tolerances on flatness, parallelism and squareness of working faces

1	2	3	4	5	6	7
	Tolerance					
	Grade A			**Grade B**		
Size of square	**Flatness**	**Parallelism**	**Squareness over length**	**Flatness**	**Parallelism**	**Squareness over length**
mm	μm	μm	μm	μm	μm	μm
150 × 100	2.5	4	5	5	8	10
200 × 125	3	4.5	6	6	9	12
250 × 150	4	6	8	8	12	16
300 × 200	4.5	7	9	9	14	18
450 × 300	7	10	14	14	20	28
600 × 400	9	14	18	18	28	36

32. Accuracy

32.1 General. All working faces of open form block squares shall be accurate to within the tolerances specified in tables 14 and 15.

32.2 Open form block squares supplied in matched pairs. When open form block squares are supplied in matched pairs, their mean lengths and widths shall be respectively equal to within the following tolerances.

Grade A: 5 μm for sizes up to 300 mm × 200 mm and 10 μm for larger sizes.

Grade B: 10 μm for sizes up to 300 mm × 200 mm and 20 μm for larger sizes.

Table 15. Tolerances on flatness and squareness of front and back surfaces

1	2	3	4	5
	Tolerances			
	Grade A		**Grade B**	
Size of square	**Overall flatness of front and back surfaces**	**Squareness over depth of working face**	**Overall flatness of front and back surfaces**	**Squareness over depth of working face**
mm	μm	μm	μm	μm
Up to 450 × 300	16	6	32	12
Above 450 × 300	24	20	48	40

33. Marking

Each open form block square shall be legibly and permanently marked on the front surface with the following particulars:

(a) the number of this British Standard, i.e. BS 939;
(b) the designating size;
(c) grade A or B, as appropriate;
(d) an identification number*;
(e) the manufacturer's name or trademark.

In addition, the four corners on the front surface shall be marked with the letters A, B, C and D respectively.

. . . .

Appendix B

Notes on the manufacture and testing of squares

B.1 General. There are numerous recognized methods of testing squares, and further information may be found in appropriate text books. The recommendations given in this appendix are offered for general guidance in the hope that they may prove helpful to users of the standard.

This appendix deals with:

(a) cylindrical and block squares;
(b) testing of engineers' try squares;
(c) the rigidity and design of engineers' try squares

B.2 Cylindrical and block squares

B.2.1 *General.* Cylindrical and block squares possess advantages in that they can be manufactured and tested initially from first principles by simple methods; the parallelism of their sides renders them, in a sense, self-checking both during manufacture and during use at a later stage; finally, their form is such as to inspire confidence that their accuracy will be maintained with reasonable use provided the material of which they are made has been stress relieved.

It therefore follows that one of the most satisfactory methods of testing the accuracy of an engineers' try square is by comparison with a known square of the cylindrical or block type.

B.2.2 *Cylindrical squares.* In manufacturing this type of square, the end faces and the cylindrical surface of the square should be ground at one setting with the cylinder mounted between centres. Care has to be taken to grind the cylindrical surface truly parallel. If true parallelism is achieved (and this may be readily checked with a comparator) it follows

*Matched pairs should have the same identification number followed by a suffix a or b, e.g. 312a, 312b.

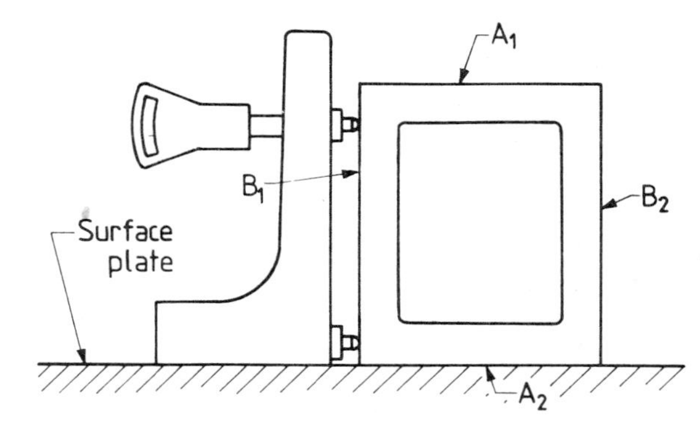

Figure 10. Method of testing squareness of block squares

that the cylindrical surface should be truly square with the two end faces. However, should a small residual error in squareness remain, the effect of such an error may be eliminated in use by making observations on diametrically opposite generators of the cylinder.

B.2.3 *Block squares.* (see figure 10). During manufacture, a pair of opposite faces of block squares, e.g. A_1 and A_2, can be 'spot ground'* so as to achieve accurate parallelism between the opposite faces. The accuracy of parallelism is checked by sliding the block square on a grade AA surface plate under a sensitive indicator.

During the process of grinding the second pair of faces, B_1 and B_2, their squareness to faces A_1 and A_2 is tested on a grade A surface plate by means of a simple form of 'squareness tester' (see figure 10). This consists of a rigid angle block, the vertical arm of which is fitted with a transverse straightedge near the bottom and a sensitive indicator towards the top. By this means, any out-of-squareness of face B_1 with respect to faces A_1 and A_2 can be determined by taking the mean of readings on B_1 with the block standing first on A_1 and then on A_2. After correcting B_1 for squareness, B_2 can readily be made parallel to B_1.

It may be mentioned that any residual error in the squareness of the block is revealed two-fold by the indicator of the squareness tester.

B.3 Engineers' try squares

B.3.1 *Testing an engineers' try square against a reference cylindrical square or block square.* When testing an engineers' try square the reference cylinder (or block) and the square are stood side by side on a grade A surface plate. The square is slid gently into contact with one side of the cylinder (or block) and the fit between the latter and the outer edge of the square is sighted against a well-illuminated background. If a tapering slit of light is seen, the magnitude of the error present in

*The process of 'spot grinding' consists of sliding the workpiece about by hand on a surface plate under the edge of a grinding wheel.

the square can be ascertained by tilting it with gauge blocks inserted under the two ends of the stock until a light-tight fit is achieved from top to bottom of the blade. The error can then be found from the difference between the two gauge blocks and their distance apart.

If the square under test has a knife-edge blade, it is preferable to use a block square with flat sides for a reference, as shown in figure 10. With flat-edged blades, however, the sensitivity of the test would be increased by using a cylindrical square for reference, as in figure 6.

Whichever type of square is used as a reference, it is important to set the plane of the blade of the engineers' try square normal to the surface of the reference (in plan view) in order to eliminate errors that might arise owing to the blade of the square being out-of-square laterally (see figure 4).

Assuming accurate parallelism between the opposite faces of the block square, the effect of any small residual error in its squareness may be eliminated by offering up in turn the opposite faces of the block square to the square under test and taking a mean of the results. This procedure is also applicable when a cylindrical square is used as the reference.

Besides their use for testing engineers' try squares, cylindrical and block squares can be used directly with advantage in assembling and testing the squareness of machine tool members and their movements and for checking squareness of tools and components in inspection rooms and workshops.

B.3.2 *Tilting square-tester.* A method of testing an engineers' try square without reference to a block square or master square of any sort is by means of a truly parallel-sided straightedge, A (see figure 11), held upright on a grade AA or A surface plate in a tiltable support that permits the straightedge to be slightly inclined in its own plane on either side of the perpendicular. The square, S, to be tested is stood alongside the straightedge and the perpendicularity of the latter is adjusted by a vertical micrometer screw, B, which forms one of the three feet of the support, until there is a light-tight fit between the outer edge of the blade and the adjacent side of the straightedge. The square is then transferred to the other edge of the straightedge. If a light-tight fit is also obtained against this edge of the straightedge, the square is quite true. On the other hand, if a tapered slit of light is seen against the second edge, the direction of the error in the square becomes immediately apparent. The magnitude of the error is shown twofold. It can be measured very accurately by readjusting the micrometer, B, so as to obtain a light-tight fit against the second edge of the straightedge and noting the difference between the micrometer reading in this position and that obtaining previously with the square in contact with the first edge. One half of this difference divided by the horizontal normal distance between the axis of the micrometer and the axis of the tilt of the ball-feet, C, of the support gives the angular error of the square per unit of length.

B.3.3 *Rigidity of engineers' try squares.* When testing or using an engineers' try square, the blade should not be pressed heavily against the reference surface or the surface under test. If the blade should happen to make contact at its tip, there is a risk that undue pressure would bend the blade and so give rise to a false impression from the test.

B.3.4 *Notes on design of engineers' try squares.* If a force of P newtons is applied to the tip of the blade of a square, as shown in figure 12, a deflection is produced equal to

$$2 \times 10^{-5} \times \frac{PL^3}{bh^3} \text{ mm}$$

This deflection may be particularly noticeable in larger-sized squares where the blade dimensions ordinarily met with in practice may often not provide adequate stiffness in relation to the length of the blade.

For the reason just stated, engineers' try squares of ordinary design, with blade lengths above about

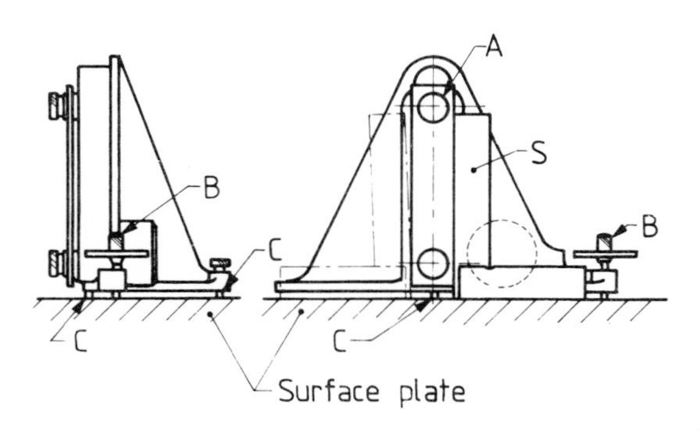

Figure 11. Tilting square tester

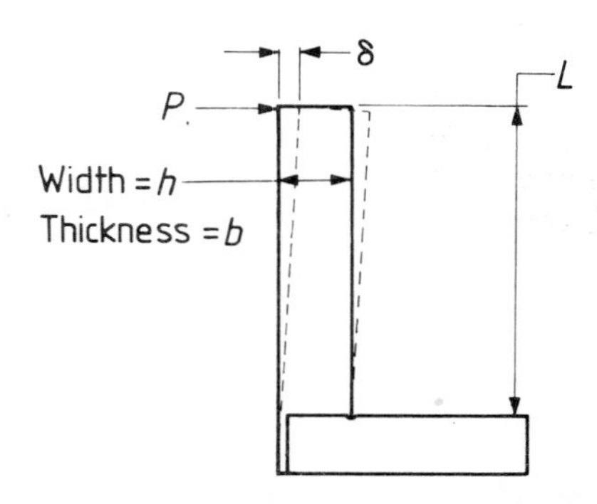

Figure 12. Testing for rigidity

300 mm, are not best suited for accurate reference purposes. For preference, larger-sized reference squares should take the form of cylindrical or block squares, or else the ordinary design should be so modified as to provide the necessary rigidity.

One obvious suggestion towards improving the design of engineers' try squares is to make the blades of larger squares wider and of I-section. Another means of increasing the rigidity of the blade would be to add a stiffening member across the hypotenuse of the triangle. The addition of this member would tend to render the inside right-angle of the square useless, but in the case of squares above 300 mm this would probably not be a serious disadvantage, since they are used almost exclusively on the outside angle.

Engineers' parallels

BS 906 'Engineers' parallels', Part 1 'Metric units', provides for grades A and B.

BS 906

Engineers' parallels

Part 1. Metric units

. . . .

Specification

1. Scope

This British Standard applies to parallel sided steel blocks of six sizes in grade A and eight sizes in grade B, namely:

Grade A

5 mm × 10 mm × 100 mm
10 mm × 20 mm × 125 mm
15 mm × 30 mm × 150 mm
20 mm × 40 mm × 200 mm
25 mm × 50 mm × 250 mm
30 mm × 60 mm × 300 mm

Grade B

5 mm × 10 mm × 100 mm
10 mm × 20 mm × 125 mm
15 mm × 30 mm × 150 mm
20 mm × 40 mm × 200 mm
25 mm × 50 mm × 250 mm
30 mm × 60 mm × 300 mm
40 mm × 80 mm × 350 mm
50 mm × 100 mm × 400 mm

. . . .

3. Finish

The sides of grades A and B parallels shall have a finely ground, or lapped finish. The ends of both grades of parallels shall be smoothly finished by milling or grinding.

All sharp edges shall be removed.

4. Accuracy

4.1 Each parallel and pair of parallels shall comply with the requirements for accuracy for the appropriate grade stated in tables 1 and 2.

4.2 Additionally, the maximum errors in squareness of adjacent width and thickness faces shall not exceed 0.005 mm per 25 mm for both grade A and grade B parallels. (For widths less than 25 mm the tolerance is 0.005 mm over the width *W* for both grades.) (See tables 1 and 2.)

5. Marking

Each parallel shall have legibly and permanently marked on it the following information:

the nominal dimensions of *T* and *W* (see tables 1 and 2);
the grade of accuracy;
the manufacturer's name or trademark;
the serial number common to each parallel of the pair.

e.g.
20 mm × 40 mm
A
X & Co.
1234

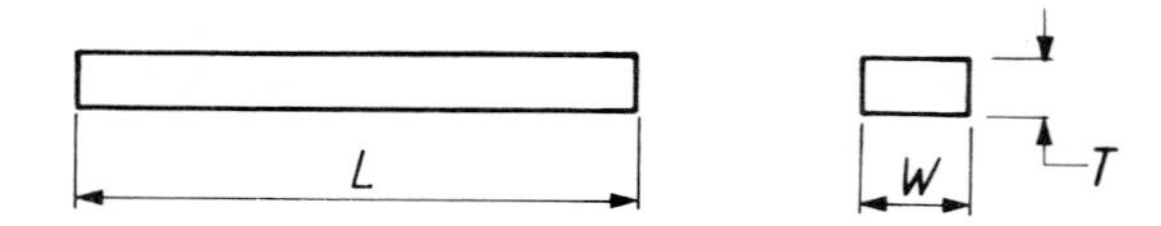

Table 1. Tolerances on grade A parallels

1	2	3	4	5	6	7
			Calliper measurements for individual parallels (see figure 1)		**Functional requirements (see figure 2)**	
Nominal size					**for individual parallels**	**for a pair**
T	***W***	***L***	**Departure from nominal *T* and *W***	**Variations in calliper measurements of *T* and *W***	**parallelism of *T* and *W***	**matching**
mm	mm	mm	mm	mm	mm	mm
5 10 15 20	10 20 30 40	100 125 150 200	±0.005	0.002	0.004	0.006
25 30	50 60	250 300	±0.010	0.003	0.006	0.010

NOTE. Tolerance on *L* is ±1 mm throughout.

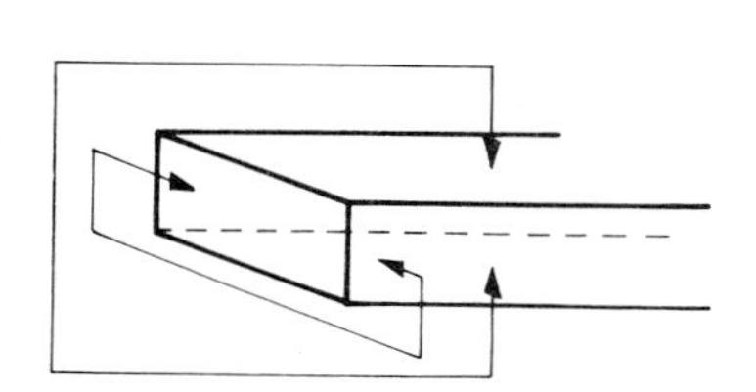

Figure 1.

Calliper measurement is the perpendicular distance from any point on one surface to the corresponding point on the opposite surface.

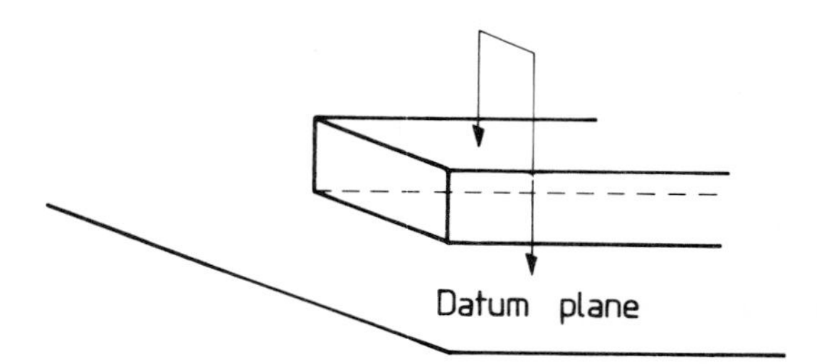

Figure 2.

When resting on a true plane, either way up in turn, the variation at any point shall be within the values shown in column 6 for a single parallel. For matching, the maximum variation over the two parallels, however associated, shall not exceed the tolerance in column 7.

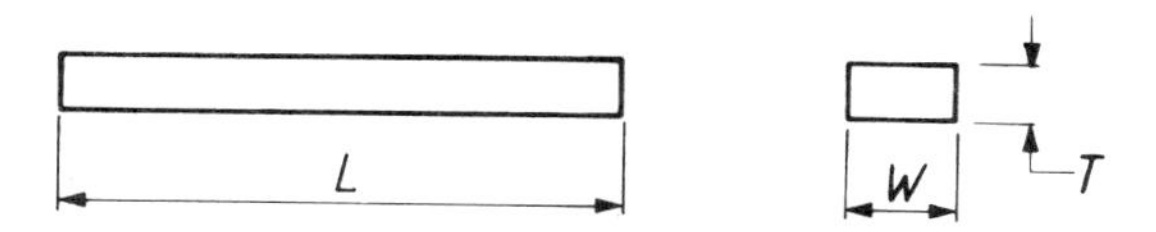

Table 2. Tolerances on grade B parallels

1	2	3	4	5	6	7
Nominal size			Calliper measurements for individual parallels (see figure 1)		Functional requirements (see figure 2)	
					for individual parallels	for a pair
T	*W*	*L*	Departure from nominal *T* and *W*	Variations in calliper measurements of *T* and *W*	parallelism of *T* and *W*	matching
mm	mm	mm	mm	mm	mm	mm
5	10	100	±0.010	0.004	0.008	0.012
10	20	125				
15	30	150				
20	40	200				
25	50	250	±0.020	0.006	0.012	0.020
30	60	300				
40	80	350	±0.025	0.007	0.014	0.020
50	100	400	±0.030	0.008	0.016	0.025

NOTE. Tolerance on *L* is ±1 mm throughout.

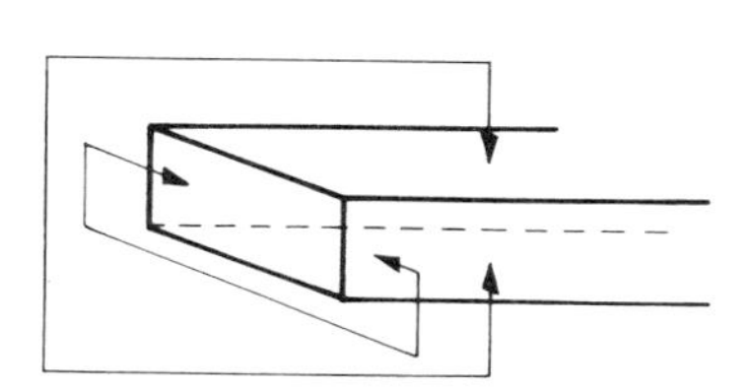

Figure 1.

Calliper measurements is the perpendicular distance from any point on one surface to the corresponding point on the opposite surface.

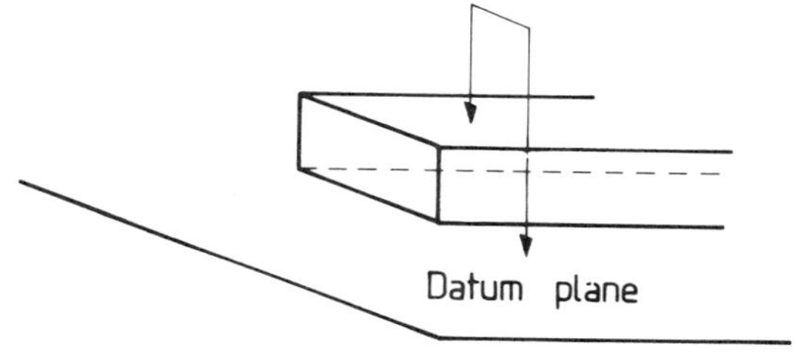

Figure 2.

When resting on a true plane, either way up in turn, the variation at any point shall be within the values shown in column 6 for a single parallel. For matching, the maximum variation over the two parallels, however associated, shall not exceed the tolerance in column 7.

. . . .

Straightedges

The two parts of BS 5204 'Straightedges' cover cast iron bow and I-section straightedges (Part 1), and steel or granite straightedges of rectangular section (Part 2). All types of straightedge are provided for in two grades, namely A and B.

BS 5204

Straightedges

Part 1. Cast iron straightedges (bow shaped and I-section)

. . . .

3. Nomenclature and definitions

For the purposes of this Part of this British Standard the nomenclature indicated in figures 1 and 2 has been adopted and the following definitions apply.

3.1 deviation from flatness of the working surface. The minimum distance between two parallel planes which just envelope the working surface.

NOTE. It may be necessary to control the maximum slope of the surface deviations with respect to the enclosing planes.

3.2 flatness tolerance. The maximum permissible value of the deviation from flatness.

NOTE. The working surfaces of straightedges covered by this standard are not strictly speaking straight edges; they are elongated surfaces for which it is appropriate to specify flatness tolerances.

3.3 deviation from squareness of two surfaces. The minimum distance between two parallel planes which just enclose one surface and are perpendicular to a datum plane in contact with the other surface.

NOTE. See figure 3.

3.4 squareness tolerance. The maximum permissible value of the deviation from squareness.

. . . .

5. Lengths of straightedges

The recommended lengths of straightedges are as follows:

Bow shaped: 300, 500, 1000, 2000, 4000, 6000 and 8000 mm;
I-section: 300, 500, 1000, 2000, 3000, 4000 and 5000 mm.

Straightedges of intermediate lengths should be ordered only when it is not practicable to adopt one of the recommended lengths. Such straightedges shall conform to the same tolerances as those given for the next shorter recommended length.

6. General features

6.1 Bow shaped. The general design of the straightedge is left to the manufacturer (see appendix A for recommended dimensions). Provision shall be made for two feet (see figure 1) of the same width as the working surface and so located as to allow the straightedge to be used in the following positions (shown in figure 4):

(a) resting upon the feet;
(b) with the working surface downward;
(c) lying on either side.

In any of these positions the accuracy of the straightedge as specified in clauses **8** to **11** inclusive shall be maintained.

. . . .

7. Finish

The working surfaces and supporting feet of grade A straightedges shall be finished by scraping or any other process which produces a similar type of surface to that obtained by scraping.

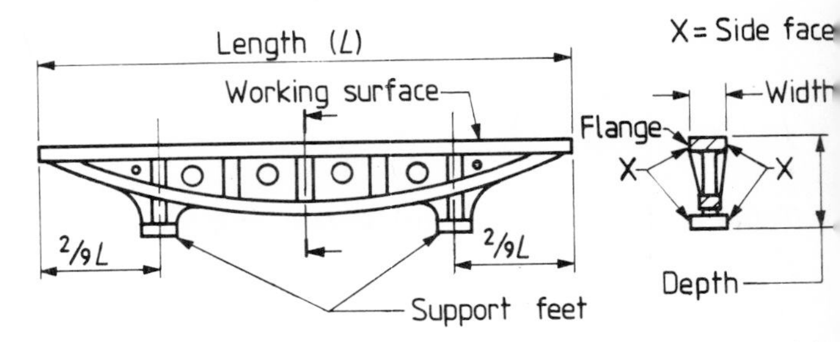

Figure 1. Bow-shaped straightedge

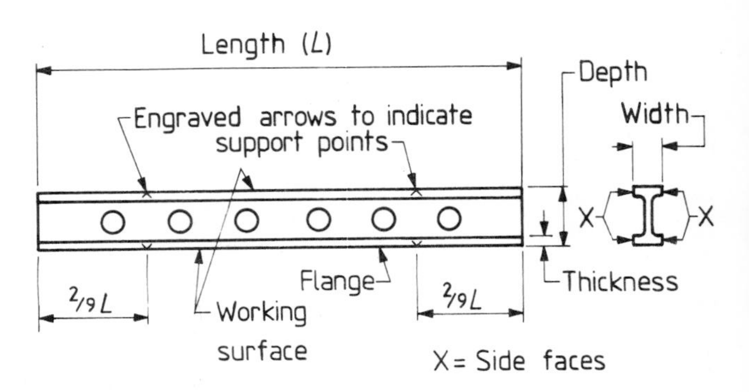

Figure 2. I-section straightedge

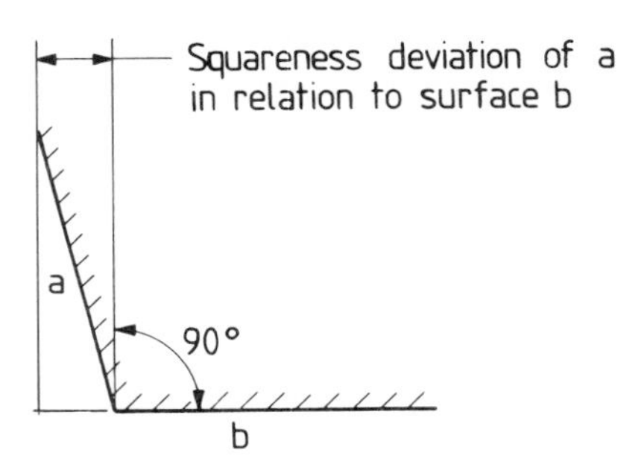

Figure 3. Exaggerated illustration of squareness deviation

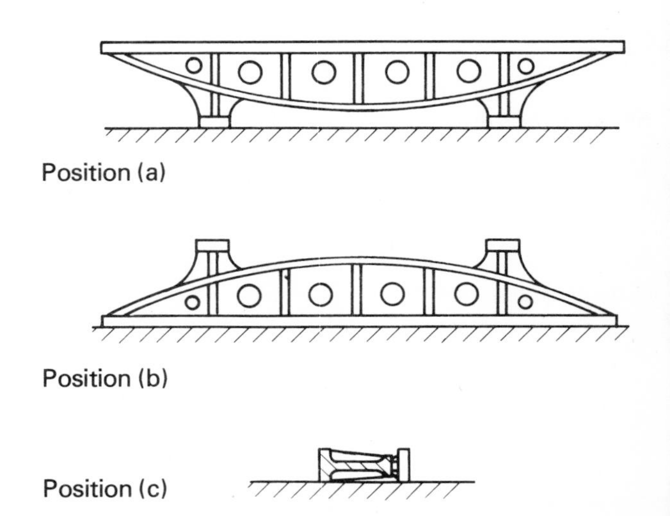

Figure 4. Alternative positions

The working surfaces of grade B straightedges and the side faces of all straightedges shall be finished by similar processes or by smooth machining.

The proportion of bearing area of the working surface shall be not less than 20 % for grade A straightedges and not less than 10 % for grade B straightedges. High spots shall be uniformly distributed and the percentage of bearing area should not be so high as to cause wringing.

NOTE. Recommended methods of testing applicable to straightedges with scraped surfaces are given in appendix B.

All sharp edges shall be removed.

All unmachined parts shall be painted.

8. Flatness of working surfaces

The bearing area of the working surfaces shall be flat within the tolerances given in table 1.

In addition to conforming to the tolerances given in table 1, the deviation from flatness over any local length on straightedges 1000 mm and larger shall not exceed the following amounts:

Grade A: 0.003 mm per 300 mm
Grade B: 0.006 mm per 300 mm.

9. Equality of depths (working surface to faces of support feet) (see figure 1)

When a bow shaped straightedge rests on a flat base (see position (a) of figure 4), the depths from the working surface of the straightedge to this base, measured immediately over each support foot, shall be equal to each other to within the tolerances given in table 2.

10. Flatness of side faces

Any 300 mm length of each side face of a grade A straightedge shall be flat to within 0.025 mm and any 300 mm length of each side face of a grade B straightedge shall be flat to within 0.05 mm.

The maximum tolerance on flatness over the whole length of straightedges up to and including 4000 mm shall be 0.05 mm for grade A and 0.1 mm for grade B; for longer straightedges, these tolerances shall be doubled for both grades.

11. Squareness of working surface and side faces

When a straightedge is laid on its side on a grade A surface plate (as specified in BS 817 : 1972*) the working surface shall be square to the surface plate to within 0.008 mm per 25 mm for grade A straightedges and 0.015 mm per 25 mm for grade B straightedges. (See also figure 4.)

* The extract of BS 817 at the beginning of this chapter is from the 1983 edition.

Table 1. Tolerances on flatness of working surfaces

Length of straightedge	Flatness tolerance	
	Grade A	**Grade B**
mm	mm	mm
300	0.003	0.006
500	0.005	0.010
1000	0.010	0.020
2000	0.020	0.040
3000	0.030	0.060
4000	0.040	0.080
5000	0.050	0.100
6000	0.060	0.120
8000	0.080	0.160

NOTE 1. These tolerances apply to the straightedge supported in any of the three positions as specified in clause **6**.
NOTE 2. Tolerances on flatness of working surfaces for straightedges with lengths other than those listed in table 1 can be calculated at the rate of 0.001 mm per 100 mm for grade A and 0.002 mm per 100 mm for grade B, with minimum of 0.003 mm for grade A and 0.006 mm for grade B.
Grade A tolerances are to be rounded up to the nearest 0.001 mm and grade B tolerances expressed to the nearest 0.002 mm.

Table 2. Tolerances on equality of depth between working surface and faces of support feet (bow shaped): tolerances on uniformity of depth between working surfaces (I-section)

Length of straightedge	Tolerances on equality or uniformity of depth	
	Grade A	**Grade B**
mm	mm	mm
300	0.003	0.006
500	0.005	0.010
1000	0.010	0.020
2000	0.020	0.040
3000	0.030	0.060
4000	0.040	0.080
5000	0.050	0.100
6000	0.060	0.120
8000	0.080	0.160

NOTE. These tolerances apply to the straightedge supported in any of the three positions as specified in clause **6**.

NOTE. If squareness of the individual side faces adjacent to working surfaces is required, the purchaser should specify the accuracy.

12. Marking

Each straightedge shall be legibly and permanently marked with the following particulars.

(a) The manufacturer's name or trademark.
(b) The number of this British Standard, i.e. BS 5204/1.

(c) An identification number (serial number).
(d) Grade A or B, as appropriate.
(e) The year of manufacture.

. . . .

Appendix A

Minimum general dimensions for straightedges

The minimum dimensions for straightedges in tables 3 and 4 are for guidance only.

Appendix B

Recommended methods of testing cast iron straightedges

B.1 General. The whole surface of a straightedge may be used as a datum flat surface; alternatively, a central longitudinal band or track may be used for establishing a datum straight line. Although tests for straightness may often be made by comparison with a datum straight line it is necessary to use the inclination method to determine the errors in straightness of the datum straight line and in all instances where the highest possible accuracy is required.

B.2 Inclination method. This method of test requires a type of instrument which will measure very small angular variation of a carriage or block as it is moved step by step along the centre line of the working surface of the straightedge. It may be a spirit level, an electronic level, an autocollimator or, in fact, any instrument which will measure small angular variations, but it is important to note that the weight of the equipment should be kept to a minimum in order to avoid distortion of the surface under test. The measurements should also be made in a controlled environment to ensure that rapid temperature variations and temperature gradients are kept to a minimum.

The principle of the inclination method is illustrated in figures 7 and 8. The carriage may consist of a plate or block to the underside of which may be fitted (by wringing or otherwise) parallel blocks of equal thickness. These blocks are placed at a separation suited to the step chosen for measurement and the carriage is moved along the line in steps of this value (see figure 7). It is also quite satisfactory to use a carriage with feet at a fixed distance; a block level with fixed feet, for example, provided the pitch covered by the feet is suitable for the measurement in hand. In either case, the span may be marked on the feet of the carriage.

To carry out a test, the straightedge is stood on a firm support and the working surface under consideration is approximately levelled in both longitudinal and transverse directions. The carriage to be used for making the flatness test is placed at one end of the straightedge with its length in the same direction as the latter. After making the first reading, the carriage is advanced along the straightedge through a distance equal to the span of its feet and the second reading is made. In this way, the carriage is advanced step by step along the straightedge until the other end of the working surface is reached, when it is usual to move the carriage backwards over the same path and to obtain a check series of readings terminating at the starting position. The inclination of each span relative to the first setting can

Table 3. Minimum general dimensions for bow shaped straightedges

Length of straightedge	Minimum width of working surface and feet	Minimum overall depth	Minimum flange thickness
mm	mm	mm	mm
300	30	80	10
500	35	130	12
1000	45	180	16
2000	65	300	24
3000	90	400	32
4000	100	500	38
6000	100	600	50
8000	100	800	55

Table 4. Minimum general dimensions for I-section straightedges

Length of straightedge	Minimum width of working surface and feet	Minimum overall depth	Minimum flange thickness
mm	mm	mm	mm
300	25	75	8
500	30	75	10
1000	35	100	12
2000	50	150	14
3000	55	250	16
4000	60	300	18
5000	65	350	20

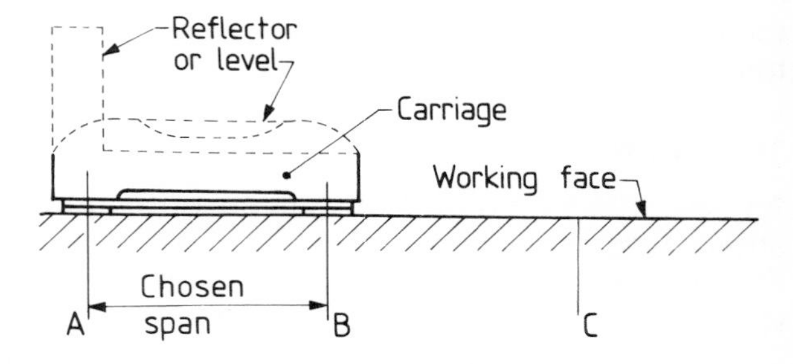

Figure 7. Test by inclination method

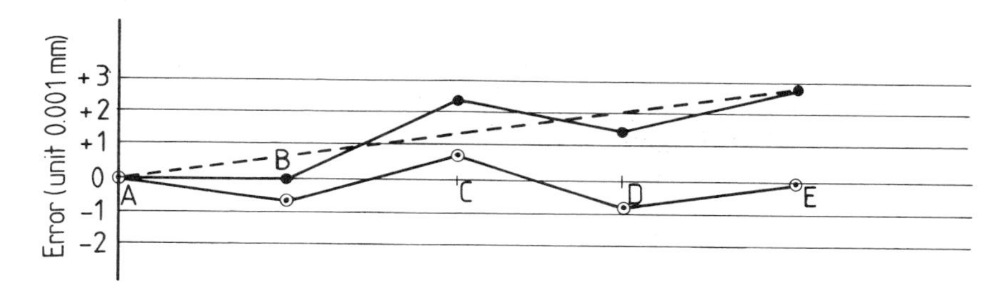

Figure 8. Plotting of readings

therefore be found and the continuous angular contour recorded. No absolute datum, such as true level, is necessary. The datum is the first angular reading, all other inclinations being initially related to this.

Assume, for example, positions A, B, C, D, etc. in figure 8 are laid out, 250 mm apart, along the centre line of the carriage. A 250 mm span will also be marked out on the feet or base of the carriage. The first span will be from A to B and this will be taken as the initial datum from which the other inclinations will be measured. Assume also that the span B–C shows an upward tilt of 5 seconds and C–D a downward tilt of 2 seconds when the carriage is moved to these positions. On a 250 mm base 1 second of arc represents a vertical displacement of 0.001 21 mm; C will therefore be 0.0060 mm above the datum line through A–B, and D will be 0.0024 mm below C or 0.0036 mm above the datum, and so on. The end point of the line will generally not be zero although it is usually convenient to take as a datum line the one between the extreme points. The final graphs must therefore be tilted about the origin, either arithmetically or graphically, to bring the end point to zero (see figure 8).

The sample table below shows the method of recording observations and working up the results. It will be noted that an additional zero is placed above the first reading in column 5; this is because the first reading, although taken as datum zero, represents the relative levels of two points A and B. In this case the error at B will not therefore be zero in column 7 if the graph has to be swung about the origin.

Although the span used may be any value within reasonable limits, it is convenient to make it about one-tenth of the length of the straightedge, giving ten steps of measurement.

In addition to the above test, which is usually carried out along the centre line of the working surface, it is necessary to test the face for a transverse twist (wind). This is most conveniently done with a level, an autocollimator being unsuitable for this test without certain special equipment. The level is turned so as to lie transversely across the surface of the straightedge and any variations in its readings are noted as it is moved from one end of the surface to the other.

B.3 Comparison methods. These methods of test require a datum reference surface of known accuracy against which a cast iron straightedge can be compared. For this purpose either a rectangular steel straightedge complying with the requirements of BS 863 or a calibrated reference track of a grade AA surface plate (as specified in BS 817 : 1972*) is suitable.

*The extract of BS 817 at the beginning of this chapter is from the 1983 edition.

Example of recording observations

1	2	3	4	5	6	7	8
Position of carriage	**Angular reading**	**Difference from first reading at position A–B**	**Tilt over 250 mm span**	**Cumulative deviations from plane A–B**	**Proportional adjustments to bring E to zero**	**Error (algebraic sum of columns 5 and 6)**	**Position along straightedge**
			Unit 1 × 10^{-3} mm				
	Second	Second					
—	—	—	—	0	0	0	A
AB	15	0	0	0	−1.8	−1.8	B
BC	20	+5	+6.0	+6.0	−3.6	+2.4	C
CD	13	−2	−2.4	+3.6	−5.4	−1.8	D
DE	18	+3	+3.6	+7.2	−7.2	0	E

B.3.1 *Comparison with a rectangular steel straightedge.* The steel straightedge is stood on its edge and supported on the working surface of the cast iron straightedge by two equal block gauge combinations placed at the positions of support marked on the steel straightedge. A suggested combination is 2.005 mm + 2.05 mm, i.e. 4.055 mm.

A gap is thus formed between the two straightedges. The width of this gap is then measured at any desired number of positions by fitting gauge blocks into it.

To obtain the true errors of the working surface of the cast iron straightedge from the results thus obtained, it is necessary to allow for the errors in the steel straightedge.

If the actual errors of the steel straightedge are not known, this method of test can still be used, but it becomes necessary to repeat the measurements of the gap between the two straightedges after inverting the steel straightedge and to measure the width of the latter at a number of positions so as to determine any errors of parallelism between its two working edges. Details of the way in which the errors in the straightness not only of the cast iron straightedge but also of both edges of the steel straightedge can be obtained from the results of these measurements will be found in appendix C of BS 863: 1939*.

B.3.2 *Comparison with a reference track of a grade AA surface plate.* It is necessary first to establish the errors in straightness of a track of the surface of grade AA surface plate using the inclination method as described in **B.2**. Comparison between the reference track and the working surfaces of bow shaped and I-section cast iron straightedges may then be made by means of a test indicator complying with the requirements of BS 2795 : Part 1, and mounted on a suitable block. The test indicator block is moved along the reference track and the difference between the known errors of the reference track and the errors of the straightedge are registered on the test indicator. To ensure that the readings obtained relate only to the bearing area of the scraped surface of the straightedge, a block gauge must be interposed between the scraped surface and the stylus of the test indicator. Where it is required to measure a large number of straightedges with their measuring surfaces uppermost, i.e. position (a) of figure 4, it is advantageous to replace the test indicator block with a carriage which just straddles the straightedge and bears on two tracks located close to either longitudinal side of the straightedge. In this instance, in addition to measuring the straightness of each track, it would be necessary to verify that the transverse twist (wind) relationship between the two tracks will not significantly influence the test indicator reading.

The method of comparison described in **B.3.1** can also be used for checking straightedges when they are positioned with their surfaces facing downward, i.e. position (b) of figure 4.

B.4 Local deviations from straightness. Clause **11** states tolerances for local deviations from straightness and the methods described above may be used for checking compliance.

B.5 Flatness. Twist or wind may be checked by observing the variations of reading of a precision level placed transversely on the straightedge at various positions in its length. A small level of sensitivity, 5 seconds or 10 seconds per division and mounted in a light frame, may be used.

B.6 Tests in three positions. Clause **9** requires that the straightedges conform to the tolerances in positions (a), (b) and (c) of figure 4. The tests for straightness described in **B.2** and **B.3** cater for positions (a) and (b). The straightedge may be tested in position (c) by the inclination method (see **B.2**), using an autocollimator or, by comparison with a reference straightedge, a test indicator or gauge blocks. The indicator must contact on a gauge block bearing on the high spots of the surface being inspected.

. . . .

BS 5204

Straightedges

Part 2. Steel or granite straightedges of rectangular section

---

. . . .

3. Definitions

For the purposes of this Part of this British Standard the following definitions apply.

straightness tolerance of working face. The straightness tolerance is the maximum permissible separation of two parallel planes between which the face lies.

tolerance on parallelism of working face. The tolerance on parallelism is the maximum permissible variation in width between the surfaces under consideration.

. . . .

5. Recommended sizes and design

Recommended lengths are 300, 500, 1000, 1500 and 2000 mm.

* BS 863: 1939 was withdrawn June 1977 and was superseded by BS 5204: Part 2 (see pages 70–3).

Minimum sections shall be as shown in table 1.

Straightedges of intermediate lengths should be ordered only when it is not practicable to adopt one of the recommended lengths. Such straightedges shall comply with the minimum depths and thicknesses given in table 1 for the next longer recommended length.

The side surfaces may be slightly relieved (I-section) to reduce weight, and long straightedges may be provided with two slots for ease in handling; the centres of these slots should be at points two-ninths of the length from each end of straightedges (see figure 1).

Table 1. Minimum section

Dimensions in millimetres

	Steel				Granite	
	Grade A		Grade B		Grades A and B	
Length *L*	Depth *d*	Thickness *t*	Depth *d*	Thickness *t*	Depth *d*	Thickness *t*
300	40	6	30	5	75	25
500	50	10	45	6	100	35
1000	75	10	60	8	125	40
1500	100	12	70	10	150	50
2000	125	13	80	12	150	50

6. Finish

The working faces shall be finished by grinding or lapping. All sharp edges shall be removed.

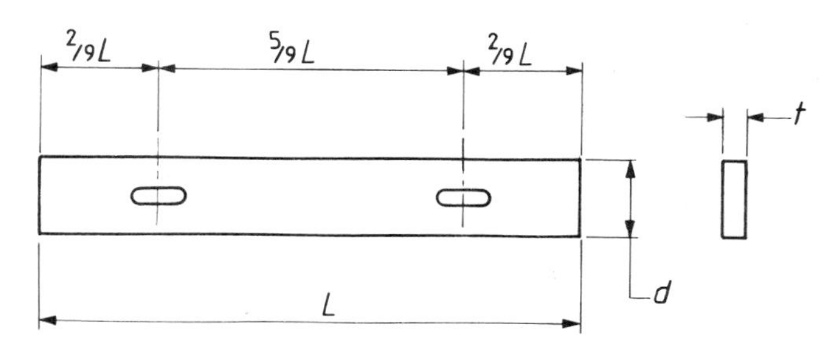

Figure 1. Position of handling slots

7. Points of support for straightedges of 1 m length and over

To ensure minimum deflection on straightedges of 1 m length and over, the positions for supporting the straightedges when used on edge should be located two-ninths of the length from each end (see appendix A) and shall be marked on one or both side faces by arrows and the word 'support' engraved on the face, as shown in figure 2.

NOTE. The attention of manufacturers is drawn to the recommendation in **A.2**.

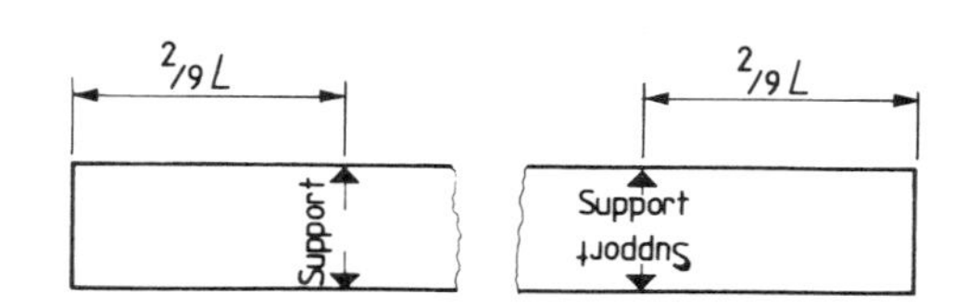

Figure 2. Alternative methods of indicating the positions of support

8. Accuracy

Each straightedge shall comply with the requirements of **8.1** to **8.4** for grade A or grade B accuracy.

8.1 Straightness tolerance of working face. When the straightedge is lying flat on a grade AA surface plate (as specified in BS 817 : 1972*) or standing 'on edge' supported at two points two-ninths of the length from each end, the permitted tolerances on straightness are given in table 2.

8.2 Tolerance on parallelism of working face. The tolerances on parallelism are given in table 3.

8.3 Combined flatness and parallelism of side faces. When the straightedge is laid flat with either of its side faces on a true surface (see **8.1**) the upper face shall be parallel to the plate to within the tolerances given in table 4.

8.4 Squareness of side faces to working faces. The squareness of the side faces to the working faces shall be within 0.001 mm per millimetre of depth for grade A and within 0.002 mm per millimetre of depth for grade B.

*The extract of BS 817 at the beginning of this chapter is from the 1983 edition.

Table 2. Tolerance on straightness of working face

Dimensions in millimetres

Length of straightedge or length under test	Tolerance	
	Grade A	Grade B
300	0.005	0.010
500	0.005	0.010
1000	0.008	0.015
1500	0.012	0.025
2000	0.015	0.030

NOTE 1. The tolerances above apply to the overall length and to intermediate lengths of a straightedge.

NOTE 2. Recommended methods of testing straightness are dealt with in appendix B.

NOTE 3. Tolerances on straightness of the working face for straightedges with lengths other than those listed in table 2 can be calculated at the rate of 0.00075 mm per 100 mm for grade A and 0.0015 mm per 100 mm for grade B, with a minimum of 0.005 mm for grade A and 0.010 mm for grade B.

Grade A tolerances are to be rounded up to the nearest 0.001 mm and grade B tolerances expressed to the nearest 0.005 mm.

11. Marking

Each straightedge shall be legibly and permanently marked with the following particulars.

(a) The manufacturer's name or trademark.
(b) The number of this British Standard, i.e. BS 5204/2.
(c) Grade A or B, as appropriate.
(d) When hardened in accordance with **4.1**, the word 'Hardened'.
(e) The year of manufacture.

Table 3. Tolerance on parallelism of working face

Dimensions in millimetres

Length of straightedge	Tolerance	
	Grade A	Grade B
300	0.005	0.010
500	0.005	0.010
1000	0.008	0.015
1500	0.012	0.025
2000	0.015	0.030

Table 4. Combined tolerance on flatness and parallelism of side faces

Dimensions in millimetres

Length of straightedge	Tolerance	
	Grade A	Grade B
300	0.05	0.07
500	0.05	0.07
1000	0.07	0.10
1500	0.09	0.14
2000	0.11	0.17

Appendix A

Note on supporting straightedges of rectangular section

A.1 General. When straightedges are used on edge, it should be realized that although they may be of comparatively deep section, they are liable to deflect under their own mass. The amount of deflection will vary according to the number and the positions of the supports along the length of the straightedge. Consider, for example, the case of a grade A steel straightedge, 1500 mm long with a section of 100 mm × 12 mm. If this rests on edge on two supports placed at its extremities, as in figure 3, it will sag at the middle by 0.025 mm. Alternatively, if the supports are brought close together at the middle of the bar, as in figure 4, the ends will droop by 0.012 mm. Thus, if such a straightedge were rested on edge on a surface which happened to be concave by as much as 0.025 mm, or convex by as much as 0.012 mm, over 1500 mm, the straightedge would rest in contact with the surface in both cases over its whole length, and one might be deceived into considering the surface as being free from any appreciable errors in straightness.

However it is evident from figures 3 and 4 that there is an intermediate position for the two supports at which the distortion of the straightedge under its own mass is reduced to a minimum. This position is shown in figure 5. With the supports in this optimum position, the deflection of the straightedge under consideration is reduced to as low as 0.0007 mm.

The disposition of the supports shown in figure 5 should always be adopted when using steel straightedges of rectangular section on edge. For example, when testing the straightness of a horizontal surface with such a straightedge, the latter should be supported at the positions indicated in figure 5 on two gauge blocks of equal height (say 2.5 mm) and the parallelism of the gap between the surface and the lower edge of the straightedge should be tested by means of other gauge blocks.

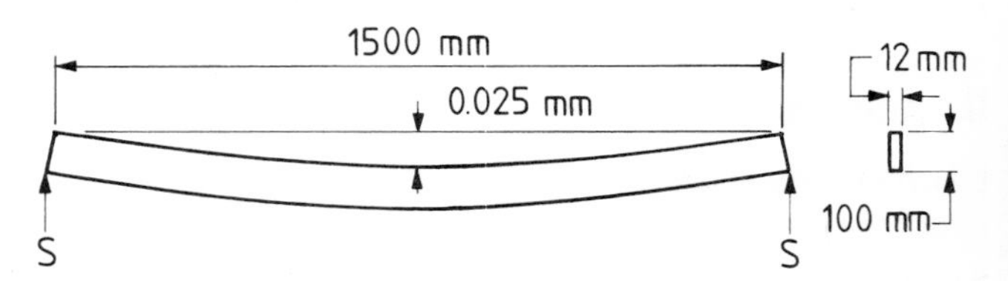

Figure 3. Deflection with supports at ends

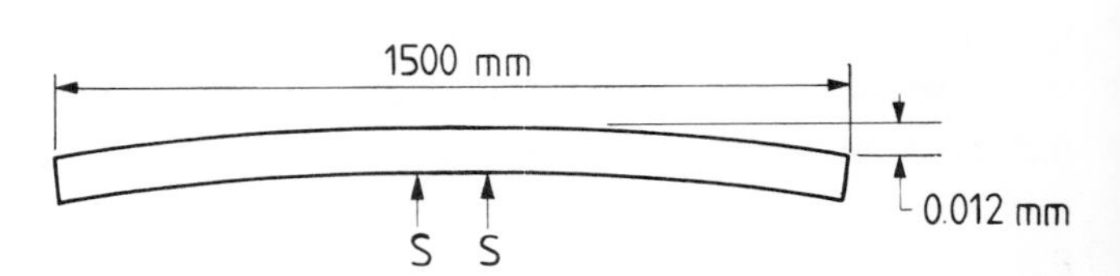

Figure 4. Deflection with supports near the centre

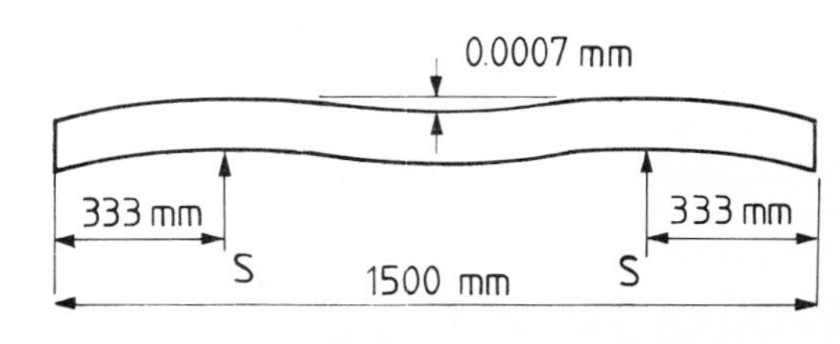

Figure 5. Optimum position for supports

Table 5. Best positions for supporting straightedges

Dimensions in millimetres

Length of straightedge	300	500	1000	1500	2000
Distance from each support from end	67	111	222	333	444

Table 5 shows the best positions for supporting straightedges of various lengths from 300 mm upwards.

. . . .

Appendix B

Recommended methods of testing straightness of a straightedge

NOTE. Same as for BS 5204: Part 1 with the exception of clause **B.6** which is as follows:

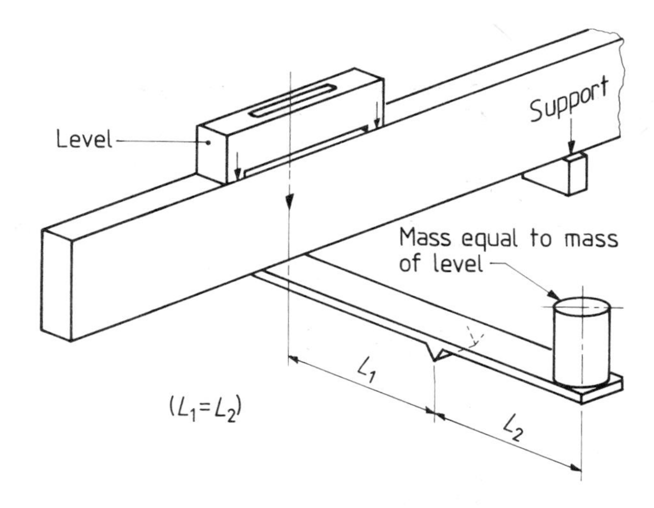

Figure 8. Method of compensation for mass of level

B.6 Deflection compensation. Deflection of the straightedge due to the mass of the level may be compensated for by application of an equal force acting vertically upwards on to the bottom edge of the straightedge. This compensating force should be applied in the vertical plane which passes through the centre of gravity of the level (see figure 8).

Vee blocks

Hollow, solid and double vee blocks having a nominal included vee angle of 90 ° are provided for in BS 3731 'Metric vee blocks'. The range of sizes recommended will accommodate cylinders having diameters from:

5 mm up to 200 mm in the case of hollow blocks;
5 mm up to 85 mm in the case of solid blocks;
5 mm up to 40 mm for double vee blocks; and
5 mm up to 125 mm for granite blocks.

BS 3731

Metric vee blocks

. . . .

3. Nomenclature and definitions

For the purposes of this standard, the nomenclature indicated in figure 1 has been adopted and the following definitions apply.

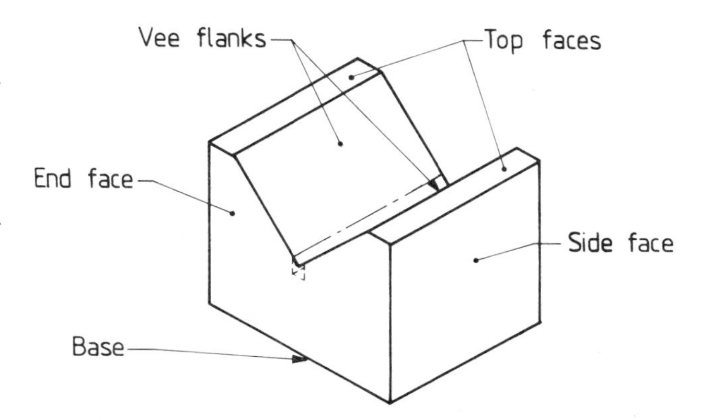

Figure 1. Nomenclature for vee blocks

3.1 designating size. The diameter of the maximum cylinder that the block will accommodate.

NOTE. But see **5.1**.

3.2 working faces. The vee flanks, the base and the end, top and side faces.

3.3 deviation from flatness. The minimum distance between two parallel planes which just enclose the measuring face.

NOTE. It may be necessary to control the maximum slope of the surface deviations with respect to the enclosing planes.

3.4 deviation from squareness of two surfaces. The minimum distance between two parallel planes which just enclose one surface and are perpendicular to a datum plane in contact with the other surface.

3.5 deviation from parallelism of two surfaces. The variation in distance between the surfaces under consideration.

3.6 tolerances. The maximum permissible deviations from flatness, squareness and parallelism are listed as tolerances.

. . . .

5. General features of design

5.1 Details of design are not specified but there should be a minimum length of vee flank of 1 mm beyond the points of contact of the maximum and minimum cylinders.

. . . .

6. Finish

6.1 General. Grade A vee blocks shall have a ground and/or lapped finish on all working faces.

Grade B vee blocks shall be finished by a similar process or by planing, shaping, milling or scraping.

The vee flanks shall be finished between the limiting positions defined in **5.1**. An undercut or radius shall not commence before the lower limiting position.

6.2 Edges. All sharp edges shall be removed.

6.3 Bearing area. The percentage area of high spots (bearing area) of each working face shall be not less than 20 %.

7. Clamps

When clamps are supplied they should be sufficiently robust to secure cylinders within the capacities shown in table 4. In order to facilitate the use of vee blocks when located on their side faces, clamps may be designed so that the body does not extend beyond the side faces of the block.

8. Recommended dimensions

Recommended general dimensions for steel and cast iron vee blocks are given in tables 1, 2 and 4. Because of the nature of the material, special consideration has been given to the size of section required for granite vee blocks and recommended general dimensions are given in table 3.

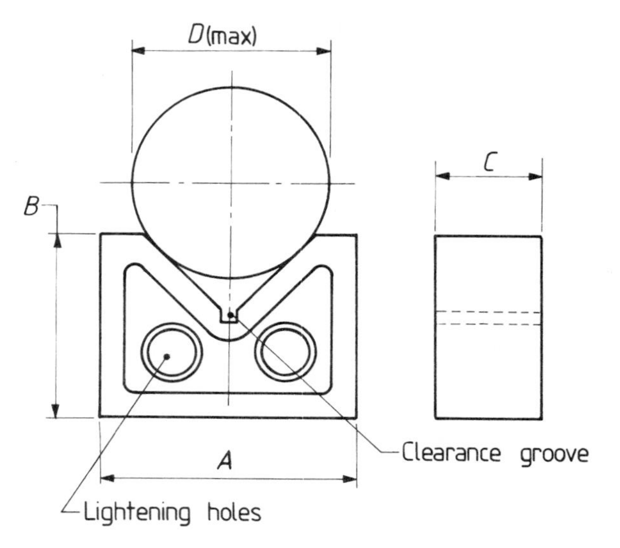

Figure 2. Vee blocks (hollow type)

Table 1. Recommended general dimensions: hollow type (see figure 2)

Designating size *D* (max.)	Width *A*	Height *B*	Length *C*
mm	mm	mm	mm
63	80	60	35
80	100	75	40
100	130	90	45
125	150	100	50
160	180	130	60
200	220	160	70

Table 2. Recommended general dimensions: solid type, steel or cast iron (see figure 3)

Designating size *D* (max.)	Width *A*	Height *B*	Length *C*
mm	mm	mm	mm
40	50	40	25
85	100	70	50

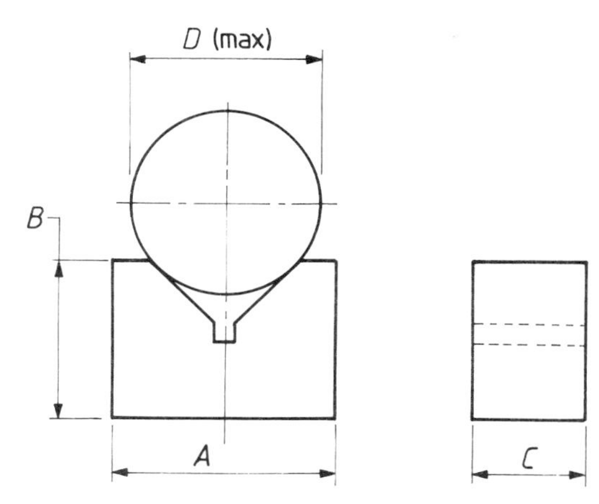

Figure 3. Vee blocks (solid type)

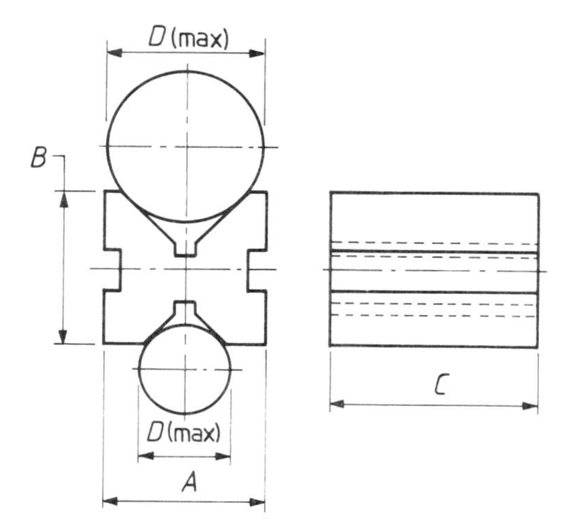

Figure 4. Vee blocks (double type)

Table 3. Recommended general dimensions: solid type, granite (see figure 4)

Designating size *D* (max.)	**Width *A***	**Height *B***	**Length *C***
mm	mm	mm	mm
50	75	75	75
75	100	100	100
85	150	150	150
125	200	200	200

Table 4. Recommended general dimensions (see figure 4)

Designating size *D* (max.)	**Width *A***	**Height *B***	**Length *C***
mm	mm	mm	mm
20/25	32	32	42
20/40	40	40	50

9. Accuracy

9.1 Grade A vee blocks. Grade A vee blocks shall be geometrically accurate within the tolerances given in columns 2 to 7 of table 5.

Table 5. Geometrical accuracy for grade A vee Blocks

1	2	3	4	5	6	7	8
Designating size	**Flatness of working faces**	**Squareness of adjacent exterior faces (measured over height)**	**Parallelism of opposite faces**	**Parallelism of vee axis to base and to side faces over length of vee***	**Equality of semi-angles of vee flanks***	**Centrality of vee†**	**Matching tolerance (see 9.3)**
mm	μm	μm	μm	μm	minute	μm	μm
20	2	4	4	2	1	4	2
25	2	4	4	2	1	4	2
40	3	6	6	4	1	6	3
50	3	6	6	4	1	6	3
63	4	8	8	6	1	8	4
75	5	10	10	8	1	10	5
80	5	10	10	8	1	10	5
85	5	10	10	8	1	10	5
100	6	12	12	10	1	12	6
125	8	16	16	14	1	16	8
160	10	20	20	18	1	20	10
200	12	24	24	22	1	24	12

* For a recommended method of checking these features, see appendix A.
† Displacement of vee axis from centre of block.

NOTE. Intermediate sizes should be made to the accuracy specified for the next smaller designating size.

9.2 Grade B vee blocks. Grade B vee blocks shall be geometrically accurate within the tolerances given in columns 2 to 7 of table 6.

Table 6. Geometrical accuracy for grade B vee blocks

1	2	3	4	5	6	7	8
Designating size	**Flatness of working faces**	**Squareness of adjacent exterior faces (measured over height)**	**Parallelism of opposite faces**	**Parallelism of vee axis to base and to side faces over length of vee***	**Equality of semi-angles of vee flanks***	**Centrality of vee†**	**Matching tolerance (see 9.3)**
mm	μm	μm	μm	μm	minute	μm	μm
20	6	12	12	12	2	6	4
25	6	12	12	12	2	6	4
40	8	16	16	16	2	8	6
63	12	24	24	24	2	16	8
80	16	32	32	32	2	16	10
85	16	32	32	32	2	16	10
100	18	36	36	36	2	18	12
125	24	48	48	48	2	24	16
160	30	60	60	60	2	30	20
200	36	72	72	72	2	36	24

* For a recommended method of checking these features, see appendix A.
† Displacement of vee axis from centre of block

NOTE. Intermediate sizes should be made to the accuracy specified for the next smaller designating size.

9.3 Matched pairs. Vee blocks in either grade may be manufactured or selected to form matched pairs. For matched vee blocks, the tolerances in columns 2 to 7 of table 5 or 6, as appropriate, shall apply and, in addition, any difference between the heights of the vee axes above the base and any difference in width between the vee axes and side faces of the two blocks comprising the pair shall not exceed the values given in column 8 of the respective tables.

Attention is drawn to the fact that if this latter requirement is to be met, full advantage cannot be taken of the whole amount of the tolerances for the individual blocks.

. . . .

12. Marking

Each vee block shall be legibly and permanently marked with the following particulars in a manner that does not affect the accuracy of the vee block.

(a) The manufacturer's name or trademark.
(b) The number of this British Standard, i.e. BS 3731.
(c) The grade (grade A or grade B).
(d) Matched pairs of vee blocks shall bear a suitable pairing symbol, e.g. A and B.

All vee blocks which are to be certified by a testing authority are required to bear a serial number.

NOTE. It is recommended that the designating size be marked on the block.

Appendix A

Recommended methods of testing vee blocks

Most methods of checking vee blocks are too well known to need description, and recommendations in this appendix are confined to four particular features.

A.1 Vee axis

A.1.1 Parallelism of the vee axis to the base can be determined from indicator readings x taken over the ends of a precision cylinder placed in the vee (see figure 5*a*). The cylinder should extend beyond the end faces of the block.

A.1.2 Parallelism of the vee axis to the side faces can be determined in a similar manner, but using a side face as a datum (see figure 5*b*).

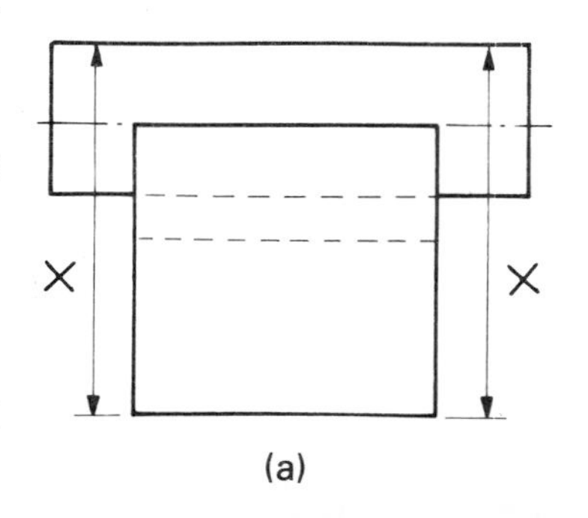

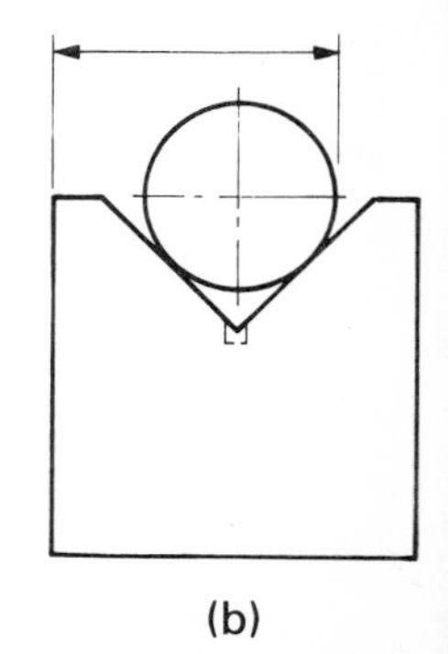

Figure 5. Checking the vee axis

A.2 Equality of matched pairs. A method similar to that described in **A.1.1** is also recommended for checking the heights of the vee axes above the base in matched pairs of vee blocks (see **9.3**), but in this case height readings should be taken over the 'minimum' and 'maximum' cylinders that can be accommodated by the blocks in question. Measurement of the widths between the vee axes and side faces of the blocks can also be effected by means of cylinders.

A.3 Equality of semi-angles. Equality of the semi-angles of the vee is conveniently checked by means of a sine table and test indicator.

Standards publications referred to in this chapter

BS 817 Surface plates
BS 869 Toolmakers' flats and high precision surface plates
BS 906 Engineers' parallels
  Part 1 Metric units
BS 939 Engineers' squares (including cylindrical and block squares)
BS 3731 Metric vee blocks
BS 5204 Straightedges
  Part 1 Cast iron straightedges (bow shaped and I-section)
  Part 2 Steel or granite straightedges of rectangular section
BS 5535 Right angle and box angle plates

Note: The above list does not include all the standards referred to in the extracts.

4 Linear measurement

Introduction

Chapter 2 is concerned with why a system of measurement is needed and how, via the NPL and BCS, the primary standards are maintained and disseminated in the United Kingdom. While this is of great importance to the engineering metrologist, and an activity with which he should be familiar, it is more the province of the physicist.

For most applications the engineering metrologist uses a range of equipment from steel rules to comparators. This is the range covered in this chapter.

Steel rules

BS 4372 'Engineers' steel measuring rules' recommends that the figuring on rules be in millimetres. However, in those cases where practical considerations make it preferable to use centimetre figuring, this is permissible as a second choice. The requirements for steel rules and folding steel rules in preferred lengths up to one metre nominal length are specified.

BS 4372

Engineers' steel measuring rules

. . . .

1.2 Definitions. For the purposes of this British Standard, the following definitions apply:

steel rule. A graduated continuous length of steel with one or two square datum ends.

folding steel rule. Two or more graduated lengths of steel connected together by swivel joints, with two square datum ends.

1.3 Marking. The rule shall be marked with at least the manufacturer's name or trademark and the number of this British Standard, i.e. BS 4372.

. . . .

2.2 Recommended lengths. It is recommended that the lengths given below be regarded as preferred and that those in the first choice series be selected wherever possible:

First choice: 150, 300, 500, 1000 mm

Second choice: 100, 200, 600 mm

. . . .

2.5 Accuracy. When referred to the standard reference temperature of 20 °C, the tolerances in **2.5.1** to **2.5.4** shall apply.

2.5.1 Rule edges shall be straight to within 0.1 mm on any length up to 300 mm.

2.5.2 Rule edges shall be parallel to within 0.1 mm on any length up to 300 mm.

2.5.3 Flat ends shall be square to the rule edges to within 0.05 mm over the width of the rule.

2.5.4 Graduation lines shall be accurate to within the tolerances specified in table 2. These tolerances apply to the distances between the centre lines of the graduation lines.

Table 2. Accuracy of graduation

Rule length	Departure from nominal		
	Up to and inc. 300 mm	Over 300 mm up to and inc. 500 mm	Over 500 mm up to and inc. 1 m
	mm	mm	mm
Distance between any two graduation lines on a single scale	0.1	0.2	0.25
Distance between any two adjacent graduation lines	0.05	0.05	0.05
Position of the 10 mm graduation line from its flat end datum	0.08	0.08	0.08

3. Folding steel rules

3.1 Construction

3.1.1 The rule shall consist of 'legs' of approximately equal length, connected end to end by permanent joints to form a straight line. Both extreme ends of the rule shall be flat and square to the rule edges.

3.1.2 The joints shall locate the legs in a straight line during service and also allow the rule to fold into a convenient length.

Both ends of the rule may be fitted with tips if so desired; such tips shall be within the graduated length. The tip shall be of the same width as the rule and shall fit flush with the rule end and edges. Tips shall be securely fastened to the outer face only of the appropriate legs.

. . . .

3.5 Accuracy. When referred to the standard reference temperature of 20 °C, the tolerances in **3.5.1** to **3.5.4** shall apply.

3.5.1 Flat ends shall be square with the rule edges to within 0.05 mm over the width of the rule.

3.5.2 Each joint shall ensure the alignment of the graduated edges of its two associated legs to within 0.25 mm. It shall also locate the positional continuity of the graduation to within 0.2 mm.

3.5.3 Each individual leg of the rule shall be accurately graduated with a maximum permissible error between the centre lines of any two graduation lines on a single scale in accordance with the tolerances given in table 2.

3.5.4 The positional error of the centre line of the first and last 10 mm or 1 cm graduation lines relative to its nearest square end shall not exceed 0.12 mm.

Feeler gauges

BS 957 'Feeler gauges', Part 2 'Metric units' specifies the requirements for feeler gauges with blades of thickness from 0.03 mm to 1.00 mm. The recommended combinations of blades are intended to enable the user to obtain as wide a range as possible with the minimum number of blades.

BS 957

Feeler gauges

Part 2. Metric units

. . . .

4. Dimensions of blades

4.1 Length. The following lengths are recommended:
75, 100, 150 and 300 mm.

4.2 Width. The blades shall be approximately 12 mm wide at the heel and may be parallel or tapered. Blades 300 mm or more in length may be wider. The outer ends of the blades shall be approximately semicircular.

. . . .

6. Accuracy

The thickness of a blade shall not depart from its nominal thickness by more than the amount given in column 2 of table 1 and any variation in the thickness of a blade shall not exceed the amount given in column 3.

Table 1. Tolerances on thickness of blades

1	2	3
Nominal thickness of blade	Permissible departure from nominal thickness	Permissible variation in thickness of blade
mm	mm	mm
0.03 up to and including 0.04	±0.004	0.004
above 0.04 up to and including 0.35	±0.005	0.005
above 0.35 up to and including 0.65	±0.008	0.008
above 0.65 up to and including 1.00	±0.010	0.010

. . . .

8. Marking

8.1 Blades. Each blade shall be legibly and permanently marked with its nominal thickness in millimetres, e.g. 0.04.

8.2 Sheath. The sheath shall be legibly and permanently marked 'mm' and with the manufacturer's or vendor's name or trademark.

. . . .

Appendix A

Recommended combinations of blades

The past practice of manufacturers in regard to the various combinations of feeler blades assembled together in sets has been very diverse. With a view to simplifying the considerable number of combinations at present listed, 4 recommended series are given here. These series are so devised as to furnish sets of the greatest utility with a minimum number of blades.

The order in which the blades are given in the series below is NOT that most suitable for assembly. It is desirable that each thin blade should be given the maximum protection by being interleaved between 2 thicker blades.

Set No. 1	Set No. 2	Set No. 3	Set No. 4
mm	mm	mm	mm
0.05	0.05	0.05	0.03*
0.10	0.10	0.10	0.04*
0.15	0.15	0.15	0.05
0.20	0.20	0.20	0.06
0.25	0.25	0.25	0.07
0.30	0.30	0.30	0.08
0.40	0.40	0.35	0.09
0.50	0.50	0.40	0.10
	0.60	0.45	0.15
	0.70	0.50	0.20
	0.80	0.55	0.30
	0.90	0.60	0.40
	1.00	0.65	0.50
		0.70	
		0.75	
		0.80	
		0.85	
		0.90	
		0.95	
		1.00	

* In view of the delicate nature of the 0.03 mm and 0.04 mm blades it is recommended that these blades be included in duplicate.

Dial gauges

BS 907 'Dial gauges for linear measurement' relates to gauges with the plunger movement parallel to the plane of the dial and scale divisions of 0.001 inch (sometimes sub-divided into half divisions of 0.0005 inch in which case they are commonly designated '0.0005 inch gauges'), 0.0001 inch or 0.01 mm.

BS 907

Dial gauges for linear measurement

. . . .

2. Nomenclature and definitions

For the purposes of this British Standard the nomenclature is as given in figure 1 and the following definitions apply:

scale mark. One of the marks constituting a scale.

minimum scale value. The smallest value of the measured quantity which the scale is graduated to indicate.

scale division. A part of a scale delineated by two adjacent scale marks.

discrimination. The sensitiveness of an instrument, the smallest change in the quantity measured which produces a perceptible movement of the pointer.

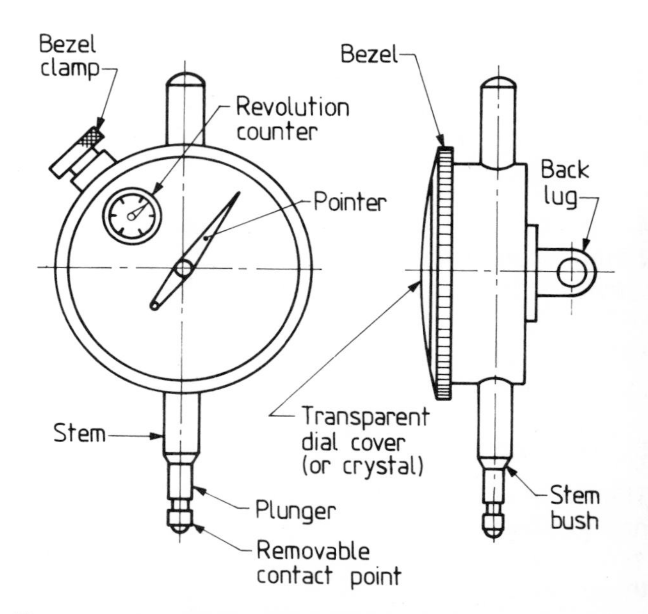

Figure 1. Nomenclature of dial gauges

repeatability. The reproducibility of the readings of an instrument when a series of tests is carried out in a short interval of time under fixed conditions of use.

limits of error. The positive and/or negative values of the errors which must not be exceeded under test.

. . . .

10. Performance

(a) General. All measurements of the accuracy of performance specified in this clause shall be referred to the standard reference temperature of 20 °C.

(b) Repeatability of reading. The gauge shall repeat its readings under all ordinary methods of operation to within 0.0002 inch on 0.001 inch and 0.0005 inch gauges, 0.000 02 inch on 0.0001 inch gauges and 0.002 mm on metric gauges. (See appendix C. Use of plunger lifting lever.)

(c) Discrimination. 0.001 inch and 0.0005 inch gauges shall be capable of indicating small gradual changes of the order of 0.001 inch in either direction of the plunger to within 0.0003 inch.

The 0.0001 inch gauge shall be capable of indicating small gradual changes of the order of 0.0002 inch in either direction of the plunger to within 0.000 04 inch.

The metric gauge shall be capable of indicating small gradual changes of the order of 0.025 mm in either direction of the plunger to within 0.003 mm.

The degree of discrimination specified above shall be achieved without resort to tapping the gauge.

(d) Limits of error. Errors permissible in dial gauges differ according to the type of instrument and the magnitude of the difference measured.

Tables 1 to 3 give the errors allowed for different types of dial gauges in common use. Since dial gauges are primarily intended for the measuring of small differences in dimensions, it will be noted that the same order of accuracy of the instrument when used for the measuring of small differences in dimensions will not be realized when larger differences are concerned.

The error over an interval of reading may be either plus or minus and the permissible errors shall not be exceeded for either direction of movement of the pointer.

Calibration shall take place from the first 12 o'clock position of the pointer (see figure 4) unless otherwise specified by the purchaser.

The permissible errors stated in tables 1 to 3 apply to any intervals and are not restricted to those commencing at zero; they shall also embrace any eccentricity of the dial in relation to the axis of rotation of the pointer and shall not be exceeded for any position of the bezel, when this is rotatable.

. . . .

Table 1. Limits of error for dials with scale divisions of 0.001 inch and 0.0005 inch

Interval of reading	Error in reading over stated interval
	inch
Any 0.01 inch	0.000 25
Any half revolution	0.000 5
Any one revolution	0.000 5
Any two revolutions	0.000 75
Any large interval	0.001

Table 2. Limits of error for dials with scale divisions of 0.0001 inch

Interval of reading	Error in reading over stated interval
	inch
Any half revolution	0.000 1
Any one revolution	0.000 15
Any two revolutions	0.000 25
Total travel	0.000 7

Table 3. Limits of error for dials with scale divisions of 0.01 mm

Interval of reading	Error in reading over stated interval
	mm
Any 0.1 mm	0.005
Any half revolution	0.007 5
Any one revolution	0.01
Any two revolutions	0.015
Any larger interval	0.020

12. Marking

Each gauge shall be legibly and permanently marked with the number of this British Standard, viz BS 907, and with the manufacturer's or vendor's name or trademark.

. . . .

Appendix B

Notes on methods of testing dial gauges

B.1 Repeatability of reading. For limits of error see subclause 10(b). The accuracy to which a dial gauge will repeat its readings is usually tested in one or two ways. Having clamped the gauge firmly in a suitable rigid fixture over a flat steel base, a true cylinder is rolled under the contact-point a number of times from various directions and the readings noted. This test is repeated at two or three positions along the range of the gauge. The cylinder is also passed through both slowly and rather abruptly.

A further test is to allow the contact-point to rest directly upon the flat base below and to note the constancy of the reading obtained when the plunger is lowered on to the base both slowly and abruptly, and when an attempt is made to rotate the plunger first in one direction and then in the other.

B.2 Discrimination. It is important that a dial gauge should be free from any trace of 'stickiness' or backlash, particularly when it is used for detecting small errors in alignments, or for testing the accuracy of the centring of work mounted in a machine. The most practical way of testing a dial gauge for this type of defect is to mount it in a *rigid* fixture with the end of its plunger in contact with the surface of a slightly eccentric precision mandrel mounted between centres. The actual amount of the eccentricity of the mandrel can be determined beforehand by means of a known sensitive indicator set to read on its surface. For preference the eccentricity should be of the order of only 0.001 inch for 0.001 inch gauges, 0.0002 inch for 0.0001 inch gauges and 0.025 mm for 0.01 mm gauges. When a dial gauge is tested against such a mandrel it should indicate the amount of eccentricity present to within the accuracy stated in subclause 10(c).

B.3 Calibration. The calibration of a dial gauge is usually carried out in a fixture in which the gauge is held rigidly opposite, and in line with, a calibrated micrometer head; or above and normal to a base plate on to which slip gauges can be wrung. In either case a series of readings is taken at suitable intervals throughout the range of the gauge. If the gauge has a limited range of only two or three turns of its pointer, the readings can be taken at intervals of a tenth of a turn throughout the whole range. This is hardly practicable, however, in the case of gauges with longer ranges. To keep the number of readings on such instruments within reasonable bounds, only a limited number are taken at the small interval of one-tenth of a turn during each revolution of the pointer. For example, in the case of a gauge having 100 divisions of 0.001 inch and a range of 0.5 inch (five turns of the pointer), the readings could be distributed in the manner shown in table 4. Each of the readings shown in the body of the table represents the mean error, found on the instrument tested, at that particular point of observation, the unit being 0.0001 inch i.e. 0.1 of a division.

Table 4. Error of reading of dial gauge given in units of 0.0001 inch

Divisions of dial	0	10	20	30	40	50	60	70	80	90
Turn I	0	−1	+1	—	—	—	+6	+4	+2	—
Turn II	—	—	0	+2	+4	—	—	—	+3	+2
Turn III	0	—	—	—	+4	+6	+8	—	—	—
Turn IV	+1	−1	0	—	—	—	+7	+5	+3	—
Turn V	—	—	0	+2	+4	—	—	—	+5	+3

NOTE. Large errors, as shown in the table, are not typical of dial gauges but have been chosen for purposes of clear illustration.

By setting out the readings in this manner, it is a simple matter to compare the results with the limits of error given in table 1 of the standard for this particular type of gauge. Thus it will be seen that the error over none of the 0.01 inch intervals tested, i.e. between any two adjacent readings, exceeds the tolerance of 0.000 25 inch specified for that interval. On the other hand, over intervals of half a turn, i.e. 50 divisions, the corresponding tolerance of 0.0005 inch is exceeded over the following tested positions:

Turn I	10 to 60	error +0.0007 inch
Turn III	0 to 50	error +0.0006 inch
Turn III	60 to Turn IV, 10	error −0.0009 inch
Turn IV	10 to 60	error +0.0008 inch

Over one turn or two turns or any longer interval, the errors are all within the tolerances allowed.

This gauge would be regarded as being unsatisfactory on account of the large errors found over some of the half-turn intervals tested.

It is interesting to note that this particular gauge possesses a type of periodic error which is not infrequently met with in this class of measuring tool. To illustrate this point, the results of a test carried out on this gauge at every 10 divisions throughout its range are shown in figure 6. It will be noticed from this graph that the errors in each successive turn are of similar undulatory character. This type of error usually arises from centring errors in the teeth of the pinion by which the pointer is attached, or from the fact that the axis of rotation of the pointer is not truly central with the dial.

The abbreviated test described above is sufficient to detect this type of periodic error by reason of the

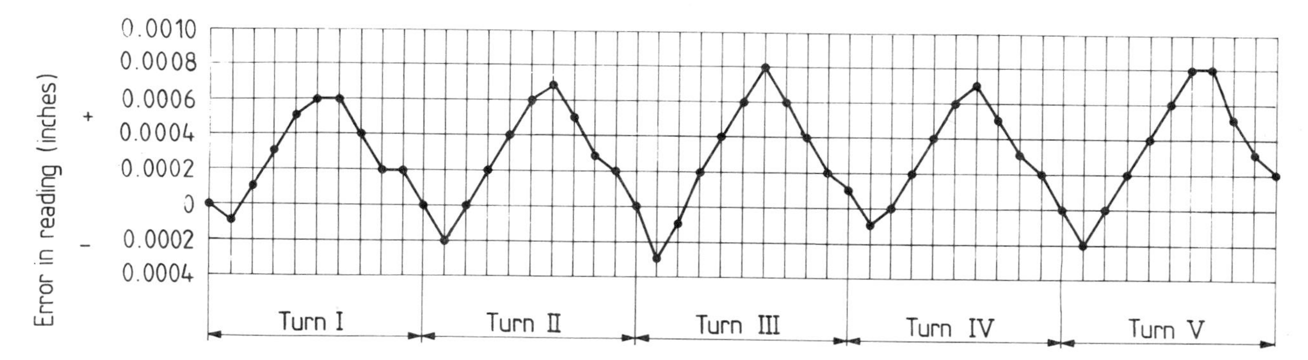

Figure 6. Calibration curve of a dial gauge illustrating periodic error

large errors found in some of the readings of half-turn intervals.

In case of any dispute regarding accuracy, resort should be had to the slip gauge method, in which case the arithmetical mean of a series of at least 5 measurements over each of the points in dispute should be taken.

Appendix C

Notes on the care and use of dial gauges

C.1 General care. Although dial gauges are made robust for workshop use it should nevertheless be realized that their mechanism is similar to that of a clock and that any fall or sharp blow may damage the pivots or upset the alignment of the bearings. If it is found at any time that the movement is not smooth, the gauge should be checked by a competent instrument maker or be returned to the makers for adjustment.

C.2 Oiling. It is important that the plunger should not be oiled or greased. It is intended to work dry. The plunger is made quite a close fit in its bush and is returned to its zero position by means of a comparatively light spring. Even a thin oil will pick up dirt which may cause the plunger to stick.

C.3 General use. Dial gauges should preferably be used as comparators and not as direct measuring instruments. For example, a piece of work 0.25 inch diameter might be tested with a dial gauge by setting the gauge to zero, raising the spindle to admit the work and then taking a reading. This method, however, would give an accuracy of measurement only within the tolerances allowed in the specification for the longer intervals of reading. On the other hand, if the gauge is used as a comparator by observing, first of all the reading when set on a 0.25 inch standard plug or slip gauge and comparing this reading with the one obtained on the work, thus determining the size of the latter by difference, the accuracy of measurement will be higher than that obtained by the first method.

C.4 Clamping. Dial gauges are usually supplied with means for alternative methods of fixing, namely (a) by means of a back lug or (b) by means of the steel stem. When the first method is used it is very necessary for the clamp to be quite tight, since slip may otherwise occur and false readings result. When the second method is adopted care must be taken not to apply too great a clamping force as this may cause the plunger to jam in the stem; the hole in the fixture which is to take the stem must be smoothly finished and truly circular and a good fit on the stem. The stem is purposely made cylindrical to fine limits of accuracy, and the hole in the fixture should be correspondingly accurate. Should the hole not be true and smooth, then there may be a tendency to nip the stem across a diameter, thus distorting it and causing the plunger to bind or even lock.

C.5 Stands. It is essential that any stand in which a dial gauge is used should be very rigid if accurate readings are to be obtained.

C.6 Use of plunger lifting lever. Where the plunger is raised by means of a lever to allow work to be inserted under the contact-point it must not be released so that the point hits the work with force. The use of a dial gauge in this way for any length of time will bring about a distortion of the teeth of the rack of the plunger at the point where the mechanism is suddenly arrested in its movement. Furthermore, such use may cause the pointer to slip on its spindle, with consequent inaccuracy of reading, particularly if the gauge is one of high magnification, i.e. 0.0001 inch or 0.01 mm.

C.7 Contact points. A variety of types of contact-points are supplied by the makers for use according to the nature of the work to be tested. Ball points are those most commonly used. It must be realized that, with constant use, a ball point will wear and that a flat on the ball may cause errors of contact resulting in inaccurate readings.

Particular care should be exercised when using flat contacts of diameters of say $\frac{1}{4}$ inch, $\frac{3}{8}$ inch or $\frac{1}{2}$ inch. It is essential that the flat surface of such contacts should be truly square to the plunger and parallel to the measuring surface of the fixture if accurate readings are to be obtained.

C.8 Choice of an appropriate type of gauge. In choosing a dial gauge for a certain purpose, one should be selected with a magnification which is ample in relation to the accuracy required on the work under test. For example, if the tolerance on the work is ±0.001 inch then a 0.001 inch gauge is not the most suitable, the movement of the pointer would be too small for accurate reading. For such a case a 0.0001 inch gauge should be selected.

Dial test indicators

A lever type dial test indicator is particularly suitable for making linear measurements where the plunger type of dial gauge would prove unsuitable. BS 2795 'Dial test indicators (lever type) for linear measurement' relates to indicators fitted with an adjustable stylus lever which may be displaced in either of two opposite directions. Unless otherwise stated, the indicators are intended for applications to the work in a direction normal to the stylus level. Requirements are specified for indicators with metric graduations in scale divisions of 0.01 mm with a minimum magnification of × 70 or 0.02 mm with a minimum magnification of × 350, each having a measuring range of at least one revolution of the pointer.

BS 2795

Dial test indicators (lever type) for linear measurement

. . . .

3. Nomenclature and definitions

For the purposes of this British Standard, the nomenclature given in figure 1, together with the following definitions apply.

NOTE. The general definitions given in BS 5233 have been modified as applicable to dial test indicators.

3.1 repeatability. The ability to give the same reading when a dial test indicator is used by the same person and in the same manner and under the same conditions.

3.2 discrimination. Change of reading for small displacements of the stylus contact. The displacement may be caused gradually, cyclically or abruptly.

3.3 error of reading. The difference between a repeated known displacement applied normal to the stylus lever and the mean of the readings of the dial test indicator.

. . . .

9. Performance

9.1 General. A dial test indicator shall comply with the requirements for accuracy within the tolerances given in table 1:

(a) for either direction of operation;

(b) at any position within the range of the pointer

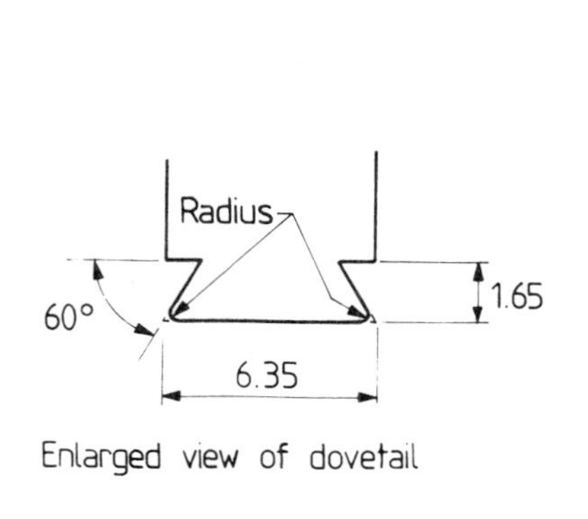

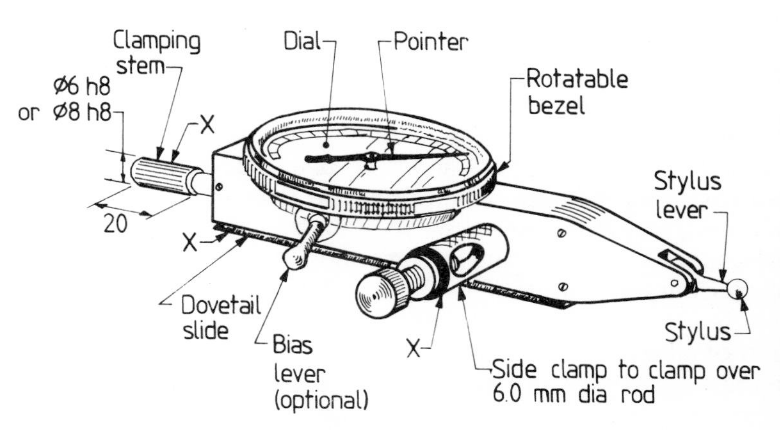

All dimensions are in millimetres

Figure 1. Nomenclature and dimensions (see clause **3**)

Table 1. Tolerances for performance (metric units)

Requirement	Tolerance	
	0.01 mm dial	**0.002 mm dial**
	mm	mm
(a) Repeatability	0.001	0.0005
(b) Discrimination		
Displacement of 0.01 mm	0.003	—
Displacement of 0.002 mm	—	0.0006
(c) Error of reading over an interval of:		
any one division	0.002	0.0005
any five divisions	0.005	0.001
any half revolution	0.01	0.002

movement excluding two divisions at the beginning and end of the travel;
(c) without resort to tapping the indicators;
(d) for an involute form of stylus, at any angle of application up to the maximum specified by the manufacturer.

9.2 Repeatability of reading. The indicator shall repeat its readings under all ordinary methods of operation to within the tolerance given for (a) in table 1.

9.3 Discrimination. The indicator shall be capable of indicating gradual, cyclical or abrupt changes of the order of one division to within the tolerance given for (b) in table 1.

9.4 Accuracy of readings. The errors of readings taken over various intervals of the scale shall not exceed the tolerance given for (c) in table 1.

The permissible error over an interval of reading may be either plus or minus and the tolerance shall not be exceeded for either direction of movement of the pointer.

Indicators with the involute form of stylus shall satisfy the above requirements of all angles of application up to the maximum specified by the manufacturer.

. . . .

11. Marking

Each dial test indicator shall be legibly and permanently marked with the manufacturer's or vendor's name or trademark.

Appendix A

Methods of test

A.1 General. Subclauses **A.2** and **A.3** describe some of the methods of test that may be used. Dial test indicators have two directions of operation and are usable under different conditions of displacement of the stylus lever. Although their versatility poses the testing of several variables, the tests can be organized economically. For example, the tests described for discrimination and error of reading can also inform about repeatability; the test on gauge blocks for repeatability checks discrimination.

A.2 Repeatability. The dial test indicator is securely fixed to a rigid stand that can be moved over a precision surface plate. The stylus on the dial test indicator is then applied, in turn to (1) a semicircular block B and (2) two gauge blocks C of different sizes, all of which are wrung to a steel block A, the surfaces D, E and F of which are lapped (see figure 3). The parts B and C are intended to simulate conditions of contact met with in practice. The stylus contact should approach B and C from various directions and the variation of reading compared with the prescribed tolerance.

It is recommended that the difference in size between two gauge blocks C should be the equivalent of one division movement of the pointer.

In order to test the indicator operating in the reverse direction, the jig can be inverted so that the lapped surface D is wrung to the surface plate, as illustrated in figure 3*b*.

A.3 Discrimination. A practical way of testing a dial test indicator for discrimination is to mount it on a rigid fixture with the stylus contact on the surface of an eccentric precision mandrel* mounted on centres. The throw or total indicator reading (TIR) of the mandrel can be determined beforehand by means of a more sensitive indicator (preferably a high quality electronic instrument known to have a good discrimination) set to read on the mandrel surface. For this test the TIR should be of the order of one division of the dial test indicator under test.

When the dial test indicator is tested against such a mandrel it should indicate the known amount of throw present within the prescribed tolerance.

Another test is to move the dial test indicator along a sine bar tilted at a small angle and to observe the difference of the gradual displacement of the stylus lever from the change in indicator reading.

Lack of discrimination may be caused by stickiness, backlash or hysteresis. It is not usual to isolate each of these effects.

A.4 Errors of readings. The calibration of a dial test indicator can be carried out by using the jig illustrated in figure 3*a* and *b*. Gauge blocks chosen to cover the measuring range of the indicator at

*A convenient form of mandrel for this purpose can be produced by slightly off-setting the centre at one end only. It is then possible to select the required eccentricity by contacting the mandrel at the appropriate position along its length. The surface finish of the mandrel should not exceed 0.05 μm R_a at 0.25 mm cut-off.

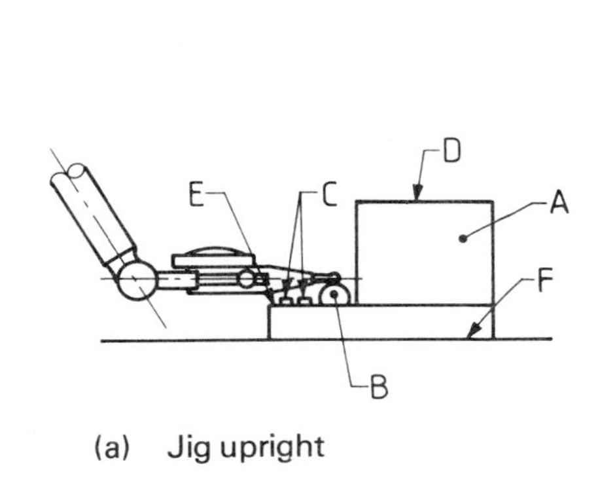

(a) Jig upright

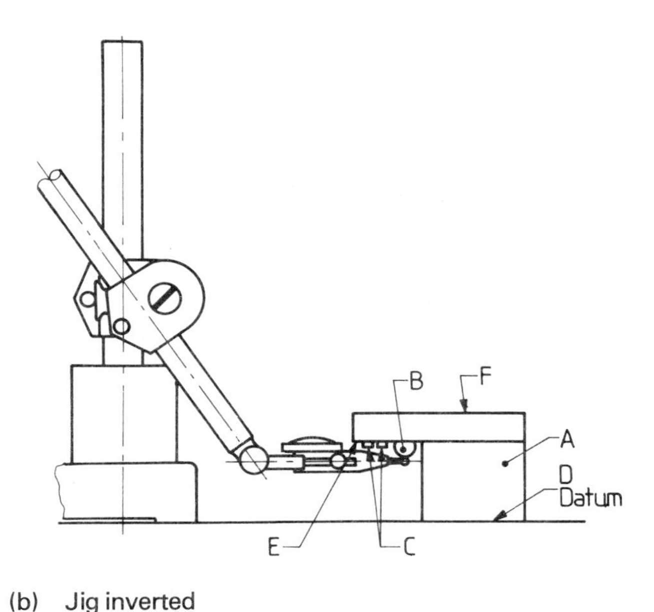

(b) Jig inverted

Figure 3. Tests for repeatability of readings

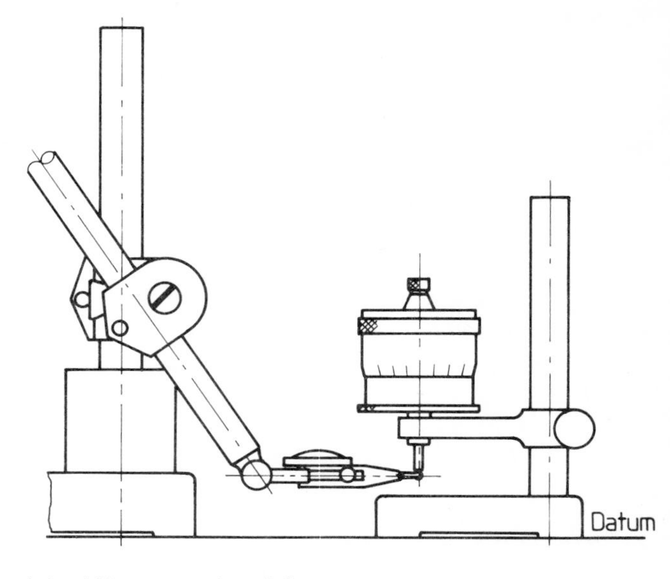

(a) Micrometer head down

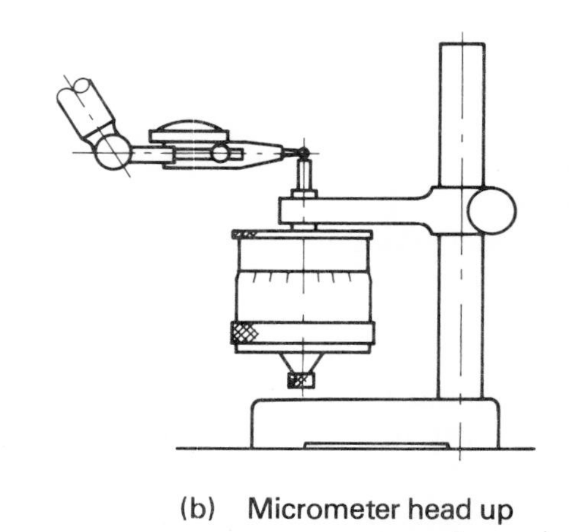

(b) Micrometer head up

Figure 4. Tests for accuracy of readings

suitable intervals are wrung to face E of the block and readings taken.

The errors of reading may be checked against a calibrated micrometer head as shown in figure 4*a* and *b*. In this arrangement the dial test indicator is supported rigidly with the stylus lever nominally at right angles to the axis of the micrometer head.

In both methods of test the mean of several readings at each point should be used to compare with the prescribed tolerance.

Appendix B

Notes on care and use

B.1 General care. In order to retain the accuracy of reading of a dial test indicator, care should be taken to avoid its receiving a sharp blow or any similar mishap which might damage the pivots or bearings.

When not in use, the indicator should be kept in a closed case to avoid the entry of dust into the movement.

The manufacturer's instructions regarding care of the indicator should be carried out. If at any time the movement of an indicator is suspect, it is advisable for the indicator to be returned to the maker for attention.

B.2 Stands: clamping. An indicator should be attached to a robust stand having a mass that is large in proportion to the measuring force of the indicator. If the conditions of measurement require the stand to be moved, it should have a well finished base.

Overhang of the indicator from the clamping knuckles and from the base should be kept to a minimum.

B.3 Principles of use. When comparing an indicator reading on a standard with a reading on a workpiece, it is desirable to use a length standard made up as near as possible to the size of the workpiece to be determined and thus cause a minimum movement of the pointer.

As a general principle, the indicator stand should be kept stationary on a datum plane and the workpiece, brought to the indicator. Where, however, the indicator has to be brought on to the workpiece, on the whole more representative readings and a more reliable mean is obtained when the test indicator is successively withdrawn and re-offered to the surface under measurement.

It should be emphasized that the measurements described above depend upon the flatness of the datum plane on which the workpiece, length standard and stand are moved. Grade A surface plates complying with the requirements of BS 817 offer a suitable order of flatness for the 0.01 mm indicator in general; selected areas of the same plate may be used with the 0.002 mm indicator.

To avoid inaccuracies, dial test indicators should be applied to the work with the stylus lever normal to the direction of measurement. When this is not possible, and a spherical stylus is used, the observed measurement has to be multiplied by the cosine of the angle of inclination of the stylus lever to the correct position. However, if the stylus contact has involute form, the cosine correction is not to be applied, provided that the angle of inclination does not exceed the limit specified by the manufacturer.

. . . .

Precision vernier callipers

The principal features of callipers are shown in Figure 1 of BS 887 'Precision vernier callipers', but details of design may vary between manufacturers.

The accuracy of performance of a calliper is based upon the quality and precision of the various features, for example the accuracy of the dividing of the scales, the quality of the graduation marks, the straightness of the guiding edge and the flatness of the side. These are all features which the manufacturer controls in order that measurements made with the instrument may be reliable within the tolerances given in BS 887.

The standard specifies requirements for the construction, the accuracy (at 20 °C) and the protection of precision vernier callipers with internal and external contact faces. A maximum of 1000 mm measuring length with vernier scale graduated to read to 0.02 mm are covered.

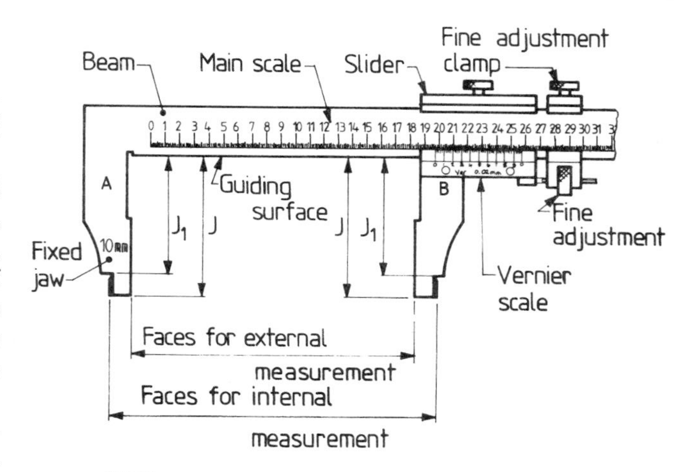

Figure 1. Example and nomenclature of a precision vernier calliper
NOTE. The illustrations are not intended to specify details of design.

BS 887

Precision vernier callipers

. . . .

3. Nomenclature and definitions

For the purposes of this standard, the nomenclature shown in figure 1, together with the following definitions, apply.

3.1 deviation of reading. The difference between the vernier scale reading and the length separating the external measuring faces.

3.2 measuring range. The range of lengths that the external jaws may be used to measure without the vernier scale extending beyond the main scale.

4. Construction

. . . .

4.2 Measuring ranges. The measuring ranges shall be as shown in table 1.

Table 1. Measuring ranges

Metric
mm
0 to 150
0 to 300
0 to 600
0 to 1000

Table 2. Projection of jaws *J* from guiding edge of beam

Measuring range	Projection *J*	
	Minimum	**Maximum**
mm	mm	mm
0 to 150	30	50
0 to 300	50	70
0 to 600	55	75
0 to 1000	60	80

4.3 Beam. The beam shall be long enough to prevent overhang of the sliding jaw assembly at the limit of the nominal measuring range. A recommended minimum section for the beam for each measuring range is shown in appendix A.

4.4 Jaws

4.4.1 *Projections from beam (J).* The projections of the jaws from the edge of the beam shall be as shown in table 2.

The projections J of the two jaws shall be equal to within 0.03 mm, as shall projections J_1. (See figure 1.)

. . . .

4.4.2 *Faces for external measurement.* The lengths of the faces for external measurement shall be at least half that of projection J (see figure 1), and the remainder of the jaws shall be relieved.

The faces shall be flat to within 0.005 mm parallel to within 0.008 mm and shall have a surface finish no coarser than 0.1 μm R_a at a cut-off value of 0.8 mm. The face of the fixed jaw (A) shall be square to the guiding edge of the beam within 0.008 mm per 25 mm. These requirements shall be tested in accordance with **B.2.1** to **B.2.3**.

4.4.3 *Faces for internal measurement.* The internal measuring faces shall be of cylindrical form with a radius not exceeding half their combined width.

Table 3. Faces for internal measurement

Measuring range	Minimum length of face	Minimum combined width of jaws	Maximum error in	
			Combined size	**Parallelism of the faces**
mm	mm	mm	mm	mm
0 to 150	4	5	±0.013*	0.01
0 to 300	6	10		
0 to 600	6	10		
0 to 1000	6	10		

* This requirement does not apply to callipers having separate scales and verniers which permit direct reading of internal and external measurements.

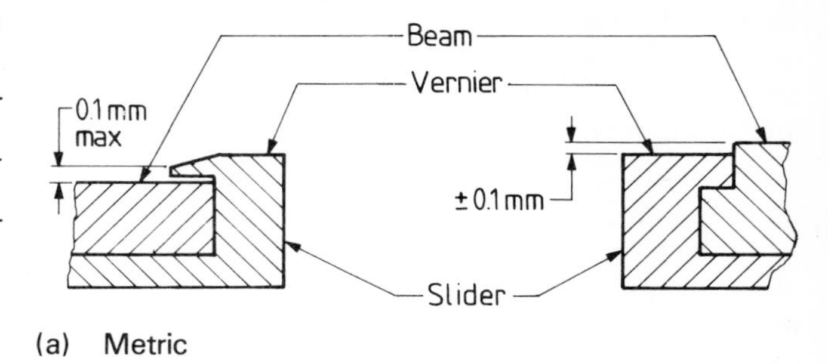

(a) Metric

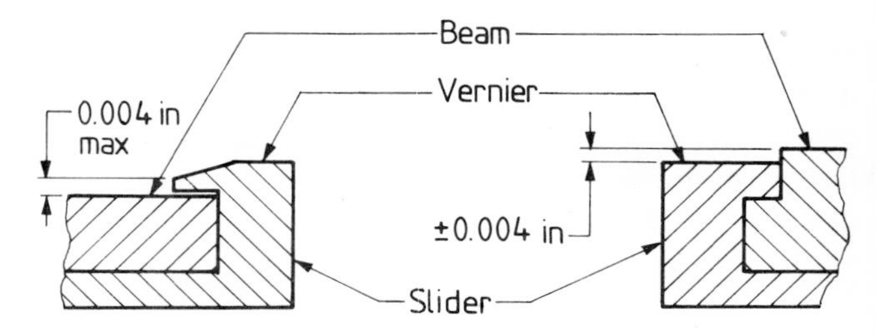

(b) Imperial

Figure 2. Distance of vernier from main scale

The faces shall have a surface finish of not less than 0.4 μm R_a at a cut-off value of 0.8 mm and shall comply with table 3 when tested in accordance with **B.2.4** and **B.2.5**.

4.4.4 *Alignment of jaws.* When the sliding jaw is clamped to the beam in the zero position the surfaces of the jaws (A and B) shall be co-planar to within 0.05 mm.

4.4.5 *Slider.* The slider shall be a good sliding fit over the full working length of the beam and a suitable fitting shall be incorporated to give fine adjustment to the slider. A clamp shall be provided on the slider so that it may be effectively locked after fine adjustment has been made and in such a manner that the setting of the vernier scale relative to the main scale is not altered. Parallelism of the faces for external measurement shall not be affected by the clamping or locking of the slider.

. . . .

6. Accuracy of reading

The deviation of reading at any position within the measuring range of the calliper shall be not greater than as shown in table 5 when tested in accordance with the method specified in **B.1**.

Table 5. Deviation of reading

Measured length	Maximum deviation of reading
mm	mm
Between 0 and 300	±0.02
Over 300 up to and including 600	±0.04
Over 600	±0.06

7. Protection

Each calliper shall be supplied in a suitable protective case and, as a protection against climatic conditions, shall be coated with a thin non-corrosive oil and securely wrapped.

. . . .

Appendix B

Methods of test

B.1 Deviation of reading. Support the calliper along the side of the beam with the jaws horizontal. Measure the separation of the external measuring faces by combinations of calibrated gauge blocks and other end gauges or by other means of similar accuracy such as a length measuring machine. Check the deviation of reading at a minimum of at least five approximately equally spaced positions which cover the measuring range of the main scale and the vernier scale.

Take care to ensure that the method of measurement of the separation of the measuring faces does not distort the calliper. Take care, also, particularly on the longer separations, to ensure that the calliper and the measuring equipment have stabilized at 20 °C and that differences in temperature are not introduced by handling while carrying out the measurements.

B.2 Measuring faces

B.2.1 *Flatness of the faces for external measurement.* The flatness of the faces for external measurement shall be checked either by applying an optical flat or a dial test indicator (see BS 2795) working from a grade A surface plate (see BS 817), or by another suitable method giving the same degree of accuracy.

B.2.2 *Parallelism of the faces for external measurement.* The parallelism of the faces for external measurement shall be checked either by inserting gauge blocks between them at different points on the jaws and at different measured lengths by varying the sizes of the gauge blocks, or by using a measuring machine or an indicating comparator.

Two positions are commended, namely in the 0 to 20 mm range of reading and in the 20 mm length close to full range.

Parallelism shall not be affected by clamping the slider.

B.2.3 *Squareness of the fixed faces for external measurement.* The squareness of the fixed faces for external measurement with the guiding edge of the beam shall be checked either by sighting against a knife-edge square of stock length comparable with the length of the slider, or by another suitable method giving the same degree of accuracy.

B.2.4 *Parallelism of the faces for internal measurement.* The parallelism of the faces for internal measurement shall be checked either by a micrometer with reduced faces or by another suitable method giving the same degree of accuracy.

B.2.5 *Combined width of the internal measuring jaws.* The combined width of the internal measuring jaws shall be checked either by means of a micrometer used in conjunction with a GO ring gauge, the diameter of which exceeds the nominal combined width of the internal measuring jaws by 0.013 mm or by another suitable method giving the same degree of accuracy.

B.3 Scale lines. The thickness of the main scale lines and vernier scale lines shall be checked either by direct measurement with a microscope fitted with a micrometric device, or by using a measuring machine.

Appendix C

Use of a precision vernier calliper

C.1 General. The following guidelines facilitate the use of a precision vernier calliper. Measuring situations may be encountered where the guidelines cannot be adopted totally and in these instances it is recommended that the principles of operation are followed as far as possible.

C.2 Preparation. At the outset the measuring faces should be inspected for freedom from damage, thoroughly cleaned and the freedom of movement of the slider confirmed.

The zero reading with the external measuring faces together should then be checked.

The dimension to be measured should be aligned parallel to the beam of the calliper.

Stress on the calliper should be avoided and, if possible, the item under measurement should not be a load on the calliper.

Wherever possible the calliper and sample should be allowed to acclimatize to the same temperature.

C.3 Application. The contact making and setting is often dictated by the geometry of the item under measurement. For internal measurements, the jaws are usually set somewhat short and the amount of possible rotation of the calliper estimated by touch. The faces are gently opened and the chordal movement reduced until it is eliminated; care is essential to keep control and to avoid undue springing of the tips of the jaws.

A similar method may be followed for external measurement, starting with the jaws set long. As the fine adjustment gently closes the jaws on to the item, slight rotation of the beam to and fro assists the flat faces of the jaws and the item to align.

Prior to any final adjustment of the slider, play should be removed by lightly tightening its locking screw.

Consistency in measuring may be gained by first using the calliper on a standard length bar assembly with accessories complying with BS 5317. This is particularly useful when the operator is required to 'feel' the jaws in contact with the item to be measured.

Vernier height gauges

BS 1643 'Vernier height gauges' relates to gauges graduated in metric units up to 1 metre, to read with the vernier to 0.02 mm (provision is also made for imperial and dual imperial/metric gauges but these are not included in this manual).

BS 1643
Vernier height gauges

. . . .

2. Nomenclature

For the purposes of this standard the nomenclature given in figure 1 has been adopted.

. . . .

9. Accuracy

The test for accuracy of reading shall be made when the height gauge is standing on a precision surface plate complying with BS 869 'Toolmakers' flats and high precision surface plates'. The test measurements, as recommended in the appendix, shall be made at a distance of 1 inch from the end of the jaw. The test shall be made by using standard gauges, and when tested at the standard reference temperature of 20 °C and with the slider finally clamped, the error in the reading at any position within the measuring range of the height gauge shall not be greater than the appropriate amount given in table 3.

. . . .

Table 3. Maximum permissible errors (metric readings)

Measuring range of height gauge	Maximum permissible error in reading
mm	mm
Up to and incl. 300	±0.02
Above 300 up to and incl. 600	±0.04
Above 600	±0.05

. . . .

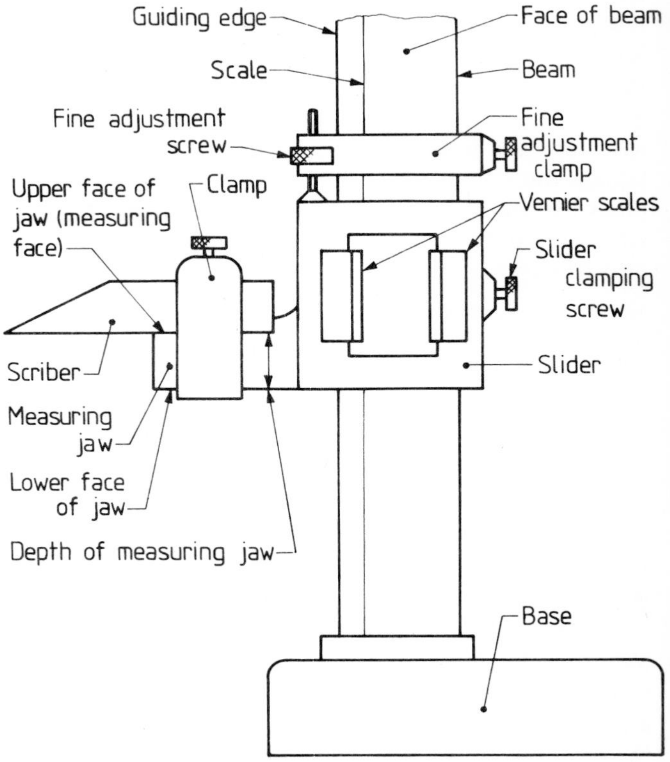

Figure 1. Nomenclature for vernier height gauges

NOTE. The illustration is diagrammatic only, and is not intended to illustrate details of design.

13. Marking

Each height gauge shall have legibly and permanently marked upon it, in characters not less than 0.04 inch high, the manufacturer's name or trademark.

. . . .

Appendix

Recommended methods of testing vernier height gauges

The most important points which require attention in the testing of vernier height gauges are the straightness, and the squareness or the parallelism of the working faces of the beam, measuring jaw and scriber, and the accuracy of the scale readings.

(a) Hardness of working faces. The hardness of the working faces may be assessed approximately by the following convenient method, namely, the careful application of a hand scriber. Such a scriber consists of a small round Swiss file, the end of which is ground to a smooth hemisphere of about 0.025 inch radius. This tool is mounted on a wooden handle and is used for comparing the surface under test with that of a 700 DPN standard of hardness by drawing the scriber across each face in turn, using as nearly the same force as possible on the standard and the face under test and comparing the resulting marks.

(b) Flatness of base. For assessing the general flatness of the base a blueing test against a known flat surface may be employed. A 'knife-edge' straightedge may afterwards be applied to the surface using an illuminated background. This visual test is sensitive to well within 0.0001 inch.

(c) Rigidity of beam. The change in parallelism between the surface of the scriber and the base when a load of 2 lb weight is applied to the tip of the scriber and the measuring jaw is in its highest position, in a plane containing the measuring jaw, may be checked with a suitably mounted indicator and a precision square block as shown in figure 5.

NOTE 1. Squareness of the precision block is not necessary for this test but it is required for the test illustrated in figure 6.

NOTE 2. For methods of producing and testing a precision square block, see appendix B of BS 939 'Engineers' squares'. This precision block will be needed for tests described under (f).

(d) Straightness and squareness of beam. The tests for straightness and squareness of the beam to the tolerances laid down in clause 5 may be made very conveniently with the aid of a precision square block used in conjunction with a precision surface plate. The method is illustrated in figure 6 and is self-explanatory.

(e) Flatness of the working surfaces of the measuring jaw and scriber. The flatness of these surfaces may be checked by applying a 'knife-edge' straightedge or, if the finish of the surface is sufficiently good, by applying an optical flat.

(f) Parallelism of the measuring jaw and scriber with the base. The test for parallelism may be carried out with an indicator, suitably mounted on a stand, used as illustrated in figure 5, standing the indicator base directly on the surface plate or on precision parallel blocks as required.

(g) Thickness of graduations. The thickness of the graduations may be checked with a microscope fitted with an eyepiece graticule scale engraved in 0.001 inch intervals, when the graduation line thickness may be read directly against the graticule scale.

(h) Dimension T. (figure 2). The dimension T may be checked with a 0.004 inch feeler gauge.

(k) Accuracy. The accuracy of reading of vernier height gauges may be checked by making settings on end-bar and slip-gauge combinations of known size, the instrument and the gauges standing side by side on a precision surface plate. The scriber should be in position for this test. The size of the standard combinations used should be chosen so as to cover a number of points over the range of the instrument and at the same time to provide a check on the accuracy of the vernier scale.

Care should be taken, particularly on longer lengths, to avoid heating the height gauge or the standard combination by warmth from the hands. Springing of the measuring jaw should also be avoided.

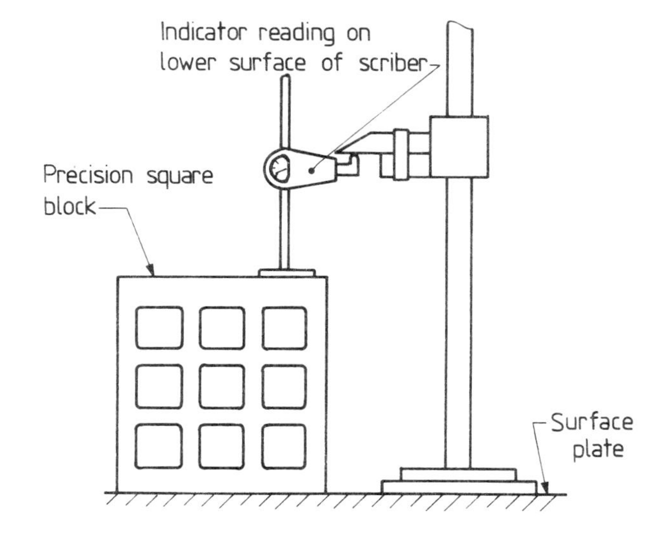

Figure 5.

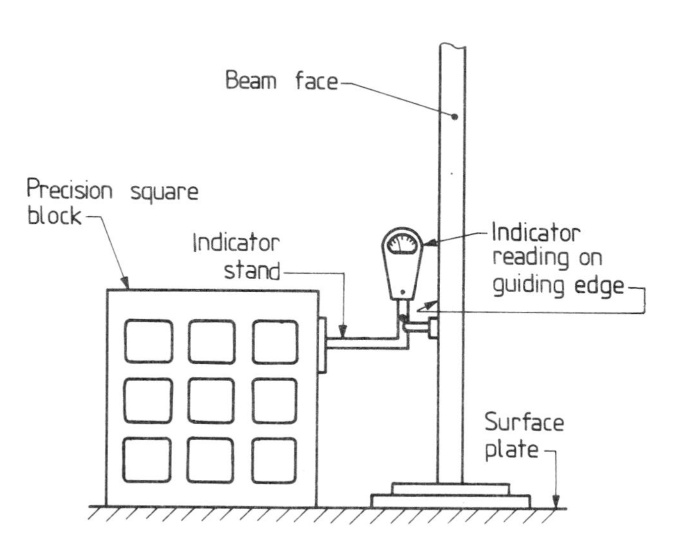

Figure 6.

Micrometers

External micrometers having both fixed and interchangeable or sliding anvils are covered in BS 870 'External micrometers'. The standard provides for metric and imperial micrometers with various measuring ranges up to 600 mm or 24 inches. Setting gauges are also specified for micrometers above 25 mm or 1 inch.

BS 870

External micrometers

....

2. Nomenclature

For the purposes of this British Standard the nomenclature given in figure 1 has been adopted.

....

10. Accuracy

When tested at the standard reference temperature of 20 °C each micrometer shall conform to the limits of error given in tables 3 to 6 according to the type of micrometer.

The zero reading (except in the case of micrometers with sliding anvils) shall be correct within the tolerance given in tables 3 and 4, column 4. In the case of a micrometer with a vernier, this requirement shall apply to the fiducial line and vernier simultaneously.

The maximum range of error in the calibration of the micrometer screw shall not exceed the tolerance given in column 5.

The alignment of the spindle and the anvil shall be such that the prolongation of the axis of the spindle shall not deviate from the centre of the anvil face by more than the amounts given in column 6.

....

11. Zero setting

A suitable gauge for setting the zero reading shall be provided with each micrometer capable of measuring a length of more than 1 inch or 25 mm.

....

14. Marking

Each micrometer shall have legibly and permanently marked upon it, in characters not less than 0.025 inch high, the manufacturer's name or trademark.

Section three. Setting gauges

....

17. Accuracy

When measured at the standard reference temperature of 20 °C, the setting gauges shall conform to the following limits of error:

(a) Disks. The disks shall be uniform in diameter to within 0.000 05 inch or 0.001 mm. The mean diameter shall be within ±0.000 05 inch or ±0.001 mm of nominal size.

(b) Rod gauges. The error in length shall not exceed the appropriate amount given in column 2 of table 8.

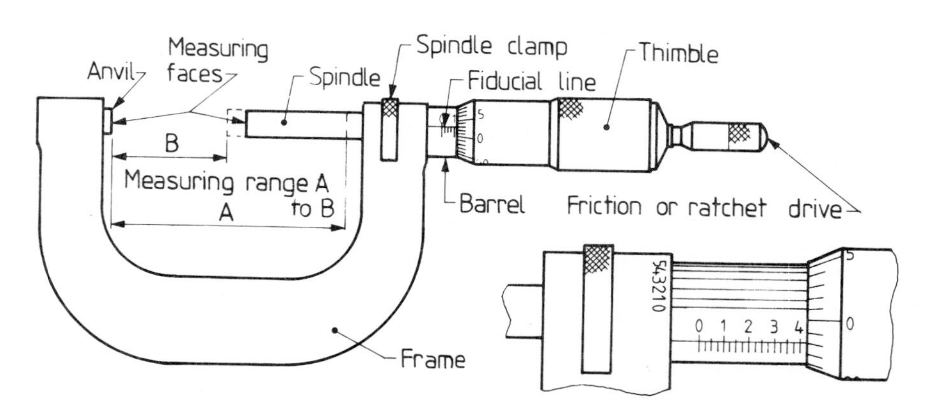

Figure 1. Nomenclature for micrometers

NOTE. The illustrations are diagrammatic only, and are not intended to illustrate details of design.

Table 4. Maximum permissible errors

Micrometers with fixed anvils (metric readings)

1	2	3	4	5	6
Measuring range of micrometer	**Measuring faces**		**Zero reading**	**Range of error of transverse of micrometer screw**	**Error in alignment**
	Flatness	**Parallelism**			
mm	mm	mm	mm	mm	mm
0–13, 0–25	0.001	0.003	±0.001	0.003	0.05
25–50	0.001	0.003	±0.001	0.003	0.08
50–75	0.001	0.005	±0.003	0.003	0.13
75–100	0.001	0.005	±0.003	0.003	0.15
100–125	0.001	0.005	±0.003	0.003	0.20
125–150	0.001	0.005	±0.003	0.003	0.23
150–175	0.001	0.005	±0.003	0.003	0.25
175–200, 200–225	0.001	0.005	±0.003	0.003	0.3
225–250, 250–275, 275–300	0.003	0.008	±0.005	0.003	0.4
300–325, 325–350, 350–375	0.003	0.013	±0.008	0.005	0.45
375–400, 400–425, 425–450	0.003	0.015	±0.008	0.005	0.5
450–475, 475–500, 500–525	0.003	0.018	±0.010	0.005	0.65
525–550, 550–575, 575–600	0.003	0.020	±0.010	0.005	0.75

See table 6, NOTE 1 and NOTE 2.

Table 6. Maximum permissible errors

Micrometers with interchangeable or sliding anvils (metric readings)

1	2	3	4	5	6
Measuring range of micrometer	**Measuring faces**		**Zero reading**	**Range of error of traverse of micrometer screw**	**Error in alignment**
	Flatness	**Parallelism**			
mm	mm	mm	mm	mm	mm
0–50	0.001	0.008	±0.003	0.003	0.13
0–100	0.001	0.010	±0.003	0.003	0.2
100–200	0.001	0.013	±0.003	0.003	0.4
200–300	0.003	0.015	±0.005	0.003	0.5
300–400	0.003	0.020	±0.008	0.005	0.6
400–500	0.003	0.025	±0.010	0.005	0.75
500–600	0.003	0.030	±0.013	0.005	1.0

NOTE 1. Where a ratchet or friction drive is fitted, this shall be made use of in taking the zero reading. The reading taken shall be that obtained as soon as the ratchet or friction drive has slipped, after the spindle has been slowly advanced to the anvil. Rapid rotation of the ratchet or friction drive will give false readings.

NOTE 2. The range of error in the calibration of the micrometer screw is independent of the zero setting; it represents the maximum difference between the ordinates of the curve of error in the readings obtained on calibrating the screw at a number of positions along its traverse. These positions are chosen so as to check not only the progressive errors in the screw but also the accuracy of the graduations round the thimble, the fiducial line, and the vernier on the barrel if present.

The error in parallelism of the faces of flat ended rods shall not exceed the appropriate amount given in column 3 of table 8.

. . . .

Table 8. Maximum permissible errors in setting rod gauges (metric readings)

1	2	3
	Maximum permissible errors	
Size of setting gauge	In length	In parallelism of flat end faces
mm	mm	mm
50, 75, 100, 125	±0.002	0.001
150, 175, 200	±0.003	0.001
225, 250, 275	±0.004	0.001
300, 325, 350, 375, 400, 425	±0.005	0.003
450, 475, 500, 525, 550, 575	±0.006	0.005

. . . .

19. Marking

Each setting gauge shall have legibly and permanently marked upon it:

(a) the manufacturer's name or trademark;
(b) the nominal size.

Appendix A

Recommended methods of testing micrometers

The most important points which require attention in the testing of micrometers are the accuracy of the measuring faces in flatness and parallelism and the accuracy of the micrometer screw.

(a) Measuring faces. The flatness of the measuring faces is best tested by means of an optical flat. Having cleaned the faces thoroughly, the glass plate is brought into contact with each one in turn. Unless the faces are perfectly flat a number of coloured interference bands will be seen on their surfaces, and the shape and number of these bands indicate the degree of flatness of the face. To comply with the specified tolerance on flatness of 0.000 05 inch, not more than five bands of the same colour should be visible on either of the faces.

The bands are rendered much more distinct if the test is carried out in the light from a mercury vapour lamp, particularly if the bands are then viewed through a green filter.

The parallelism of the measuring faces of 0–1 inch micrometers may be tested very conveniently by utilizing the same principle of optical interference. For this purpose use is made of a number of optical flats, of different thicknesses, the opposite faces of which are accurately parallel as well as flat. These optical flats are placed in turn between the measuring faces as though their thicknesses were being measured, and in each case attention is paid to the interference bands which are formed simultaneously on the two measuring faces. By carefully moving the glass plate between the faces the number of bands visible on one face is reduced to a minimum and those on the opposite face are then counted: there should not be more than 10 of the same colour for the parallelism to comply with the requirements of the standard.

The object of making the test with flats of different thicknesses is to check the parallelism of the faces for different angular positions of the face on the micrometer spindle. To do this effectively four flats are generally used, the thicknesses of which differ in succession by approximately a quarter of 0.025 inch so that the test is carried out at four positions during a complete rotation of the micrometer face.

If desired, the same method can be used for testing the parallelism of the faces of larger micrometers up to about 4 inches. Two of the optical flats are then wrung on to the ends of a combination of slip gauges and the whole combination thus formed is used as a parallel-ended test piece between the measuring faces. The test can be carried out in four positions as before by changing the length of the slip gauge combination between the glass end-pieces by 0.006 inch or 0.007 inch in succession.

For the larger sizes of micrometers the optical test for parallelism of the faces is too sensitive in relation to the tolerance. In such cases it is usual to make test pieces of suitable length for each size of micrometer by soldering $\frac{1}{8}$ inch or $\frac{3}{16}$ inch steel balls on to the ends of a length of $\frac{1}{4}$ inch diameter steel rod. These test pieces are inserted between the measuring faces of the micrometer at a number of positions round their periphery. Any resulting variation which is found to occur in the micrometer reading should not exceed the tolerance prescribed for the parallelism between the measuring faces.

Instead of making a separate test piece for each size of micrometer, a single one capable of adjustment for length may be used.

(b) Calibration of micrometer screw. The accuracy of the readings of a 0–1 inch micrometer or a 0–25 mm micrometer is usually checked by taking readings on a series of slip gauges. The sizes of these gauges should be chosen so as to test the micrometer not only at complete turns of its thimble but also at intermediate positions. This is required as a check on the accuracy of the

graduations round the thimble. The following series of gauges serve for testing an inch micrometer both for progressive errors throughout its range and for periodic errors:

0.105, 0.210, 0.315, 0.420, 0.500, 0.605, 0.710, 0.815, 0.920, and 1.000 inch.

For a metric micrometer a convenient series of gauges is:

2.5, 5.1, 7.7, 10.3, 12.9, 15.0, 17.6, 20.2, 22.8, and 25.0 mm.

Both these series give readings which work round the thimble twice over and so provide a double check on any periodic error which may be present.

Series with somewhat fewer gauges could be used, if desired, as follows:

0.130, 0.250, 0.385, 0.500, 0.615, 0.750, 0.870, and 1.000 inch.
3.1, 6.5, 9.7, 12.5, 15.8, 19.0, 21.9, and 25.0 mm.

These series would not, however, give quite so close a test of progressive and periodic errors as those first described.

As a check on the vernier graduations of an inch micrometer, readings should be taken on the following series of gauges:

0.1000, 0.1002, 0.1004, 0.1006, 0.1008, and 0.1010 inch.

In order to facilitate the accurate reading of the micrometer during such a calibration, the thimble may be viewed under a microscope of low power magnification.

In the case of micrometers of above one inch capacity, the errors in the traverse of the micrometer screw can still be checked with slip gauges of the sizes mentioned above by judiciously clamping the micrometer to a surface plate and fixing a temporary anvil, with a rounded face, close to the face of its spindle. Alternatively, an indicator having a sensitivity of at least 0.000 05 inch may be used with advantage as the temporary anvil. The errors in the traverse of the micrometer can then be read with greater certainty on the scale of the indicator instead of on the micrometer itself. When an indicator is used the micrometer is set accurately 'line-to-line' at the exact readings corresponding to the sizes of the slip gauges inserted between the micrometer face and the contact point of the indicator.

If it is not possible to fix an indicator directly opposite the face of the micrometer spindle of the smaller sizes of micrometers, such as 2 inches to 3 inches or 3 inches to 4 inches, a reading may be obtained by inserting a simple form of lever between the face and the contact point of the indicator, which is secured in some convenient position.

Appendix B

Information to be supplied by the purchaser with his enquiry

NOTE. The following items are for the guidance of purchasers in framing an enquiry, so that their requirements may be fully understood.

1. Range of micrometer, and whether inch or metric reading.

2. Any particular requirements as to type and finish of frame.

3. Whether a vernier is required.

NOTE. It is recommended that verniers should not be incorporated in micrometers with a capacity greater than 4 inches, or in micrometers of the adjustable type, or in metric micrometers of any capacity.

4. Whether a spindle clamp is required.

5. Whether a friction or ratchet drive is required, and whether a special pressure is required.

6. Whether a National Physical Laboratory certificate is required.

. . . .

BS 959 'Internal micrometers (including stick micrometers)' is in two sections and covers internal micrometers (metric and imperial) and stick micrometers (imperial only).

BS 959

Internal micrometers (including stick micrometers)

. . . .

9. Accuracy

When tested at the standard reference temperature of 20 °C each measuring head with its associated extension rods (and collars if used) shall conform to the limits of error given in table 1, when the thimble reading is zero.

Table 1. Maximum permissible errors

Internal micrometers

Measuring range	**Maximum permissible error in length**
mm	mm
25 up to and including 150	±0.005
Above 150 up to and including 300	±0.010
Above 300 up to and including 450	±0.015
Above 450 up to and including 600	±0.020
Above 600 up to and including 900	±0.025

The measuring head shall not show a range of error in the traverse of the micrometer screw greater than 0.003 mm.

NOTE. The range of error in the calibration of the micrometer screw is independent of the zero setting; it represents the maximum difference between the ordinates of the curve of error in the readings obtained on calibrating the screw at a number of positions along its traverse. These positions are chosen so as to check not only the progressive errors in the screw but also the accuracy of the graduations round the thimble and on the fiducial line.

. . . .

10. Rigidity

When the micrometer is assembled, using any combination of extension rods and spacing collars, and this assembled unit is supported at its ends as shown in figure 2, the deflection of the mid-point under an applied load of 1 lb weight shall not exceed 0.002 inch for an assembly 6 inches long with an increase of 0.002 inch for each increase in length of 1 inch (e.g. a maximum deflection of 0.050 inch for an assembly 30 inches long).

11. Handle

Each micrometer set for the measuring range 1 inch to 2 inches shall be supplied with a suitable detachable handle; the inclusion of a handle in micrometer sets for larger ranges of measurements shall be optional.

. . . .

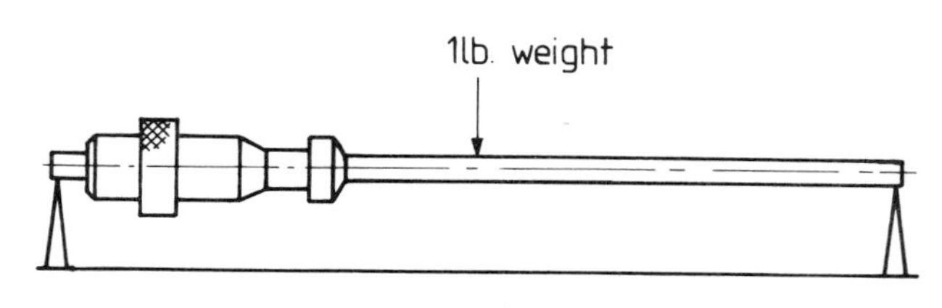

Figure 2. Application of load in testing for rigidity

14. Marking

Each measuring head and component in a set shall have legibly and permanently marked upon it, in characters not less than 0.025 inch high, a common serial number. The measuring head shall be similarly marked with the manufacturer's name or trademark.

Section two. Stick micrometers

. . . .

16. Nomenclature

For the purposes of this standard the nomenclature given in figure 3 has been adopted.

. . . .

20. Accuracy of micrometer unit

The micrometer unit, with terminal piece in position, shall record correctly throughout its range to within ±0.0002 inch. The micrometer head shall incorporate a means of adjusting the zero setting.

21. Accuracy of extension rods

The lengths of the extension rods shall agree with their nominal sizes at 20 °C within the following tolerances:

Nominal length	Tolerance on mean axial length
inches	inches
Up to and including 3	±0.0001
Above 3 up to and including 6	±0.0002
Above 6 up to and including 12	±0.0003
Above 12 up tc and including 24	±0.0005
Above 24 up to and including 36	±0.0007

* The approximate equivalent hardness numbers on the Rockwell C scale are 62 and 59 respectively.

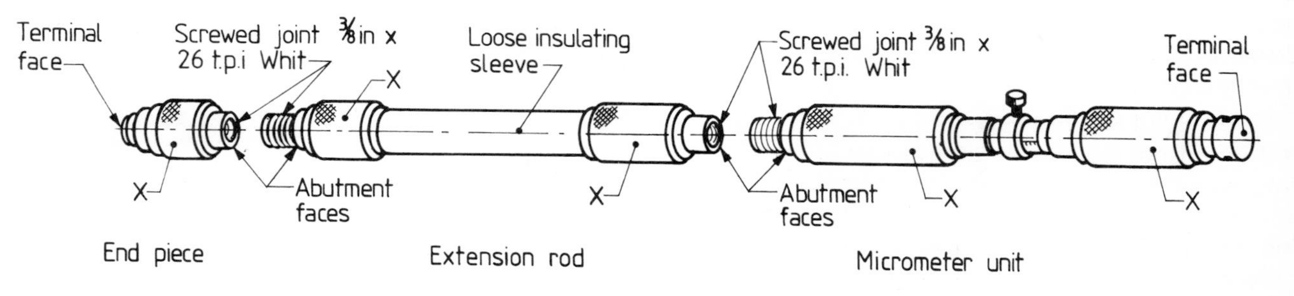

Figure 3. Nomenclature for stick micrometer

22. Rigidity

When a six foot length of the micrometer is assembled, using any combination of rods, and this assembled unit is supported at its ends as shown in figure 2, the deflection of the mid-point under an applied load of 1 lb weight shall not exceed 0.15 inch.

NOTE. When lengths of between 4 and 12 feet are assembled for use, the unit should be supported at least at the ends and in the centre.
For lengths over 12 feet, the unit should be supported at regular intervals and the maximum separation between the supports should not exceed 6 feet. Failure to support may result in damage to the instrument.

23. Marking

The micrometer unit shall be marked '6–7 inches with end piece' or '12–13 inches with end piece' and the micrometer head and each extension rod of a set shall be legibly and permanently marked with a common identification number. The micrometer unit shall be similarly marked with the manufacturer's name or trademark.

24. Gap setting gauges

Gap setting gauges, if supplied, shall have a gap of either $6\frac{1}{2}$ inches or $12\frac{1}{2}$ inches, according to the length of the micrometer unit, the permissible errors on the length of the gap being ±0.0001 and ±0.0002 inch respectively. The faces of the gap shall be hardened and shall be flat to within 0.0001 inch.

Appendix A

Method of testing internal micrometers

The checking of the accuracy of internal micrometer readings is best carried out in two parts: (1) determining the accuracy of the traverse of the measuring head and (2) determining the accuracy of the overall lengths when the measuring head, set to zero, is associated with the various extension rods in turn.

(1) To test the traverse of the measuring head it is convenient to clamp it in a vee-block with its axis in line with a sensitive indicator such as for example a 0.0001 inch dial gauge. The indicator is fitted with a flat contact face which is juxtaposed to the rounded contact face of the measuring head. The latter is set initially to read exactly zero and a slip gauge 0.1 inch longer than the travel of the measuring head is inserted between the two

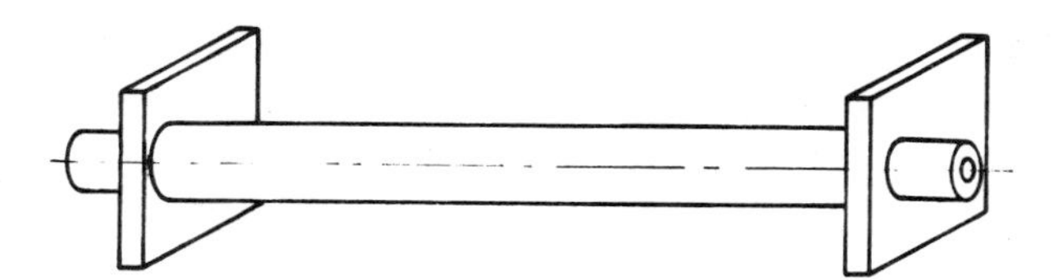

Figure 4. Gap setting gauge

contact faces. The indicator is then also set to read zero. To test the accuracy of the reading of the head at any point it is set exactly to that reading and the initial size of the slip gauge is reduced in length by the value of the reading. Any error in the reading is then revealed by a corresponding departure of the indicator pointer from its initial zero position.

It is usual to test a measuring head at a number of readings along its traverse and, in order to check not only the progressive but also the periodic errors, it is convenient to test at the following series of readings:

Travel of micrometer screw	Series of readings					
inches	inches					
$\frac{1}{4}$	0	0.045	0.090	0.135	0.180	0.250
$\frac{1}{2}$	0	0.105	0.210	0.315	0.420	0.500
1	0	0.105	0.210	0.315	0.420	0.500
	0.605	0.710	0.815	0.920	1.000	
mm	mm					
5	0	1.4	2.3	3.2	4.1	5.0
6	0	1.3	2.6	3.9	5.4	6.5
10	0	5.9	6.8	7.7	8.6	10.0
13	0	1.3	2.6	3.9	5.2	6.5
	7.8	9.1	10.4	11.7	13.0	
25	0	2.4	5.1	7.7	10.3	12.5
	14.9	17.6	20.2	22.8	25.0	

(2) The accuracies of the various overall lengths can be conveniently determined by means of a vertical comparator consisting of a flat, horizontal base plate over which is supported a sensitive indicator having a flat contact face set accurately parallel to the base plate. Having fixed the indicator at a height to suit the overall length to be measured, its reading is noted when the internal micrometer is passed under it in the maximum position.

A comparative reading is then taken on an accurate length-gauge of the same nominal size. From these two readings and the known size of the gauge the desired overall length is readily determined.

BS 1734 'Micrometer heads' relates to micrometer heads supplied for assembly with measuring tools or machine tools and having measuring ranges of:

(a) ½ inch
(b) 1 inch
(c) 13 mm
(d) 25 mm

It applies to micrometer heads with rotatable and non-rotatable spindles, classed according to the following ranges of thimble diameter (diameter D, Figure 4.1):

Type 1 below 1 inch;
Type 2 1 inch and below 2 inches;
Type 3 2 inches and over.

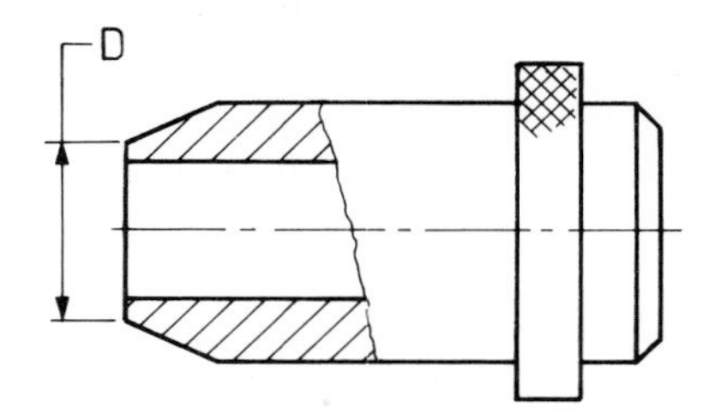

Figure 4.1 *Diameter of thimble at bevel edge*

BS 1734

Micrometer heads

. . . .

2. Nomenclature

For the purpose of this standard the nomenclature given in figure 2 has been adopted.

. . . .

10. Accuracy

When tested at the standard reference temperature of 20 °C each micrometer head shall comply with the tolerances given in tables 3 and 4.

Repetition of reading. Micrometer heads of types 1 and 2 shall be capable of repetition of reading over any portion of the scale to 0.000 05 inch (0.001 mm). Type 3 micrometer heads shall be capable of repetition of reading over any portion of the scale to 0.000 02 inch (0.0005 mm).

. . . .

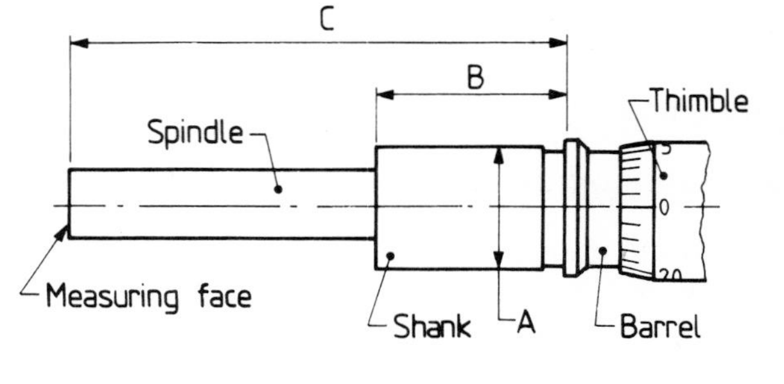

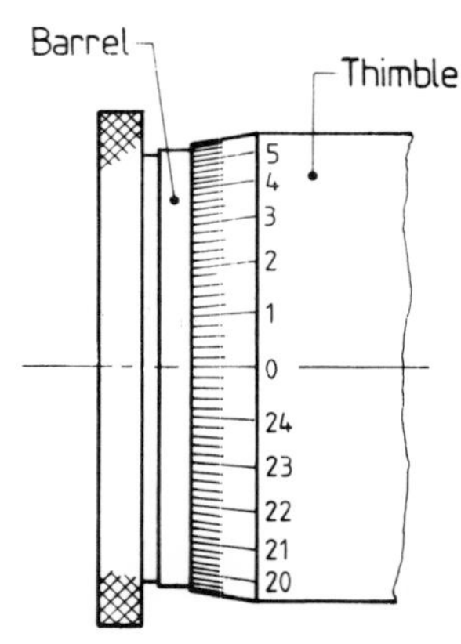

Figure 2. Nomenclature for micrometer heads

Table 3. Maximum permissible errors for type 1 micrometer heads

1	2	3	4	5
Diameter of thimble	**Flatness of measuring face**	**Squareness of measuring face to axis of spindle (measured over diameter of face)**	**Squareness of measuring face to outside diameter of shank (measured over diameter of face)**	**Maximum error of traverse of measuring face at any position**
Type 1 Below 1 in	0.000 05 in (0.001 mm)	0.000 05 in (0.001 mm)	0.0003 in (0.008 mm)	0.0001 in (0.003 mm)

Table 4. Maximum permissible errors for types 2 and 3 micrometer heads

1	2	3	4	5
Diameter of thimble	**Flatness of measuring face**	**Squareness of measuring face to axis of spindle (measured over diameter of face)**	**Squareness of measuring face to outside diameter of shank (measured over diameter of face)**	**Maximum error of traverse of measuring face at any position**
Type 2 1 in and below 2 in	0.000 03 in (0.0008 mm)	0.000 03 in (0.0008 mm)	0.0001 in (0.003 mm)	0.0001 in* (0.003 mm)*
Type 3 2 in and over	0.000 02 in (0.0005 mm)	0.000 03 in (0.0008 mm)	0.000 03 in (0.0008 mm)	0.0001 in* (0.003 mm)*

* The periodic error shall not exceed ±0.000 05 in (±0.001 mm) for type 2 micrometer heads and ±0.000 02 in (±0.0005 mm) for type 3 micrometer heads.

13. Marking

Each micrometer head shall have legibly and permanently marked upon it, in characters not less than 0.025 inch high, the manufacturer's name or trademark.

Gauge blocks

There are two distinct and well-recognized uses for gauge blocks; they are in general use for precise measurement wherever accurate work sizes are required or they may be reserved as standards of lengths to be used with very high magnification comparators to establish the sizes of gauge blocks in general use. BS 4311 'Metric gauge blocks', Part 1 'Gauge blocks' provides for three grades of gauge blocks for general use (grades 0, I and II), a special grade intended essentially for use in measuring other gauge blocks (calibration grade) and a further special grade which may be used on those rare occasions when the combination of high geometrical accuracy and close length tolerances is required (grade 00).

The magnitudes and dispositions of the tolerances for the various grades were determined with the different uses in mind and it should be noted that it is not only uneconomic but may also be disadvantageous from a functional point of view to select one of the special grades because it appears to be 'the best' when it may not be the one most suitable for the user's requirements.

When deciding which grade is required for a particular function, it may be useful to consider not only the applications envisaged for the blocks in question but also the accuracy of the equipment with which they are likely to be used. These two aspects should be taken into account since there is little point in using a grade of block for a purpose which will not take full advantage of its qualities and even less point in recognizing this fact by locking the gauges away and seldom using them at all.

The grade most commonly used in the production of components, tools and gauges is grade I; for applications such as preliminary setting up or comparatively rough checking where product tolerances are relatively wide, grade II blocks may be suitable. The tolerances on these blocks are somewhat coarser than those normally associated with precision measuring equipment but it is realized that blocks of this type are widely used and the standard provides some control over their geometry. The tolerance on the length of grade I and II blocks is arranged to allow for wear which takes place rather rapidly when these blocks are first put into use and helps to ensure that they have as long a working life as possible, consistent with their remaining within the specified limits.

For gauge inspection and higher precision work in general, grade 0 blocks may be required.

Calibration grade gauge blocks should not be used for general inspection work, even on gauges. They are intended for calibrating other gauge blocks and for checking gauges in a comparator where the tolerance is one or two micrometres or less. In such cases, the required accuracy of measurement demands that the actual size of the reference gauge should be used in the computation; that this size is very close to nominal is of less importance than that it should be accurately known from calibration of the block gauge. These block gauges are therefore required to have a high quality of geometrical form (flatness and parallelism) but relatively large tolerances on length can be allowed. Their calibrated sizes can be relied on to accuracies up to 0.05 μm or even 0.02 μm.

Where there is a genuine need to use gauge blocks to very high accuracy without recourse to a calibration chart, i.e. to assume that each block is true to its nominal size, the grade 00 may be used. These gauge blocks are made to the same tolerances for flatness and parallelism as the calibration grade but the length tolerances are much smaller. To use those gauge blocks as standards, assuming them to be of nominal size, *may* introduce systematic errors into the measurements up to the value of the total tolerance on the gauge blocks used. This could be as much as twice the uncertainty of calibration of the same gauges, i.e. the use of calibrated sizes could reduce this systematic error by half.

BS 4311 specifies requirements for gauge blocks up to 100 mm. For lengths in excess of 100 mm the use of length bars is recommended (see page 107).

BS 4311

Metric gauge blocks

Part 1. Gauge blocks

. . . .

2. Nomenclature and definitions

For the purposes of this British Standard, the nomenclature as shown in figures 1 and 2 and the definitions given below apply.

2.1 gauge length

2.1.1 *general.* The definition of gauge length takes into account the thickness of one wringing film, which is associated with a gauge block in its customary use, and is referred to the standard reference temperature of 20 °C.

2.1.2 *gauge length.* The normal distance L (see figure 1) measured between the centre of the free measuring face of the gauge block and a surface, having a finish similar to that of the gauge block itself, upon which the other measuring face of the gauge block has been wrung.

2.2 face length. The larger of the two dimensions other than the gauge length (figure 2).

2.3 face width. The smaller of the two dimensions other than the gauge length (figure 2).

2.4 measuring faces. The faces whose distance apart forms the gauge length L (figure 1).

2.5 side faces. The faces other than the measuring faces (figure 2).

2.6 flatness tolerance. The maximum permissible distance separating two parallel planes within which the measuring face can just be enclosed.

2.7 parallelism tolerance between measuring faces. The maximum permissible variation in length.

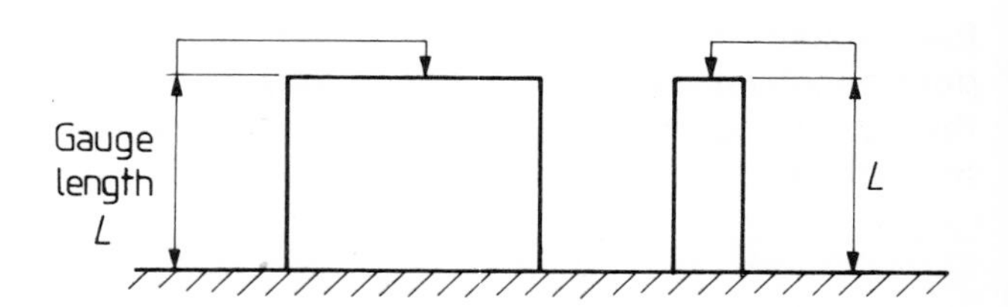

Figure 1. Gauge length L

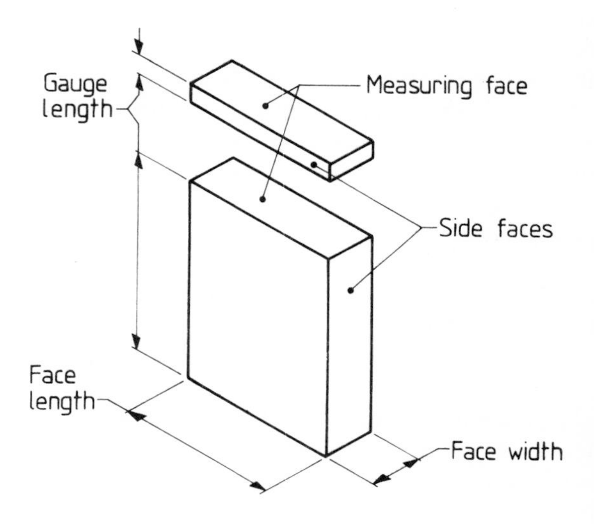

Figure 2. Nomenclature

2.8 squareness tolerance between side and measuring faces. The maximum permissible distance separating two parallel planes within which a side face can just be enclosed when the planes are perpendicular to a measuring face.

. . . .

5. Finish

5.1 General. The measuring faces shall be free from surface defects likely to affect the use of the gauge blocks adversely or, in the case of a new gauge block, to detract from its appearance. They shall be finished by high grade lapping to within the tolerances specified in clause **6** and shall have surfaces which will wring readily.

NOTE. A high mirror-like finish on the measuring faces is not essential, for it should be recognized that on some materials such a finish tends to reveal by contrast the minute scratches produced by the wringing action; in general these scratches may not affect the accuracy of the blocks but can give a false impression of its deterioration.

The side faces shall have a ground or similar quality finish.

All sharp edges shall be removed to the equivalent of a radius not exceeding 0.3 mm by a process which does not impair the wringing property of the measuring faces.

5.2 Surface texture. The quality of the finish of the lapped faces plays an important part in the firmness of adherence when wrung, but the relationship between surface texture and wringing quality has not, as yet, been fully established. A maximum value of 0.025 micrometre R_a assessed in accordance with BS 1134* shall not be exceeded; this value has proved satisfactory.

6. Accuracy

6.1 General. The gauge blocks shall conform to the requirements for accuracy appropriate to the grade specified as given below. It should be noted, however, that the requirements for accuracy apply to an area of the measuring face omitting a border zone of 1 mm in the face length direction and one of 0.5 mm in the face width direction, as measured from the side faces. In this excluded border zone the surface shall not lie above the plane of the measuring face.

*BS 1134 'Method for the assessment of surface texture'.

6.2 Flatness and parallelism of measuring faces

6.2.1 The two measuring faces of any gauge block shall be flat and parallel to each other to within the tolerances specified in table 1.

6.2.2 It is recognized that thin gauges, i.e. 2.5 mm in length and smaller, may not be flat in the unwrung state, therefore the tolerances specified for flatness of these gauges in table 1 shall be taken as applying to each face in turn when the opposite one is wrung to a lapped surface.

Gauges of all grades 2.5 mm in length and smaller shall be flat in the unwrung state to 0.005 mm.

In the case of grade 0 and grade 00 gauges longer than 2.5 mm and up to 5 mm in length, the flatness tolerance for gauges in the unwrung state may be up to four times that specified in table 1 for these grades.

Thin gauge blocks of all grades shall wring readily.

6.3 Squareness. The side faces shall be square to the measuring faces to within 0.025 mm over the length of the gauge block.

6.4 Gauge length

6.4.1 *Gauge length of calibration grades and grade 00 gauge blocks.* The gauge length *L* of a calibration grade or grade 00 gauge block, referred to the standard temperature of 20 °C, shall normally be measured by interference of light methods and include the thickness of one wringing film (see **2.1.1**) and shall not depart from the length marked on it by more than the amount specified in table 1.

6.4.2 *Gauge length of grades 0, I and II gauge blocks.* The gauge length of a grade 0, I or II gauge block, referred to the standard temperature of 20 °C, shall

Table 1. Tolerances on flatness, parallelism and length of gauge blocks

Size of gauge block		Tolerances Unit = 0.01 micrometre (0.000 01 mm)												
		Calibration grades and grade 00				Grade 0			Grade I			Grade II		
				Gauge length *L*										
Over	Up to and including	Flatness	Parallelism	Calibration grade	Grade 00	Flatness	Parallelism	Gauge length *L*	Flatness	Parallelism	Gauge length *L*	Flatness	Parallelism	Gauge length *L*
mm	mm													
—	20	5	5	±25	±5	10	10	±10	15	20	+20 −15	25	35	+50 −25
20	60	5	8	±25	±8	10	10	±15	15	20	+30 −20	25	35	+80 −50
60	80	5	10	±50	±12	10	15	±20	15	25	+50 −25	25	35	+120 −75
80	100	5	10	±50	±15	10	15	±25	15	25	+60 −30	25	35	+140 −100

be measured by comparison with a standard of similar material and of known length and shall not depart from the length marked on it by more than the amount specified for the appropriate grade in table 1.

7. Case

7.1 Each set of gauge blocks shall be supplied in a suitable case in which there is a separate compartment for each block.

7.2 The case shall be of substantial construction with substantial hinges and fastenings. The joint between the lid and the case shall be close fitting and should preferably be rabbeted or made with fillet and groove to prevent the ingress of dust.

The case shall be so designed that the gauge blocks cannot become displaced.

7.3 If the sizes of the blocks, as engraved upon them, are not visible when the blocks are in their respective compartments, each size shall be marked in the case, immediately adjacent to the appropriate compartment.

7.4 The case shall be marked with the number of this British Standard, i.e. BS 4311.

8. Marking and essential particulars

8.1 Marking

8.1.1 *Gauge blocks*

8.1.1.1 Each gauge block shall be legibly and permanently marked with its nominal size in figures at least 2 mm high.

8.1.1.2 The marking on all gauge blocks shall be applied clear of the centre of the measuring face.

8.1.1.3 Each gauge block on calibration grade and grade 00 sets shall bear some identification number or distinguishing mark common to all gauges in that set.

8.1.2 *Case.* Each case shall be legibly and permanently marked with the following particulars:

(a) the manufacturer's name or trademark and the number of this British Standard, i.e. BS 4311;
(b) the code number appropriate to the set, e.g. M 112/1, in the case of 1 mm base sets, M 112/2, in the case of 2 mm base sets, and the grade designation, i.e. calibration grade, 00, 0, I or II;
(c) an identification number, preferably the same as that for the gauge blocks where applicable;
(d) 20 °C;
(e) gauge block material.

8.2 Essential particulars. The manufacturer shall supply the following particulars with each set:

(a) a statement that the gauge blocks contained in the set have been given a treatment to promote stability of size;
(b) when the gauge blocks are made from a material other than steel, the coefficient of linear expansion for the material from which they are made;
(c) for calibration grades and grade 00 gauge blocks, a calibration chart giving the measured size of each block.

. . . .

Appendix A

Recommended sets of gauge blocks

A.1 General. The recommendations for sets of metric gauge blocks given below are offered only for the guidance of intending users. They are based on information concerning the sets at present in most popular demand, but it should be noted that other sets are also available from the manufacturers.

The economical advantages offered by the small sets, e.g. M 32 should be carefully compared with the technical ones provided by the more comprehensive but expensive larger sets, e.g. M 88 and M 112. This is particularly important when considering the purchase of additional sets when one or more large sets are already available.

The 2 mm based size series are recommended as they are less likely to suffer deterioration in flatness than similar size series based on 1 mm thin slips.

A.2 Protector gauge blocks. Grades 0, I and II sets of gauges may include a pair of extra gauges usually of 2 mm length, known as 'protector gauges'. These gauges are of the same general dimensions and accuracy as the other gauges in the set and are customarily marked with the letter 'P' on one measuring face. (See **B.7** for method of use.) Protector gauges of steel, or a wear-resisting material, such as tungsten carbide, may be obtained as separate accessories.

A.3 Typical sets – 1 mm base

	Size or series mm	Increment mm	Number of pieces
Set M 32/1	1.005	—	1
	1.01–1.09	0.001	9
	1.1 –1.9	0.1	9
	1 –9	1.0	9
	10, 20, 30	10.0	3
	60	—	1
			Total 32 pieces
Set M 41/1	1.001–1.009	0.001	9
	1.01 –1.09	0.01	9
	1.1 –1.9	0.1	9
	1 –9	1.0	9
	10, 20, 30	10.0	3
	60	—	1
	100	—	1
			Total 41 pieces

	Size or series mm	Increment mm	Number of pieces
Set M 47/1	1.005	—	1
	1.01–1.09	0.01	9
	1.1 –1.9	0.1	9
	1 –24	1.0	24
	25–100	25	4
			Total 47 pieces
Set M 88/1	1.0005	—	1
	1.001–1.009	0.001	9
	1.01 –1.49	0.01	49
	0.5 –9.5	0.5	19
	10 –100	10	10
			Total 88 pieces
Set M 112/1	1.0005	—	1
	1.001–1.009	0.001	9
	1.01 –1.49	0.01	49
	0.5 –24.5	0.5	49
	25 –100	25	4
			Total 112 pieces

Set M 46/1 As set M 41/1 plus 5 slips: 40, 50, 70, 80 and 90 mm.
The economic value of this set is dubious, particularly as the additional slips are expensive ones.

A.4 Typical sets – 2 mm base

	Size or series mm	Increment mm	Number of pieces
Set M 33/2	2.005	—	1
	2.01–2.09	0.01	9
	2.1 –2.9	0.1	9
	1 –9	1.0	9
	10, 20, 30	10.0	3
	60	—	1
	100	—	1
			Total 33 pieces
Set M 46/2	2.001–2.009	0.001	9
	2.01 –2.09	0.01	9
	2.1 –2.9	0.1	9
	1 –9	1.0	9
	10 –100	10.0	10
			Total 46 pieces
Set M 88/2	1.005	—	1
	2.001–2.009	0.001	9
	2.01 –2.49	0.01	49
	0.5 –9.5	0.5	19
	10 –100	10.0	10
			Total 88 pieces

A.5 Auxiliary sets

	Size or series mm	Increment mm	Number of pieces
Set M 009	1.001–1.009	0.001	9
Set M 09	1.01 –1.09	0.01	9
Set M 9	1.1 –1.9	0.1	9

Appendix B

Notes on the care and use of gauge blocks

B.1 General care. The greatest care should be exercised in protecting the gauges and their case from dust, dirt and moisture.

When not in actual use, the gauges should always be kept in their case and the case should be kept closed.

The gauges should be used as far as possible in an atmosphere free from dust. Care should be taken that the gauges do not become magnetized or they will attract ferrous dust.

B.2 Preparation before use. If the gauges are new or have been covered with a protective coating after being last used, most of this coating may be removed with an appropriate solvent. Gauges should finally be wiped with a clean chamois leather or soft linen cloth. This wiping should be carried out in every instance before a gauge is used, irrespective of whether it has been stored, coated or merely returned temporarily to the case uncoated.

It is, however, undesirable to aim at removing all traces of grease since a very slight film of grease is an aid to satisfactory wringing.

B.3 Care in use. Fingering of the lapped faces should be avoided to preclude the risk of tarnishing and unnecessary handling of the gauges in use should be avoided lest they take up the heat of the hand. If the gauges have been handled for some time they should be allowed to assume the prevailing temperature of the room before being used for test purposes. This is particularly important in the case of the larger sizes.

When the highest accuracy is required, a test room thermostatically controlled at the standard temperature of 20 °C becomes necessary, but for ordinary purposes a sufficient degree of accuracy can be obtained if the following precautions are taken.

The work to be tested and the gauge blocks which have to be used should both be allowed to assume the prevailing temperature of the room. Thus, a piece of work should not be tested directly after a cutting, grinding or other operation has just been completed nor should large combinations of gauge blocks be used immediately after they have been wrung together.

B.4 Wringing. Gauges should not be held above the open case when being wrung together lest one be accidentally dropped. The gauges required should be selected and the case then closed.

Before wringing gauges together, their faces should be wiped free from dust and examined for burrs.

B.5 Damaged gauges. Damage to the gauges is most likely to occur on the edges, resulting from the gauge being knocked or dropped. Such slight burrs may be removed with care by drawing an Arkansas type stone lightly across the damaged edge in a direction away from the measuring face of the gauge. Any gauges so treated should be thoroughly cleaned before wringing.

A gauge with a damaged measuring face should preferably be returned to the manufacturer for the surface to be restored.

B.6 Care after use. The gauges should never be left wrung together for any length of time. Slide the gauges apart, do not break the wringing joint between them.

Immediately after use each gauge should be wiped clean and be replaced in its proper compartment in the case. It is particularly important to remove any finger marks from the measuring faces.

If the gauges are used infrequently they should be coated with a suitable corrosion preventive before being put away. The preparation should be applied to the faces with a clean piece of soft linen. A brush should not be used as this may aerate the preparation and moisture in the air bubbles so formed can cause rusting of the faces.

B.7 Building up a size combination. The gauges to be used for building up a size combination should be determined by the method shown below. Where protector gauge blocks are provided (see appendix A), these should be wrung to the two end gauges of the size combination and must of course be allowed for in determining the blocks to be used.

The micrometre (0.001 mm) gauge should be taken first, followed by the hundredth, tenth and millimetre gauges as in the examples below.

B.7.1 *Example*

To build up 58.343 mm with the M 46/2 set (2 mm base)

1st gauge	2.003
2nd gauge	2.04
3rd gauge	2.30
4th gauge	2
5th gauge	50

B.7.2 *Example*

To build up 58.343 mm with the M 112/1 set (1 mm base)

1st gauge	1.003
2nd gauge	1.34
3rd gauge	6
4th gauge	50

To build up 58.843 mm with the M 112/1 set (1 mm base)

1st gauge	1.003
2nd gauge	1.34
3rd gauge	6.5
4th gauge	50

The use of gauge blocks for measuring and for calibrating measuring instruments may be extended by accessories of the type specified in BS 4311 'Metric gauge blocks', Part 2 'Accessories'. For instance, pairs of jaws may be combined with gauge blocks to form an external or an internal calliper. Other accessories are a robust base for converting a gauge block combination into a height gauge; and a centre point and an end tool for scribing areas of precise radius. The various items are made with wringing surfaces for combining with gauge blocks, and holders are available for supporting the combinations in use.

The precision obtainable in using the accessories is influenced by the manner of use, for example reduction of springiness in calliper jaws, flatness of the datum surface on which the base is located, and avoidance of local wear, particularly on the scribing tool.

BS 4311

Metric gauge blocks

Part 2. Accessories

. . . .

1. Scope

. . . .

Examples of typical assemblies of accessories are shown in figure 1 and notes on their use are given in appendix A.

. . . .

4. Type A measuring jaws

4.1 General. These jaws are customarily supplied in pairs for internal or external measurements: the general design and dimensions shall be as given in figure 2 and table 1.

. . . .

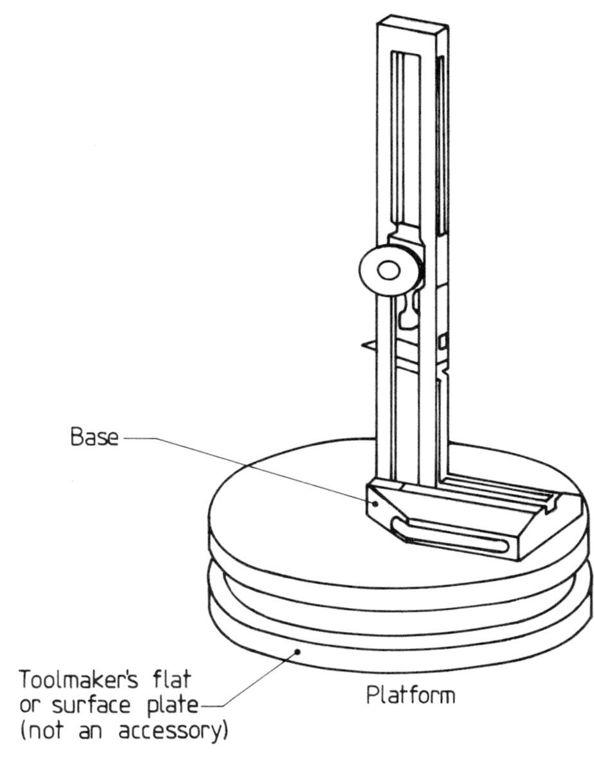

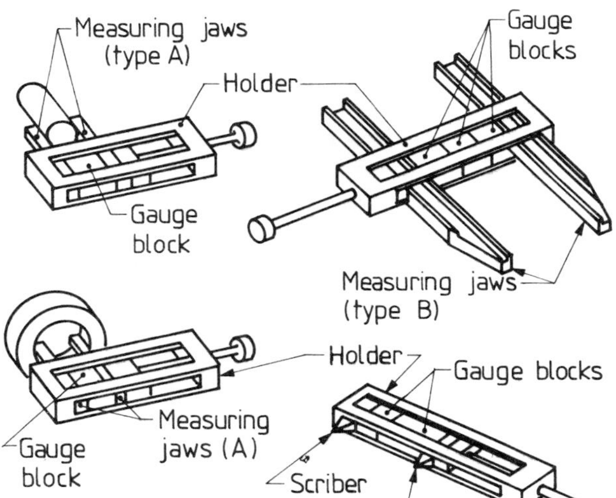

Figure 1. Examples of assemblies of gauge block accessories

Table 1. Dimensions of type A jaws

1	2	3	4	5	6
Recommended nominal dimension	**Approximate dimensions**				
A	***B***	***C***	***D* (min.)**	***E***	***F***
mm	mm	mm	mm	mm	mm
2	7.5	40	6	34	1.0*
5	7.5	45	15	30	1.5
8	12.5	50	20	30	1.5
12.5	12.5	75	45	30	20

* Radian or undercut.

5. Type B measuring jaws

5.1 General. These jaws are customarily supplied in pairs for external measurements: the general design and dimensions shall be as given in figure 3.

. . . .

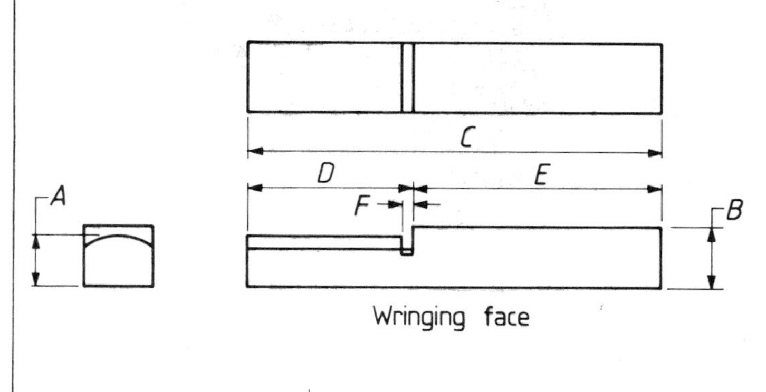

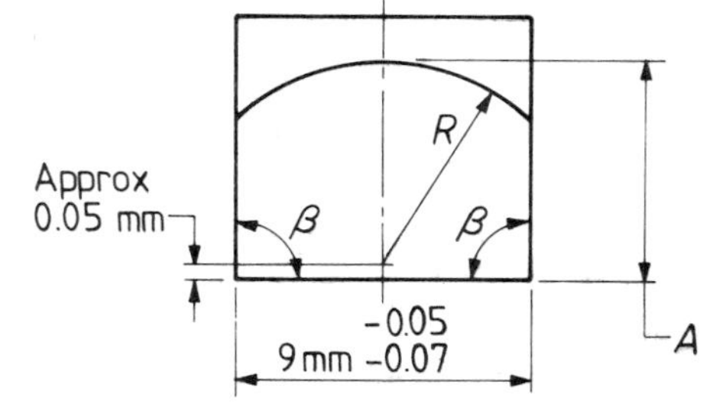

Figure 2. Type A jaws

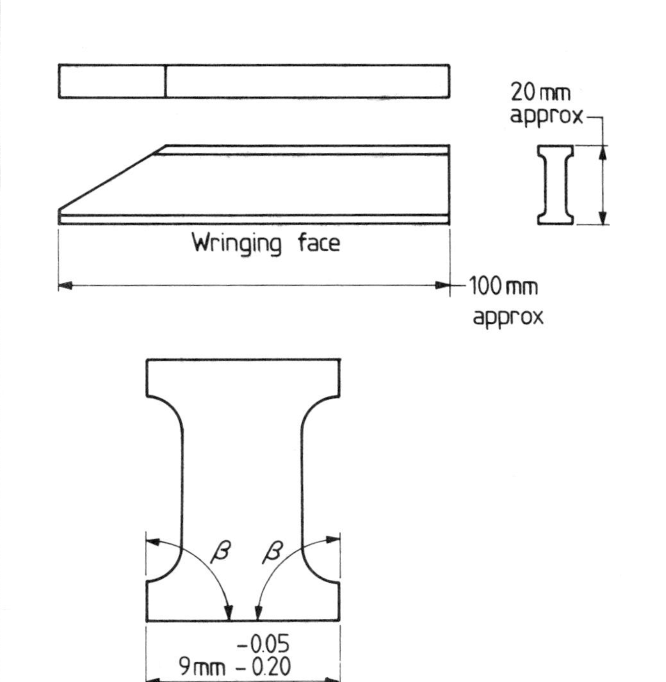

Figure 3. Type B jaws

6. Scriber point and centre point

6.1 General. These jaws are customarily supplied together as a set for scribing arcs of known radius. The general design and dimensions shall be as given in figures 4 and 5.

. . . .

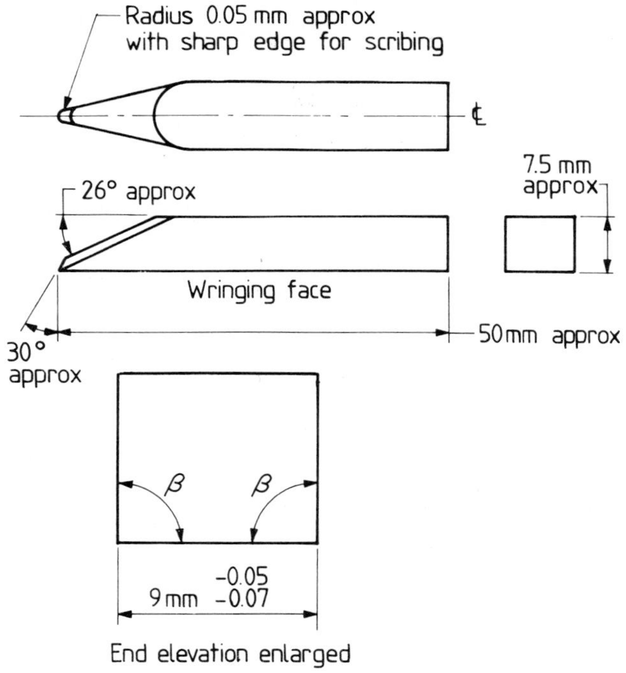

Figure 4. Scriber point

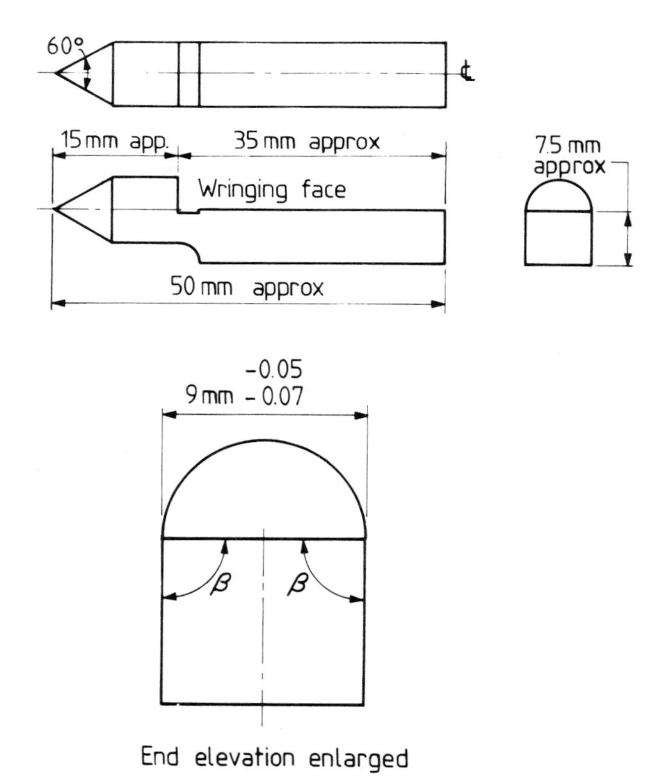

Figure 5. Centre point

7. Base

7.1 General. The base is used in combination with gauge blocks to set up a nominal dimension vertically above a datum surface in contact with the underside of the base.

The base, including a platform to which the gauge blocks are wrung, shall be sufficiently robust to provide adequate stability when used in conjunction with a combination of gauge blocks in a holder of 300 mm capacity.

Suitable means shall be provided for attaching the holder so that the combination of gauge blocks is perpendicular to the datum surface.

The underside of the base shall be relieved and an air vent shall be provided. The sides shall be relieved to form convenient finger grips.

The platform shall be at least 35 mm long and 9 mm broad.

. . . .

8. Holders

8.1 General. The holders shall be of a suitable design to accommodate combinations of gauge blocks and accessories, and to hold such combinations rigidly while they are in use. The holder shall incorporate a suitable form of swivel joint between the sliding block and the clamping screw in order to ensure an even pressure on the combination. The surfaces of the sliding and fixed blocks against which the combination abuts shall be finished by lapping and the centres of these surfaces shall be relieved (see figure 6).

The distance between the side plates shall be 9 +0.1, −0 mm.

9. Associated items

The accessories specified in this standard are often supplied accompanied by a knife edge straightedge complying with the requirements of BS 852*. In use, the base may also be associated with a grade AA surface plate as specified in BS 817 or a toolmakers' flat in accordance with BS 869.

*BS 852 has been superseded by BS 5204 'Straightedges'.

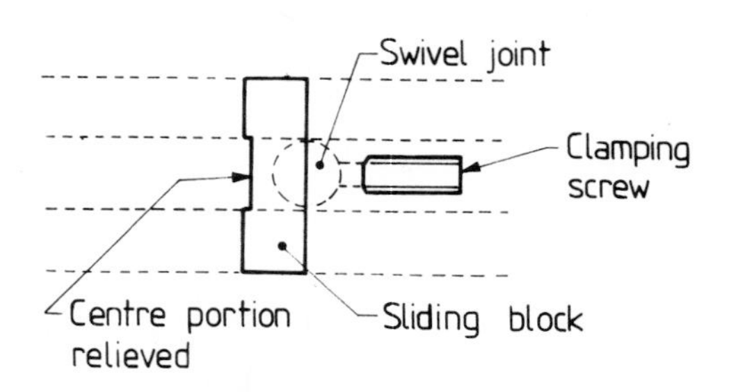

Figure 6. Detail of holder

10. Marking

Each accessory shall be marked with the manufacturer's name or initials or trademark.

Type A jaws and the base shall be marked with the nominal size.

. . . .

Appendix A

Notes on the use of gauge block accessories

A.1 Type A and type B jaws. When wringing the jaws with gauge blocks the combination should be aligned by bringing the items into contact along one side with a surface plate. The projection of the jaws beyond the gauge blocks should be minimized to avoid loss of accuracy through springing of the tips of the jaws.

A.2 Type A jaws. The user is reminded that the 2 mm jaws are not undercut at the junction of the cylindrical and rectangular sections; hence if the jaws are allowed to mate to full depth with an internal component, contact may be made with the fillet at the base of the cylindrical portion, thus falsifying the measurement.

These jaws wear most at the leading end: a sign of wear is an abraded line along the cylindrical surface.

A.3 Scriber and centre points. In all manipulation care should be taken to protect both points against damage in the form of burrs or blunting. When assembling with gauge blocks in a combination, it is advisable to align the combination along one side against a surface plate. Furthermore, the centre lines of scribers and centre points should be kept parallel with the axis of the gauge blocks.

The scribing of arcs is facilitated if the sample is polished beforehand and if it can be rotated beneath the scribing assembly. If thick lines result there is uncertainty as to the true location of the marks since the scratch may be unsymmetrical, and the innermost edge, from which measurements should be taken, may be irregular in outline.

The scriber will wear in due course and a ready check is to view its condition under optical magnification. Rectification is best entrusted to the manufacturer.

Where a travelling microscope is available, the user may prefer to scribe arcs of nominal radius and measure the diameter of the scribed circle; the interposed gauge blocks may then be adjusted if necessary, and arcs rescribed.

Length bars

Length bars are used in a similar manner to gauge blocks but for larger sizes of work. They are made from 22 mm diameter high quality tool steel which has been hardened and stabilized. BS 5317 'Specification for metric length bars and their accessories' provides for four grades of length bars, namely: reference, calibration, grade I and grade II.

The reference grade bars are intended for use as reference standards and embody the highest order of accuracy with regard to the quality of their end faces and the finishing of their actual lengths to nominal size. They should be used only with comparators of suitably high sensitivity.

The calibration grade bars are intended for use in the calibration of length measuring standards. They should be used with comparators of suitably high sensitivity. The tolerances of flatness and parallelism of the end faces of reference and calibration grades of length bars are the same but the tolerance on length of the calibration grade is approximately twice that of the reference grade. Both grades should be supplied with a certificate of accuracy giving the deviations from their nominal lengths and confirming their full accordance with BS 5317.

While the reference and calibration grades have plane end faces the grade I and II bars have internally threaded ends and can thus be used in combination with each other. The grade I bars are intended for use in inspection rooms and tool rooms, while the grade II bars can be used with gauge blocks, comparators or various accessories for measuring gauges, jigs, or workpieces.

Length bars are best used in the vertical position but when used in the horizontal position they should be supported at two points equidistant from each end and the distance between the support points being 0.577 × the length of the bar. These points are known as Airy points (after Sir George Airy) and ensure that the slope of the end faces of the bar is zero (see Figure 4.2).

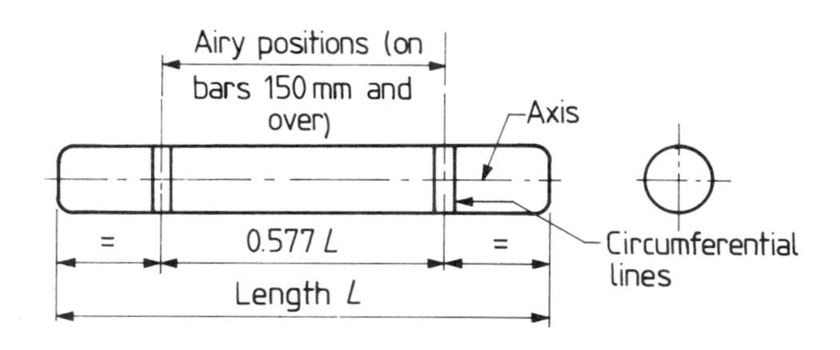

Figure 4.2 *Airy positions for length bars*

The Airy positions are marked on all bars 150 mm and over in length, but when bars are screwed together in combination and intended for use in the horizontal position, the Airy positions for the total length must be calculated and the marked positions on individual bars ignored.

In addition to specifying the requirements of the length bars, BS 5317 includes the essential features of design and accuracy of accessories for use with grade I and II bars.

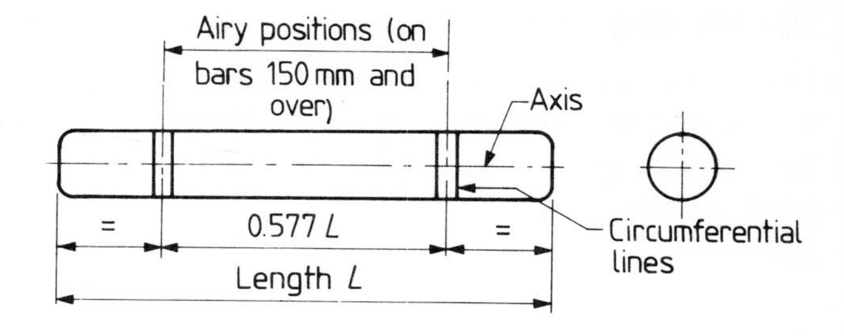

Figure 1. Indication of Airy positions and axis

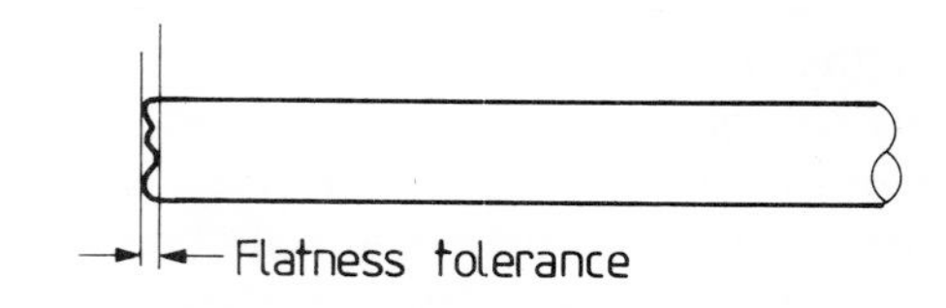

Figure 2. Exaggerated illustration of flatness tolerance

BS 5317

Metric length bars and their accesssories

Section one. General

. . . .

3. Definitions

For the purposes of this British Standard the following definitions apply.

3.1 airy positions. Those points at which a bar of uniform cross section has to be supported when used with its axis horizontal. (See figure 1.)

3.2 axis of the bar. The line passing through the centres of the cross sections at the Airy positions. (See figure 1.)

3.3 length. This is defined, with the bar mounted horizontally and referred to the standard reference temperature of 20 °C, as the distance from the centre of one of its faces to a flat surface in wringing contact with the opposite face, measured normal to the surface.

NOTE 1. To realize this definition for the grade 1 and 2 types of bar it will be necessary to wring plane parallel end pieces of known thickness to one or both annular faces.

NOTE 2. It follows that when length measurements are made by comparison with a bar whose known length, as defined above, is traceable to the primary standard of length, the length so determined always refers to the horizontal position of support regardless of whether the *comparison* is made with the bars supported vertically or horizontally.

3.4 deviation from flatness. The minimum distance between two parallel planes which just envelop the measuring face.

3.5 flatness tolerance. The maximum permissible deviation from flatness. (See figure 2.)

3.6 deviation from parallelism. The difference between the maximum and minimum lengths at any points on the measuring faces measured perpendicular to the surface to which one face is wrung.

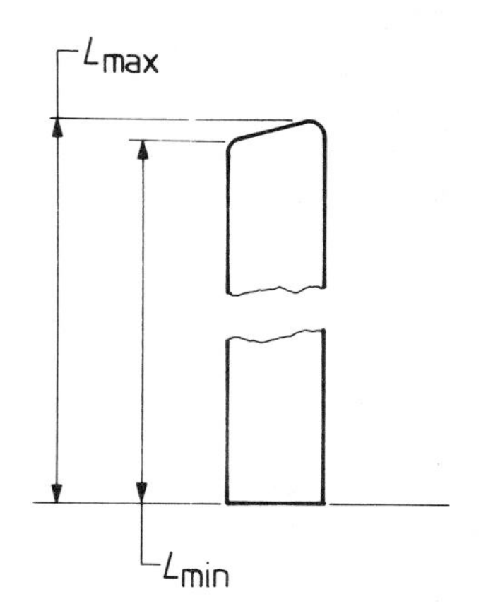

Figure 3. Exaggerated illustration of parallelism tolerance

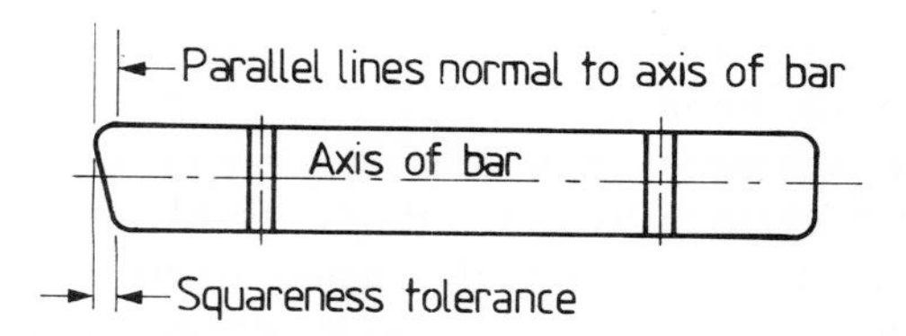

Figure 4. Exaggerated illustration of squareness tolerance

3.7 parallelism tolerance. The maximum permissible deviation from parallelism. (See figure 3.)

3.8 deviation from squareness. The minimum distance between two parallel planes normal to the axis of the bar which just envelop the measuring face under consideration. (See figure 3.)

3.9 squareness tolerance. The maximum permissible deviation from squareness. (See figure 4.)

Table 1. Tolerances on reference grade length bars

1	2	3	4
Nominal length	Tolerances on accuracy of faces		Tolerance on length at 20 °C
	Flatness	Parallelism	
mm	μm	μm	μm ±
Up to 25	0.08	0.08	0.08
50	0.08	0.10	0.12
75	0.10	0.16	0.15
100	0.10	0.16	0.20
125	0.10	0.20	0.25
150	0.10	0.20	0.30
175	0.15	0.20	0.30
200	0.15	0.20	0.35
225	0.15	0.20	0.40
250	0.15	0.30	0.40
275	0.15	0.30	0.45
300	0.15	0.30	0.50
400	0.15	0.30	0.65
500	0.15	0.30	0.80
600	0.15	0.30	0.95
700	0.15	0.30	1.10
800	0.15	0.30	1.25
900	0.15	0.30	1.40
1000	0.15	0.30	1.55
1200	0.15	0.30	1.85

NOTE. Bars of intermediate sizes should be made to the same tolerances as those for the next smaller size.

Table 2. Tolerances on calibration grade length bars

1	2	3	4
Nominal length	Tolerances on accuracy of faces		Tolerance on length at 20 °C
	Flatness	Parallelism	
mm	μm	μm	μm ±
Up to 25	0.08	0.08	0.15
50	0.08	0.10	0.20
75	0.10	0.16	0.30
100	0.10	0.16	0.35
125	0.10	0.20	0.45
150	0.10	0.20	0.50
175	0.15	0.20	0.60
200	0.15	0.20	0.65
225	0.15	0.20	0.70
250	0.15	0.30	0.80
275	0.15	0.30	0.90
300	0.15	0.30	0.95
400	0.15	0.30	1.30
500	0.15	0.30	1.60
600	0.15	0.30	1.90
700	0.15	0.30	2.20
800	0.15	0.30	2.50
900	0.15	0.30	2.80
1000	0.15	0.30	3.10
1200	0.15	0.30	3.70

NOTE. Bars of intermediate sizes should be made to the same tolerances as those for the next smaller size.

Section two. Length bars

. . . .

7. Finish

7.1 The body of each bar shall be finished all over by fine grinding.

The end faces of the bars shall be lapped to provide a finish of the highest quality both for appearance and wringing quality. They shall be free from scratches and blemishes of any noticeable character. The edges of all bars shall be radiused as shown in figure 7*a*.

Threaded holes and screws, when provided, shall be cleanly finished and all sharp edges shall be removed.

8. Accuracy

8.1 Diameter. The diameter of each bar shall be uniform within 15 μm for bars up to 300 mm in length, 25 μm for bars over 300 mm up to and including 600 mm in length, and within 50 μm for bars longer than 600 mm.

8.2 Straightness. The body shall be straight within 10 μm per 100 mm of length.

Table 3. Tolerances on grade 1 length bars

1	2	3	4
Nominal length	Tolerances on accuracy of faces		Tolerance on length at 20 °C
	Flatness	Parallelism	
mm	μm	μm	μm
Up to 25	0.15	0.16	+0.40 −0.20
50	0.15	0.18	+0.60 −0.20
75	0.15	0.18	+0.70 −0.30
100	0.18	0.20	+0.85 −0.35
125	0.18	0.20	+1.00 −0.40
150	0.18	0.20	+1.10 −0.50
175	0.20	0.25	+1.25 −0.55
200	0.20	0.25	+1.40 −0.60
375	0.20	0.35	+2.40 −1.00
400	0.20	0.35	+2.50 −1.10
575	0.20	0.40	+3.50 −1.50
600	0.20	0.40	+3.65 −1.55
775	0.20	0.50	+4.60 −2.00

NOTE. Bars of intermediate sizes should be made to the same tolerances as those for the next smaller size.

8.3 End faces

8.3.1 *Flatness and parallelism.* The tolerances on flatness and parallelism of the end faces shall be as given in tables 1 to 4.

8.3.2 *Squareness.* The end faces of all grades of bars shall be square with the axis of the bar to within 1.2 μm over the diameter of the face for bars up to and including 400 mm in length and to within 2.5 μm for bars over 400 mm in length.

8.4 Length

8.4.1 *Tolerances.* The tolerances on length shall be as given in tables 1 to 4.

. . . .

Table 4. Tolerances on grade 2 length bars

1	2	3	4
	Tolerances on accuracy of faces		
Nominal length	**Flatness**	**Parallelism**	**Tolerance on length at 20 °C**
mm	μm	μm	μm
Up to 25	0.25	0.30	+0.75 −0.35
50	0.25	0.30	+0.95 −0.45
75	0.25	0.35	+1.20 −0.50
100	0.25	0.35	+1.40 −0.60
125	0.25	0.40	+1.60 −0.70
150	0.25	0.40	+1.80 −0.80
175	0.25	0.40	+2.00 −0.90
200	0.25	0.40	+2.20 −1.00
375	0.25	0.50	+3.70 −1.60
400	0.25	0.50	+3.90 −1.70
575	0.25	0.70	+5.40 −2.30
600	0.25	0.70	+5.60 −2.40
775	0.25	0.80	+7.10 −3.00

NOTE. Bars of intermediate sizes should be made to the same tolerances as those for the next smaller size.

9. Connecting screws for grade 1 and grade 2 bars

At least six unhardened connecting screw M 10 × 1.5 − 6 g shall be provided with each set of grade 1 or grade 2 bars for holding them together. They shall screw quite freely into the threaded holes in the bars and shall allow the wringing faces to come into satisfactory contact.

10. Marking and essential particulars

10.1 Marking. Each bar shall be legibly and permanently marked with the following particulars:

(a) the nominal length and grade, e.g. 75 mm, grade 1;
(b) the number of this British Standard (BS 5317);
(c) reference and calibration bars with an identification number and grade 1 and 2 bars with their set number;
(d) 20 °C;
(e) the manufacturer's name or trademark.

10.2 Essential particulars. The manufacturer shall supply the following particulars with each set:

(a) a statement that the length bars contained in a set have been given a treatment to promote stability of size;
(b) for reference and calibration grade length bars, a calibration chart giving the measured size of each bar.

. . . .

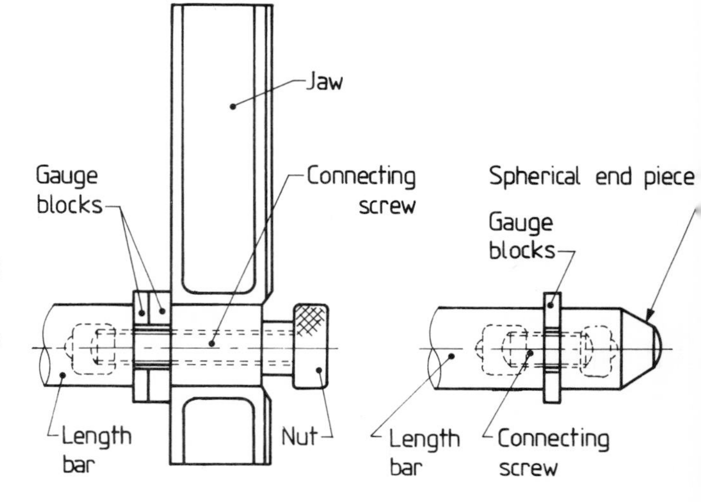

Figure 8. Assembly of accessories

Section three. Accessories

NOTE. Various accessories have been designed to extend the possible applications of length bars. However, it should be appreciated that some loss of accuracy in length measurement may result from their use due to factors such as offset from the measuring axis, the cumulative effect of tolerances, etc.

. . . .

14. Method of attachment

Accessories shall be attached to the bars by means of connecting screws provided for the purpose.

Diagrammatic illustrations of the assembly of accessories are given in figures 8 and 9.

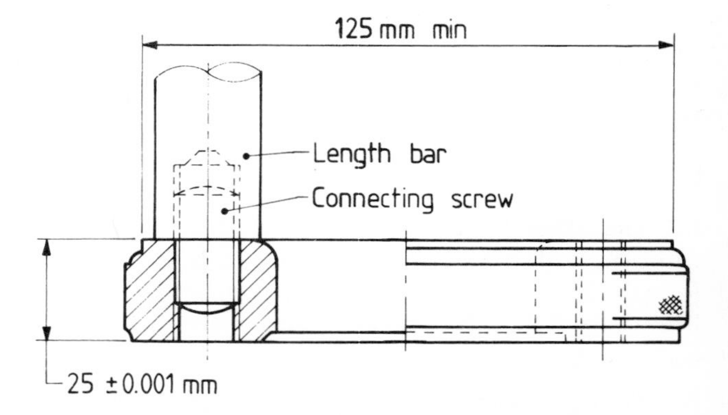

Figure 9. Base

15. Base

15.1 Design. The base (see figure 9) shall have a diameter of at least 125 mm to provide adequate stability when used in conjunction with length bars up to 1500 mm long.

. . . .

16. Large radiused jaw

16.1 General design. The design of the large radiused jaw shall be generally as shown in figure 10.

NOTE. In the case of internal measurement, this jaw should only be used for diameters greater than 150 mm.

. . . .

17. Small plane-faced jaw

17.1 General design. The general shape shall be as shown in figure 12.

. . . .

18. Spherical end piece

18.1 General design. The general form of the spherical end piece shall be as shown in figure 13.

. . . .

19. Screws

Sufficient M 10 × 1.5 − 6 g connecting screws for assembly shall be provided with each set of accessories. (See figure 8 for illustration of a typical assembly.)

The screws shall assemble the accessories quite freely with the length bars. They shall be well finished all over and all sharp edges shall be removed.

. . . .

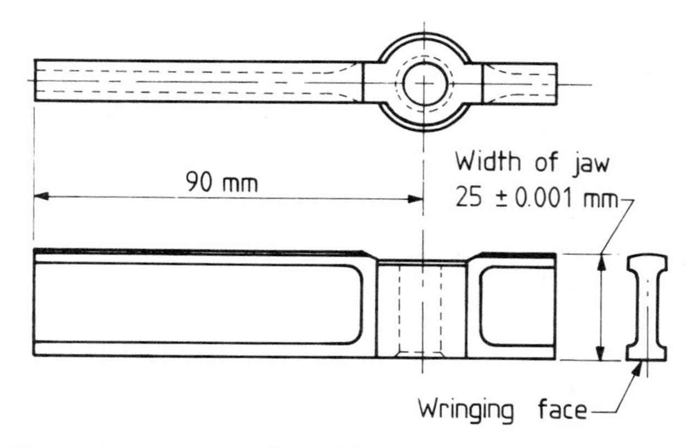

Figure 10. Large radiused jaw

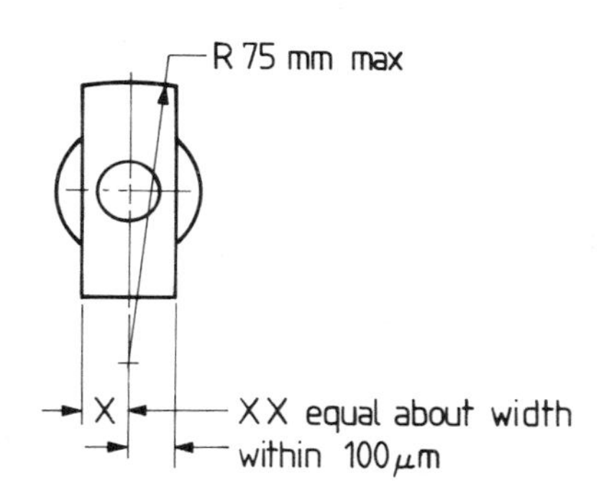

Figure 11. Centrality of radius to thickness

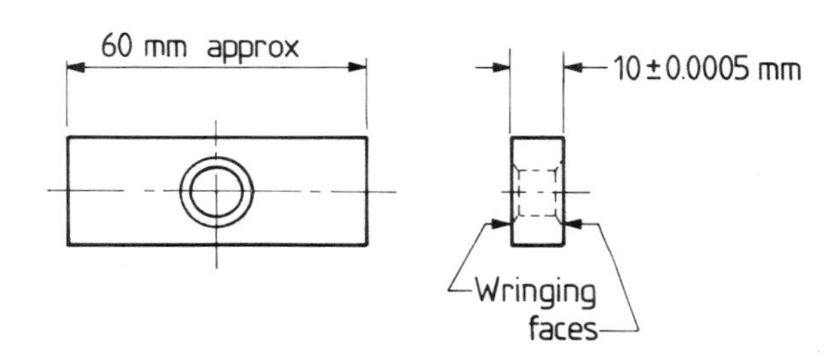

Figure 12. Small plane-faced jaw

NOTE. Width to be below minimum diameter of bar.

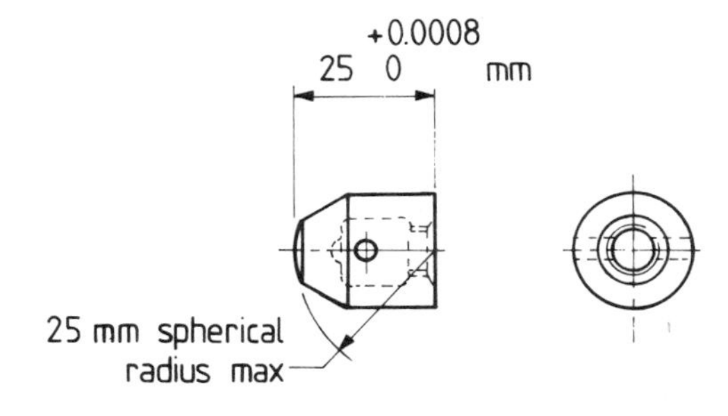

Figure 13. Spherical end piece

NOTE. Outside diameter to be smaller than minimum outside diameter of bar.

Comparators

Many different types of comparator have been developed and BS 1054 'Specification for engineers' comparators for external measurement' is restricted to the accuracy of performance and touches on features of design only where these are likely to have a direct effect on the accuracy. BS 1054 specifies the requirements for comparators having magnification factors of 250 and over, and primarily applies to instruments comprising a rigid stand supporting a measuring head over a work table. The means of amplification may be mechanical, electrical, electronic, optical, fluid or pneumatic. Only the performance requirements of BS 1054 have been selected for inclusion in this publication.

BS 1054

Engineers' comparators for external measurement

. . . .

9. Performance

9.1 Consistency of reading. The consistency of reading of the comparator shall be examined by carrying out the following series of tests; the permissible errors are stated in table 1 and are related to the magnification factor.

9.1.1 When the instrument is set to read on a workpiece and the measuring head is gently tapped, the maximum change in reading shall not exceed one half of the appropriate amount given in table 1.

9.1.2 When a true cylinder is passed under the measuring tip

from the front,
from the rear,
from the LH side,
from the RH side,

the maximum difference in reading found shall not exceed one half of the values in table 1.

9.1.3 The reading obtained when sliding a gauge block under the tip shall not differ from subsequent readings obtained by operating the lifting device by more than the appropriate amount given in table 1.

Table 1. Tolerances on consistency of reading

Magnification factor of measuring head	Tolerance on reading
	mm
From 250 up to and including 300	0.002 0
Above 300 up to and including 400	0.001 5
Above 400 up to and including 600	0.001 0
Above 600 up to and including 1 000	0.000 8
Above 1 000 up to and including 2 000	0.000 5
Above 2 000 up to and including 5 000	0.000 3
Above 5 000 up to and including 10 000	0.000 1
Above 10 000	0.000 05

9.2 Accuracy of scale. The accuracy of the scale of the instrument shall be tested by taking readings on a suitable series of calibrated gauge blocks, first over the positive half of the scale and then over the negative half of the scale, checking at each major division.

No reading shall have an error exceeding ±1 % of the reading or the appropriate amount (±) given in table 1, whichever is greater.

For example, in the case of a comparator having a magnification factor of 1000, the maximum permissible error over the range 0 to 0.05 mm would be ±0.000 5 and, beyond this range, ±1 % of the reading.

. . . .

Standards publications referred to in this chapter

BS 870 External micrometers
BS 887 Precision vernier callipers
BS 907 Dial gauges for linear measurement
BS 957 Feeler gauges
Part 2 Metric units
BS 959 Internal micrometers (including stick micrometers)
BS 1054 Engineers' comparators for external measurement
BS 1643 Vernier height gauges
BS 1734 Micrometer heads
BS 2795 Dial test indicators (level type) for linear measurement
BS 4311 Metric gauge blocks
Part 1 Gauge blocks
Part 2 Accessories
BS 4372 Engineers' steel measuring rules
BS 5317 Metric length bars and their accessories

Note: The above list does not include all the standards referred to in the extracts.

5 Angular measurement

Introduction

It was stated in Chapter 2 that since it is possible to use a combination of linear measurements to obtain an angular measurement, length measurement is of special importance to the engineering metrologist. This is reflected in two of the types of equipment covered in this chapter. Sine bars (and tables) use the length between their rollers and the height of the gauge block pile to obtain a required angle, while spirit levels depend upon the radius of their vial and the length of the base. Bevel protractors, on the other hand, are graduated in degrees and minutes of arc, and these are also covered in this chapter.

Bevel protractors

BS 1685 'Bevel protractors (mechanical and optical)' provides for three mechanical and one optical type of protractor. In the mechanical types the measurement is read direct from the scale, either with or without the assistance of a vernier. In the optical type the measurement is read on an internal circular scale by means of an optical magnifying system integral with the instruments.

The standard relates to the following types of bevel protractors:

(a) Mechanical
- Type A With vernier graduated to read to five minutes of arc; and with slow motion device, and acute angle attachment.
- Type B With vernier graduated to read to five minutes of arc, but without slow motion device or acute angle attachment.
- Type C With scale graduations in degrees, without vernier or slow motion device or acute angle attachment.

(b) Optical
With an internal circular scale which is graduated in divisions of arc and read against a fixed index or vernier by means of an optical magnifying system integral with the instrument, enabling readings to be taken to approximately two minutes of arc.

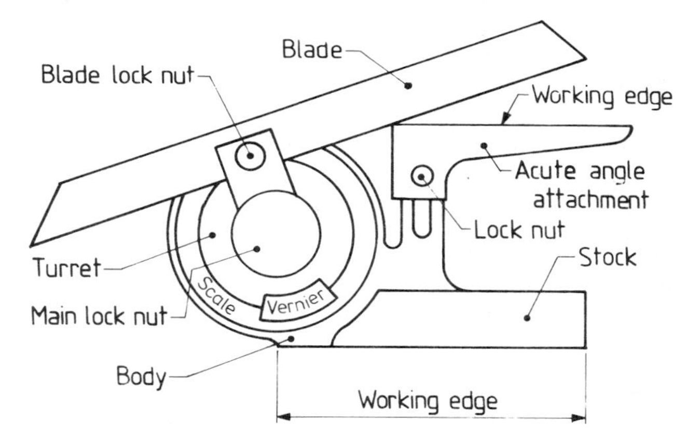

Figure 5.1 *Nomenclature for mechanical bevel protractors*

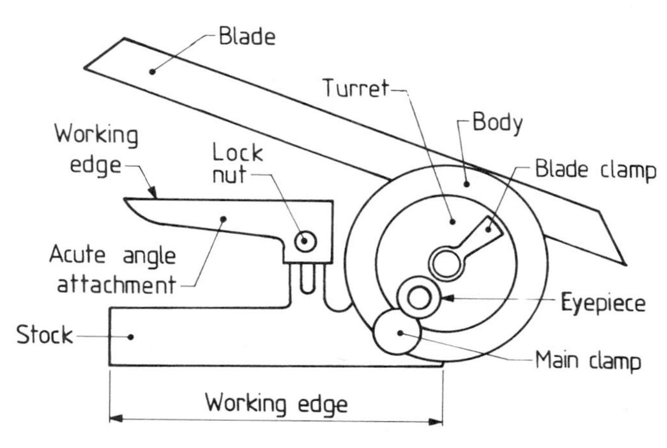

Figure 5.2 *Nomenclature for optical bevel protractors*
Note: The illustrations are diagrammatic only, and are not intended to illustrate details of design

Figures 5.1 and 5.2 show the nomenclature as used in BS 1685 for mechanical and optical bevel protractor respectively. No part of the standard has been selected for inclusion in this publication.

Sine bars and sine tables

A common type of sine bar is shown in Figure 5.3, and a typical application is shown in Figure 5.4.

In Figure 5.4, if l is the linear distance between the axes of the rollers and h is the height of the gauge blocks, then sine $\theta = h/l$.

Such applications of sine bars require:

(a) the rollers to be of equal diameter and true geometrical cylinders;
(b) the distance between the axes of the rollers must be known;
(c) the axes of the rollers must be mutually parallel;
(d) the upper surface (working surface) of the sine bar must be flat and parallel with the rollers and equidistant from each.

BS 3064 'Metric sine bars and sine tables (excluding compound tables)' provides for three sizes of sine bars and four sizes of sine tables inclinable about a single axis. Design requirements are confined to essentials but the accuracy is fully specified in order to ensure the suitability of the equipment for precision inspection.

Note: Bars of 50 mm width and over are often referred to as sine plates.

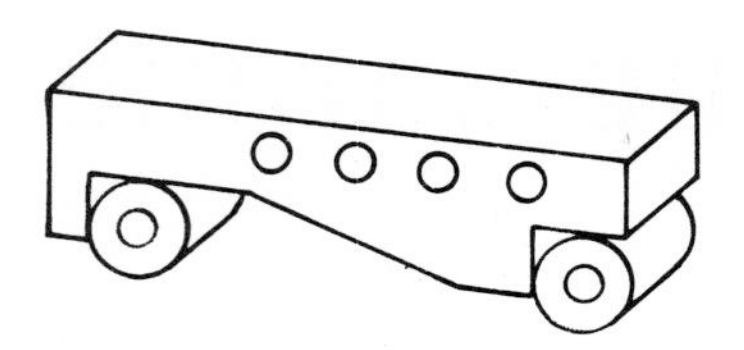

Figure 5.3 *Typical design of sine bar*

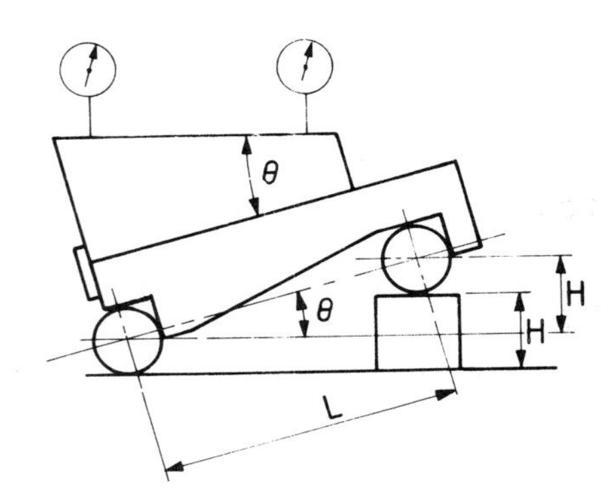

Figure 5.4 *Typical sine bar set-up*

BS 3064

Metric sine bars and sine tables (excluding compound tables)

. . . .

3. Nomenclature and definitions

. . . .

3.1 designating size. The centre distances between the rollers.

3.2 flatness tolerance. The maximum distance between two parallel planes within which the surface under consideration shall lie.

3.3 squareness tolerance of surfaces. The distance between two parallel planes that are perpendicular to the plane in contact with the datum surface and within which the surface under consideration shall lie.

3.4 parallelism tolerance of surfaces. The distance between two planes that are parallel to the plane in contact with the datum surface and within which the surface under consideration shall lie.

Section two. Sine bars

4. General design

4.1 The general form of the sine bars shall be basically in accordance with that shown for any one of the four types illustrated in figure 1.

. . . .

7. Accuracy

7.1 Upper and lower surface

7.1.1 The whole of the upper surface of the bar and its lower surface, if a working surface, shall be flat within

0.0015 mm for 100 mm bars
0.002 mm for 200 mm bars
0.003 mm for 300 mm bars

7.1.2 The lower surface, if a working surface, shall be parallel to the upper surface to within

0.0015 mm for 100 mm bars
0.002 mm for 200 mm bars
0.003 mm for 300 mm bars

7.2 Side faces. The side faces of the bars shall be:

(a) flat within
0.004 mm for 100 mm bars
0.005 mm for 200 mm bars
0.006 mm for 300 mm bars;
(b) square to the upper surface of the bar within 0.0025 mm per 25 mm;
(c) square to the axis of the hinge roller within 0.013 mm per 25 mm measured in the plane common to the axes of the rollers.

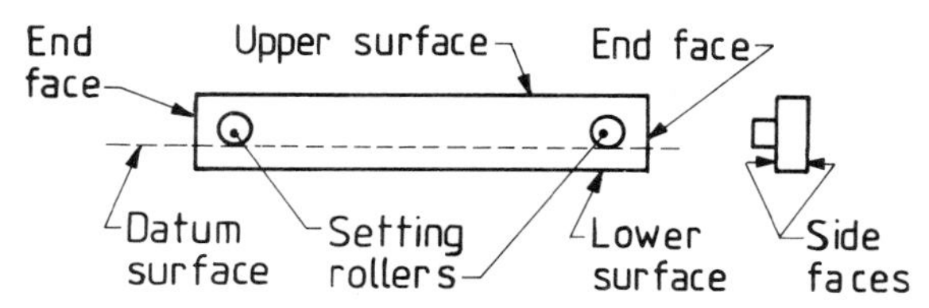

Type 1

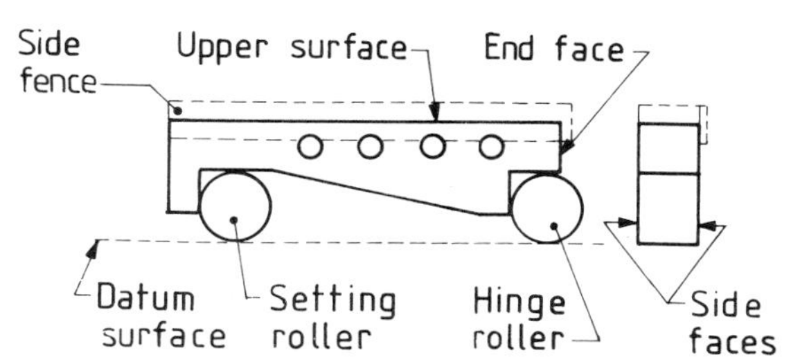

Type 2

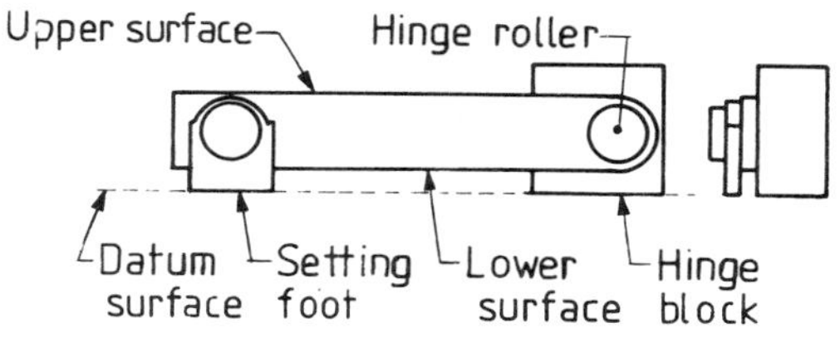

Type 3

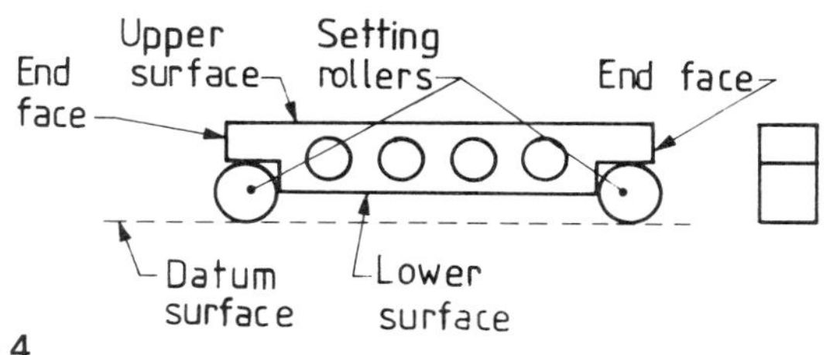

Type 4

Figure 1. Sine bars: types and nomenclature

NOTE. These illustrations show basic types of sine bars. They are diagrammatic only and are not intended to illustrate detail of design.

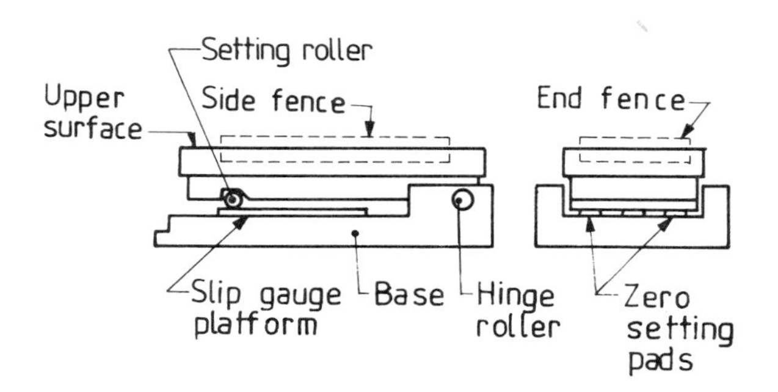

Figure 2. Sine table: nomenclature

NOTE. The illustration is diagrammatic only and is not intended to illustrate details of design.

NOTE. In the case of type 1 bars, requirement (c) applies to both rollers.

7.3 End faces. The end faces, when intended as working surfaces, shall be:

(a) flat within 0.0025 mm;

(b) square to the upper surface of the bar within 0.0025 mm per 25 mm;

(c) parallel to the axis of the hinge roller within 0.013 mm per 25 mm.

7.4 Setting and hinge rollers

7.4.1 Individual rollers shall be uniform in radius within 0.002 mm and straight over their length to the same accuracy.

7.4.2 The mean diameters of the rollers on bars of types 1, 2 and 4 shall be equal to each other within 0.0025 mm.

7.5 Roller axes

7.5.1 The distance between the roller axes shall everywhere agree with the designating size of the bars within

0.0025 mm for 100 mm bars
0.005 mm for 200 mm bars
0.008 mm for 300 mm bars

7.5.2 For all sizes of bars, the roller axes shall lie in a common plane within 0.002 mm over the length of either roller.

7.5.3 The upper surface, and the lower, if a working surface, shall be parallel to the plane tangent to the lower surface of the rollers within 0.002 mm.

7.5.4 In addition to complying with the foregoing requirements, a type 3 sine bar shall also comply with the following requirements.

(a) The bearing surface of the setting foot and the hinge block shall be flat within 0.0025 mm.

(b) When a type 3 bar is placed on a flat datum surface, it shall be free from rock and its upper surface, and lower surface, if a working surface, shall be parallel to the datum surface within

0.0015 mm for 100 mm bars
0.0020 mm for 200 mm bars
0.0030 mm for 300 mm bars

7.6 End and side fences. Sine bars may be supplied with end and side fences. The working surfaces of end and side fences shall be flat within 0.005 mm, when attached.

. . . .

9. Marking

Each bar shall have legibly and permanently marked upon it the following particulars:

(a) designating size of bar;

(b) the number of this British Standard (i.e. BS 3064);

(c) an identification number;

(d) the manufacturer's name or trademark.

Section three. Sine tables inclinable about a single axis

10. General design

10.1 The essential features of a sine table inclinable about a single axis, as illustrated in figure 2, are a base, provided with a slip gauge platform, and a work table having a setting roller at one end and a hinge roller at the other. At least one end and one side fence shall be provided.

. . . .

15. Accuracy

15.1 Gauge block platforms. The flatness of the gauge block platform and its parallelism to the under surface of the base shall lie within two parallel planes, as shown in table 2.

Table 2. Parallelism of gauge block platform

Designating size	Tolerance on parallelism
mm	mm
100	0.001
200	0.002
300	0.003
500	0.005

When two gauge platforms are provided they shall be co-planar within 0.001 mm.

15.2 Upper surface of the work table. The tolerances on flatness of the upper surface and its parallelism to the plane on which the sine table rests are given in table 3.

Table 3. Parallelism of upper surface

Designating size	Tolerance on flatness and parallelism
mm	mm
100	0.004
200	0.004
300	0.005
400	0.006

15.3 End and side faces. At least two adjacent faces of the table shall be:

(a) flat within 0.005 mm for 100, 200 and 300 mm tables; 0.006 mm for 500 mm tables;
(b) mutually square and square to the upper surface of the work table within 0.0025 mm per 25 mm;
(c) square or parallel to the axis of the hinge roller within 0.013 mm over the length of the roller.

15.4 Setting and hinge rollers. Individual rollers shall be uniform in radius within 0.002 mm and straight over their length to the same accuracy.

15.5 Roller axes

15.5.1 The distance between the roller axes shall everywhere agree with the designating size of the tables within

0.0025 mm for 100 mm tables
0.005 mm for 200 mm tables
0.006 mm for 300 mm tables
0.013 mm for 500 mm tables

15.5.2 The roller axes shall be in a common plane within 0.001 mm per 25 mm length over the length of either roller.

15.5.3 The roller axes shall be parallel to, and equidistant from, the upper surface of the work table to within 0.001 mm per 25 mm length of roller.

15.6 End and side fences. The working surfaces of end and side fences, when attached, shall be flat within 0.005 mm.

. . . .

17. Marking

Each sine table shall have legibly and permanently marked upon it the following particulars:

(a) designating size of table;
(b) the number of this British Standard (i.e. BS 3064);
(c) an identification number;
(d) the manufacturer's name or trademark.

Spirit levels

The spirit level is an angular measuring device in which the bubble settles at the highest point of a glass tube (vial), the bore of which has been produced to a large radius. The sensitivity of a spirit level is determined by the vial radius and the base length.

BS 958 'Spirit levels for use in precision engineering' provides for three types of level as follows:

Type 1 A level with an unrelieved flat base of steel, hardened and lapped, usually of base length from 100 mm to 200 mm (see Figure 5.5).

The advantage of the lapped base is that the effective length of the level can be varied by wringing two gauge blocks on to the base at the desired distance apart.

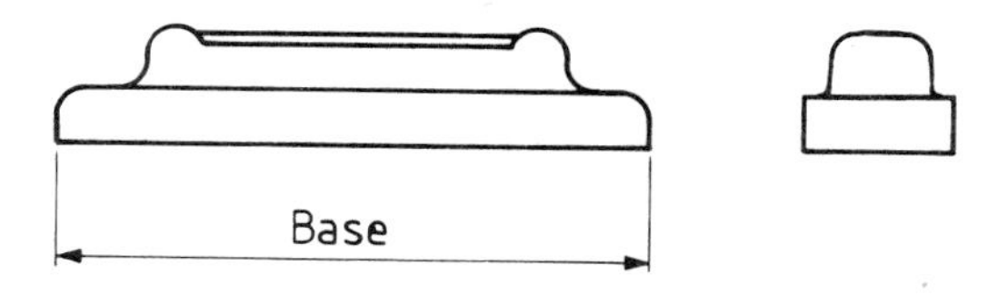

(Figure 1: BS 958: 1968)

Figure 5.5 *Type 1 level*

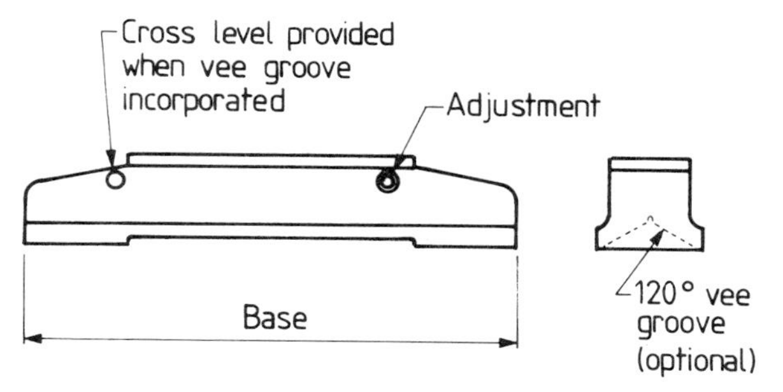

Figure 5.6 *Type 2 level* (Figure 2: BS 958: 1968)

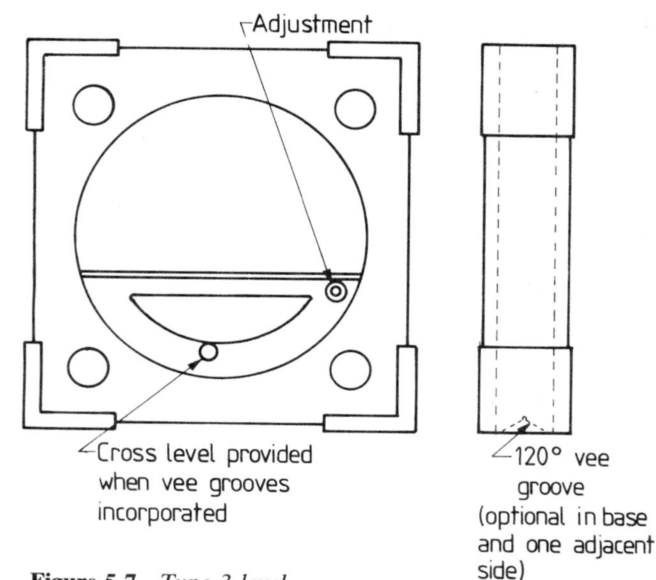

Figure 5.7 *Type 3 level*

(Figure 3: BS 958: 1968)

Type 2 A level, mounted in a body usually of cast iron or steel, having a base formed with flat bearing surfaces at the two ends, i.e. the middle portion is relieved (see Figure 5.6).

The bearing surfaces may be plain or contain a longitudinal 120 ° vee groove for use on cylindrical surfaces. In the latter case a short cross-level or small circular level is provided. This type of level is usually made with base lengths from 250 mm to 500 mm.

Type 3 A square block level, usually of cast iron and about 200 mm square. The four bearing surfaces are flat and may have the middle portions relieved; alternatively, the base and one adjacent surface may contain a longitudinal 120 ° vee groove for use on cylindrical surfaces in which case a short cross-level is provided (see Figure 5.7).

BS 958

Spirit levels for use in precision engineering

. . . .

2. Sensitivities

The sensitivities (see **3.2.(a)**) recommended for general precision engineering purposes are as follows:

Type of level	Recommended sensitivity (Value of 1 scale division)
Type 1	0.0025 mm in 100 mm i.e. approximately 5 seconds of arc* or 0.005 mm in 100 mm i.e. approximately 10 seconds of arc*
Types 2 and 3	0.005 mm in 100 mm i.e. approximately 10 seconds of arc* or 0.010 mm in 100 mm i.e. approximately 20 seconds of arc*

* 5 seconds of arc = 0.002 425 mm per 100 mm.

3. Nomenclature and definitions

. . . .

3.2 For the purposes of this British Standard, the following definitions apply:

(a) sensitivity. That angle of tilt which causes the bubble in the vial to move through one scale division.

(b) range. The total angle of tilt through which a vial can be used.

(c) roll error. The change in reading when the instrument is rotated about the axis of the vial.

(d) repeatability. The measure of the ability of the level to repeat its readings under all ordinary methods of operation.

(e) settling time. The time taken for a bubble to return to its steady position after the level has been disturbed.

(f) flatness tolerance. The maximum permissible distance between two imaginary parallel planes within which the surface under consideration can just be enclosed.

Table 1. Accuracy of performance

1	2	3	4	5	6
Sensitivity per division	**Permissible departure of average from nominal sensitivity per division**	**Maximum deviation of the value of any one scale division from average value**	**Repeatability**	**Permissible roll error over ±10 ° from normal position**	**settling time**
Per 100 mm	**Seconds of arc**	**Division**	**Seconds of arc**	**Seconds of arc**	**Seconds of time**
0.0025 (5 seconds)	$\pm\frac{1}{2}$	2/10	$\pm\frac{1}{2}$	2	10
0.005 (10 seconds)	±1	2/10	±1	2	
0.010 (20 seconds)	±2	2/10	±2	3	

(g) squareness tolerance. The maximum permissible distance separating two imaginary parallel planes which just enclose one of the surfaces under consideration and which are perpendicular to the other surface. Squareness tolerance is expressed over the total length of one of the surfaces.

(h) parallelism tolerance. The maximum permissible variation in distance between the surfaces under consideration.

. . . .

8. Accuracy of performance

8.1 Calibration. When the level is tested on a tilting-table and readings are taken at each division at both ends of the bubble, the average sensitivity shall not depart from the nominal sensitivity by more than the amount shown in column 2 of table 1. In addition, the maximum deviation of the value of any one division from the average value shall not exceed the amount shown in column 3 of table 1.

8.2 Repeatability. Identical settings of the level tilt shall be made and the departure of any one reading from the mean of six readings shall not exceed the tolerance given in column 4 of table 1.

8.3 Roll error. Roll error shall not exceed the tolerance given in column 5 of table 1.

8.4 Settling time. After the level has been disturbed to make the bubble move approximately one scale division, it shall come to within one-tenth of a scale division of its steady position within 10 seconds.

. . . .

11. Marking and manufacture

11.1 In addition to the marking referred to in **5.3**, each level shall be legibly and permanently marked with the particulars given below:

(a) the manufacturer's name and trademark;

(b) the number of this British Standard, i.e. BS 958;

(c) an identification number.

11.2 In the case of levels with hardened bases, the manufacturer shall supply a statement that the base has been suitably stabilized as recommended in **7.1**.

. . . .

Appendix A

Choice of sensitivity of a level

A.1 General. The use of unnecessarily sensitive levels is likely to introduce difficulties because ambient temperature changes and any lack of rigidity will have more noticeable effect on the readings of the instrument.

A.2 Calibration. Levels are usually calibrated on a tilting-table fitted with a pair of ball-ended feet at one end and a vertical micrometer head at the other. The tilting-table should be of substantial construction and should rest on a surface plate on a rigid foundation.

The micrometer head should be provided with a drum reading directly to 0.002 mm. It should have a ball-ended contact face.

For calibrating 5 second levels the base length of the tilting-table, i.e. the normal distance between the axis of the micrometer and the centre line passing through the two feet at the other end, should not be less than 750 mm.

A.3 Vial setting. Place the level on a rigid flat surface, which itself is approximately level, preferably a grade A surface table, rotate it until the bubble is approximately central and note the reading. Place a straightedge on the surface against the side of the level to provide a datum. Reverse the level and note the reading. Adjust the bubble with the key provided to the mean position of the two readings. Repeat this procedure until the bubble gives identical readings on reversal.

A.4 Use of spirit level vials. The application of levels to workshop practice can often be facilitated by the use of spirit level vials without fixed bases. These vials, available with or without mounting tubes, may be quickly fixed in position and adjusted to zero for temporary use by means of a suitable material. Details of standard sizes and sensitivities of such vials are given in BS 3509 'Spirit level vials'.

Standards publications referred to in this chapter

BS 958	Spirit levels for use in precision engineering
BS 1685	Bevel protractors (mechanical and optical)
BS 3064	Metric sine bars and sine tables (excluding compound tables)

Note: The above list does not include all the standards referred to in the extracts.

6 Limits and plain limit gauges

Introduction to limits and fits

It is impossible to make a part to an exact size and if, by chance, an exact size is achieved it is impossible to measure it accurately enough to prove it. Even the primary standard of length (see Chapter 2) is not exact and the error of reproduction of the metre is of the order of 1 part in 100 million. Grade 00 gauge blocks (see Chapter 4) up to 20 mm in length are accurate to within 0.05 μm, but they are not exact. Since a part cannot be made to an exact size it is necessary to specify the amount by which the size may deviate from the ideal size.

The successful functioning of most manufactured items depends not only upon the individual sizes of the parts but also upon the relationships of those parts in an assembly. If, for example, a shaft is to rotate in a hole, enough clearance must exist between the shaft and the hole to maintain an oil film, but not so much clearance that excessive radial float is permitted.

The type of fit which is required between mating parts and the acceptable deviation from the ideal size has resulted in various systems of limits and fits being developed. Such systems are particularly important where components are required to assemble together at random and much of today's manufacturing industry depends upon this interchangeability.

The ISO system of limits and fits

The ISO system is designed to provide a comprehensive range of limits and fits for engineering purposes. It is based on a series of tolerances graded to suit all classes of work from the finest to the coarsest.

BS 4500 'Specification for ISO limits and fits' relates to the ISO system of tolerances and limits of size for parts or components, and to fits obtained by their assembly. In view of their particular importance, only cylindrical parts (briefly designated as 'holes' and 'shafts') are referred to explicitly but the recommendations apply equally to other sections, and the term 'hole' or 'shaft' can be applied in the broadest sense and not restricted to round sections. For example, they may be applied to square and hexagonal sections and to lengths.

For the purpose of BS 4500 the following definitions apply and it should be noted that the terms may be used in a more restricted sense than in common use. (See also Figures 6.1, 6.2, 6.3 and 6.4.)

Size. A number expressing in a particular unit the numerical value of a length.

Actual size (*of a part*). The size of a part as obtained by measurement.

Limits of size. The maximum and minimum sizes permitted for a feature.

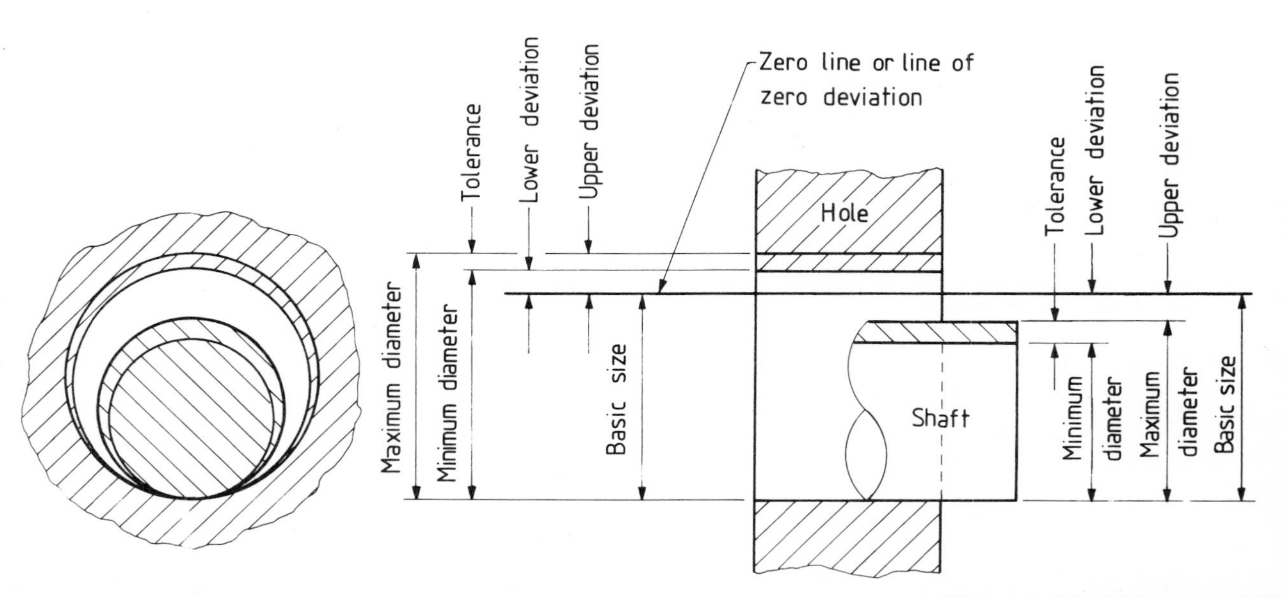

Figure 6.1 *Conventional method of illustrating the principal terms defined* (Figure 1: BS 4500: Part 1: 1969)

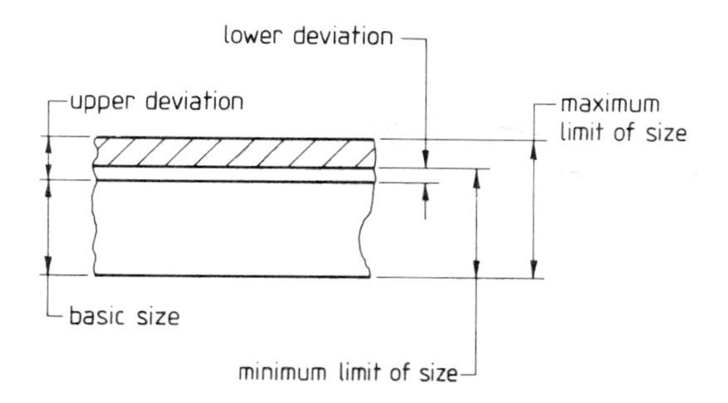

Figure 6.2 *Limit and fit terms*

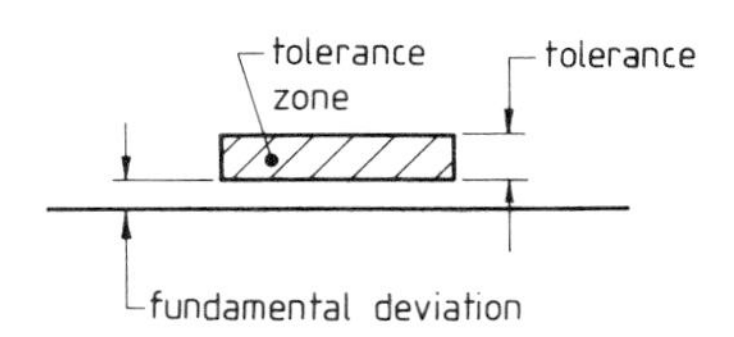

Figure 6.3 *Limit and fit terms*

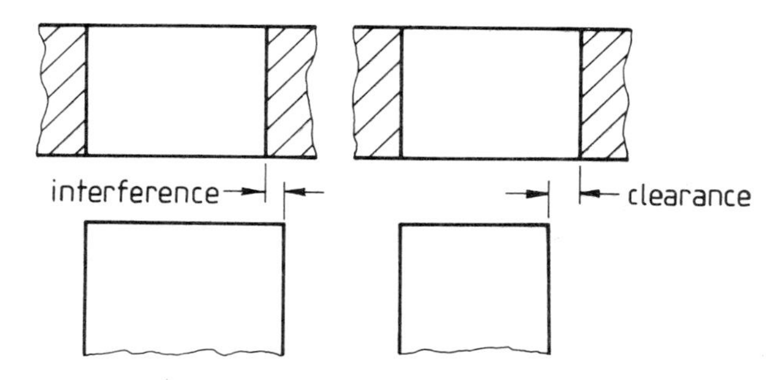

Figure 6.4 *Limit and fit terms*

Maximum limit of size. The greater of the two limits of size.

Minimum limit of size. The smaller of the two limits of size.

Basic size. The size by reference to which the limits of size are fixed. The basic size is the same for both members of a fit.

Deviation. The algebraical difference between a size (actual, maximum, etc.) and the corresponding basic size.

Actual deviation. The algebraical difference between the actual size and the corresponding basic size.

Upper deviation. The algebraical difference between the maximum limit of size and the corresponding basic size. This is designated 'ES' for a hole and 'es' for a shaft, these letters standing for the French term 'écart supérieur'.

Lower deviation. The algebraical difference between the minimum limit of size and the corresponding basic size. This is designated 'EI' for a hole and 'ei' for a shaft, these letters standing for the French term 'écart inférieur'.

Zero line. In a graphical representation of limits and fits, the straight line to which the deviations are referred. The zero line is the line of zero deviation and represents the basic size.

By convention, when the zero line is drawn horizontally, positive deviations are shown above and negative deviations below it.

Tolerance. The difference between the maximum limit of size and the minimum limit of size (or in other words the algebraical difference between the upper deviation and the lower deviation).

The tolerance is an absolute value without sign.

Tolerance zone. In a graphical representation of tolerances, the zone comprised between the two lines representing the limits of tolerance and defined by its magnitude (tolerance) and by its position in relation to the zero line.

Fundamental deviation. That one of the two deviations, being the one nearest to the zero line, which is conventionally chosen to define the position of the tolerance zone in relation to the zero line.

Grade of tolerance. In a standardized system of limits and fits, and a group of tolerances considered as corresponding to the same level of accuracy for all basic sizes.

Standard tolerance. In a standardized system of limits and fits, any tolerance belonging to the system.

Standard tolerance unit. In the ISO system of limits and fits, a factor expressed only in terms of the basic size and used as a basis for the determination of the standard tolerances of the system. (Each tolerance is equal to the product of the value of the standard tolerance unit, for the basic size in question, by a coefficient corresponding to each grade of tolerance.)

Shaft. The term used by convention to designate all external features of a part, including parts which are not cylindrical.

Hole. The term used by convention to designate all internal features of a part, including parts which are not cylindrical.

Basic shaft. In the ISO system of limits and fits, a shaft the upper deviation of which is zero.

More generally, the shaft chosen as a basis for a shaft-basis system of fits (see shaft-basis system of fits).

Basic hole. In the ISO system of limits and fits, a hole the lower deviation of which is zero.

More generally, the hole chosen as a basis for a hole-basis system of fits (see hole-basis system of fits).

GO limit. The designation applied to that one of the two limits of size which corresponds to the maximum material condition, i.e., the upper limit of a shaft or the lower limit of a hole.

(When limit gauges are used, this is the limit checked by the GO gauge.)

NOT GO limit. The designation applied to that one of the two limits of size which corresponds to the minimum material condition, i.e., the lower limit of a shaft or the upper limit of a hole.

(When limit gauges are used, this is the limit of size checked by the NOT GO gauge.)

Fit. The relationship resulting from the difference, before assembly, between the sizes of the two parts which are to be assembled.

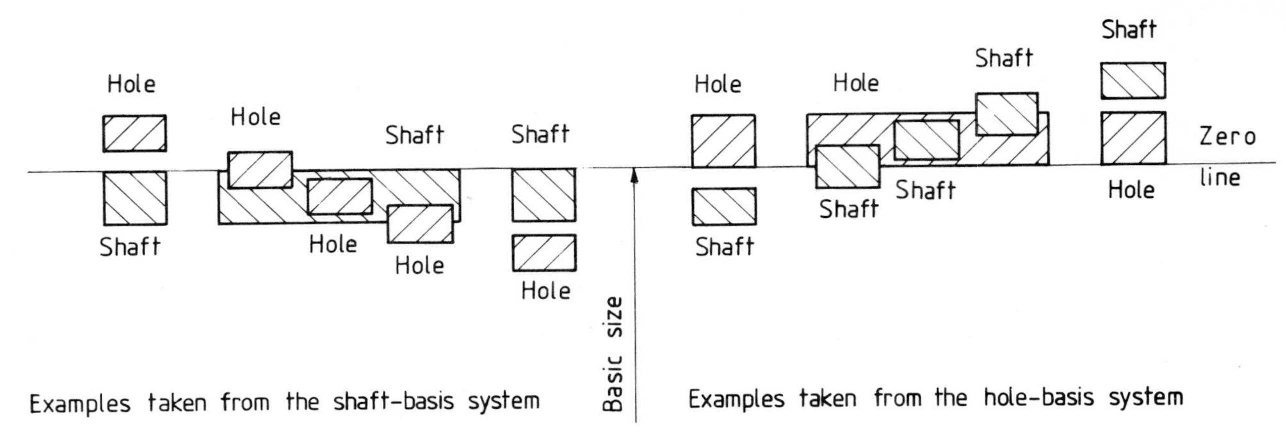

Figure 6.5 *Examples taken from the shaft-basis and the hole-basis systems* (Figure 2: BS 4500: Part 1: 1969)

Basic size (of a fit). The common value of the basic size of the two parts of a fit.

Variation of a fit. The arithmetical sum of the tolerances of the two mating parts of a fit.

Clearance. The difference between the sizes of the hole and the shaft, before assembly, when this difference is positive.

Interference. The magnitude of the difference between the sizes of the hole and the shaft, before assembly, when this difference is negative.

Clearance fit. A fit which always provides a clearance. (The tolerance zone of the hole is entirely above that of the shaft.)

Interference fit. A fit which always provides an interference. (The tolerance zone of the hole is entirely below that of the shaft.)

Transition fit. A fit which may provide either a clearance or an interference. (The tolerance zones of the hole and the shaft overlap.)

Minimum clearance. In a clearance fit, the difference between the minimum size of the hole and the maximum size of the shaft.

Maximum clearance. In a clearance or a transition fit, the difference between the maximum size of the hole and the minimum size of the shaft.

Minimum interference. In an interference fit, the magnitude of the (negative) difference between the maximum size of the hole and the minimum size of the shaft, before assembly.

Maximum interference. In an interference or a transition fit, the magnitude of the (negative) difference between the minimum size of the hole and the maximum size of the shaft, before assembly.

Limit system. A system of standardized tolerances and deviations.

Fit system. A system of fits comprising shafts and holes belonging to a limit system.

Shaft-basis system of fits. A system of fits in which the different clearances and interferences are obtained by associating various holes with a single shaft (or, possibly, with shafts of different grades, by having the same fundamental deviation).

In the ISO system, the basic shaft is the shaft the upper deviation of which is zero.

Hole-basis system of fits. A system of fits in which the different clearances and interferences are obtained by associating various shafts with a single hole (or, possibly, with holes of different grades, but always having the same fundamental deviation).

In the ISO system, the basic hole is the hole the lower deviation of which is zero.

Note
The limits and fits diagrams
A convention was developed for illustrating limits and fits, and a study of Figure 6.1 (Figure 1 in BS 4500: Part 1) will aid the understanding of this convention. It is often advantageous to use only a part of this type of diagram to show only the tolerances and their positions and this has been adopted in Figure 6.5 (Figure 2 in BS 4500: Part 1).

Two factors affect a fit: the tolerance, and the fundamental deviation of the hole and the shaft.

Tolerance

It is necessary to increase the tolerance as the basic size increases, to maintain functional conditions. For example, a tolerance of 0.025 mm applied to a basic size of 50 mm would need to be increased to 0.046 mm for a basic size of 200 mm, and to 0.063 mm for a basic size of 500 mm, to maintain a constant functional condition (with respect to size only). (See Figure 6.6.)

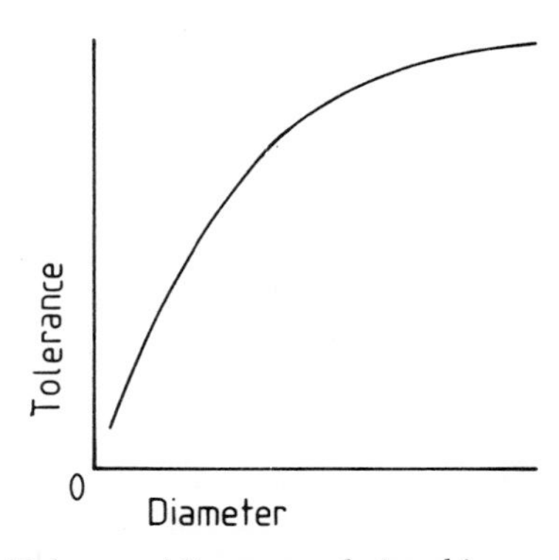

Figure 6.6 *Tolerance/diameter relationship*

Table 6.1 *Standard tolerances*

Tolerance unit 0.001 mm

Nominal sizes		*Tolerance grades*																	
Over	*To*	*IT 01*	*IT 0*	*IT 1*	*IT 2*	*IT 3*	*IT 4*	*IT 5*	*IT 6**	*IT 7*	*IT 8*	*IT 9*	*IT 10*	*IT 11*	*IT 12*	*IT 13*	*IT 14†*	*IT 15†*	*IT 16†*
mm	mm 3	0.3	0.5	0.8	1.2	2	3	4	6	10	14	25	40	60	100	140	250	400	600
3	6	0.4	0.6	1	1.5	2.5	4	5	8	12	18	30	48	75	120	180	300	480	750
6	10	0.4	0.6	1	1.5	2.5	4	6	9	15	22	36	58	90	150	220	360	580	900
10	18	0.5	0.8	1.2	2	3	5	8	11	18	27	43	70	110	180	270	430	700	1100
18	30	0.6	1	1.5	2.5	4	6	9	13	21	33	52	84	130	210	320	520	840	1300
30	50	0.6	1	1.5	2.5	4	7	11	16	25	39	62	100	160	250	390	620	1000	1600
50	80	0.8	1.2	2	3	5	8	13	19	30	46	74	120	190	300	460	740	1200	1900
80	120	1	1.5	2.5	4	6	10	15	22	35	54	87	140	220	350	540	870	1400	2200
120	180	1.2	2	3.5	5	8	12	18	25	40	63	100	160	250	400	630	1000	1600	2500
180	250	2	3	4.5	7	10	14	20	29	46	72	115	185	290	460	720	1150	1850	2900
250	315	2.5	4	6	8	12	16	23	32	52	81	130	210	320	520	810	1300	2100	3200
315	400	3	5	7	9	13	18	25	36	57	89	140	230	360	570	890	1400	2300	3600
400	500	4	6	8	10	15	20	27	40	63	97	155	250	400	630	970	1550	2500	4000
500	630	—	—	—	—	—	—	—	44	70	110	175	280	440	700	1100	1750	2800	4400
630	800	—	—	—	—	—	—	—	50	80	125	200	320	500	800	1250	2000	3200	5000
800	1000	—	—	—	—	—	—	—	56	90	140	230	360	560	900	1400	2300	3600	5600
1000	1250	—	—	—	—	—	—	—	66	105	165	260	420	660	1050	1650	2600	4200	6600
1250	1600	—	—	—	—	—	—	—	78	125	195	310	500	780	1250	1950	3100	5000	7800
1600	2000	—	—	—	—	—	—	—	92	150	230	370	600	920	1500	2300	3700	6000	9200
2000	2500	—	—	—	—	—	—	—	110	175	280	440	700	1100	1750	2800	4400	7000	11000
2500	3150	—	—	—	—	—	—	—	135	210	330	540	860	1350	2100	3300	5400	8600	13500

* Not recommended for fits in sizes above 500 mm.
† Not applicable to sizes below 1 mm.

Table 6.2 *Standard tolerances*

Tolerance unit 0.001 mm

Nominal sizes		*Tolerance grades*																	
Over	*To*	*IT 01*	*IT 0*	*IT 1*	*IT 2*	*IT 3*	*IT 4*	*IT 5*	*IT 6*	*IT 7*	*IT 8*	*IT 9*	*IT 10*	*IT 11*	*IT 12*	*IT 13*	*IT 14*	*IT 15*	*IT 16*
mm —	mm 3	0.3	0.5	0.8	1.2	2.5	3	4	6	10	14	25	40	60	100	140	250	400	600
3	6	0.4	0.6	1	1.5	2.5	4	5	8	12	18	30	48	75	120	180	300	480	750
6	10	0.4	0.6	1	1.5	2.5	4	6	9	15	22	36	58	90	150	220	360	580	900
10	18	0.5	0.8	1.2	2	3	5	8	11	18	27	43	70	110	180	270	430	700	1100
18	30	0.6	1	1.5	2.5	4	6	9	13	21	33	52	84	130	210	330	520	840	1300
30	50	0.6	1	1.5	2.5	4	7	11	16	25	39	62	100	160	250	390	620	1000	1600
50	80	0.8	1.2	2	3	5	8	13	19	30	46	74	120	190	300	460	740	1200	1900
80	120	1	1.5	2.5	4	6	10	15	22	35	54	87	140	220	350	540	870	1400	2200

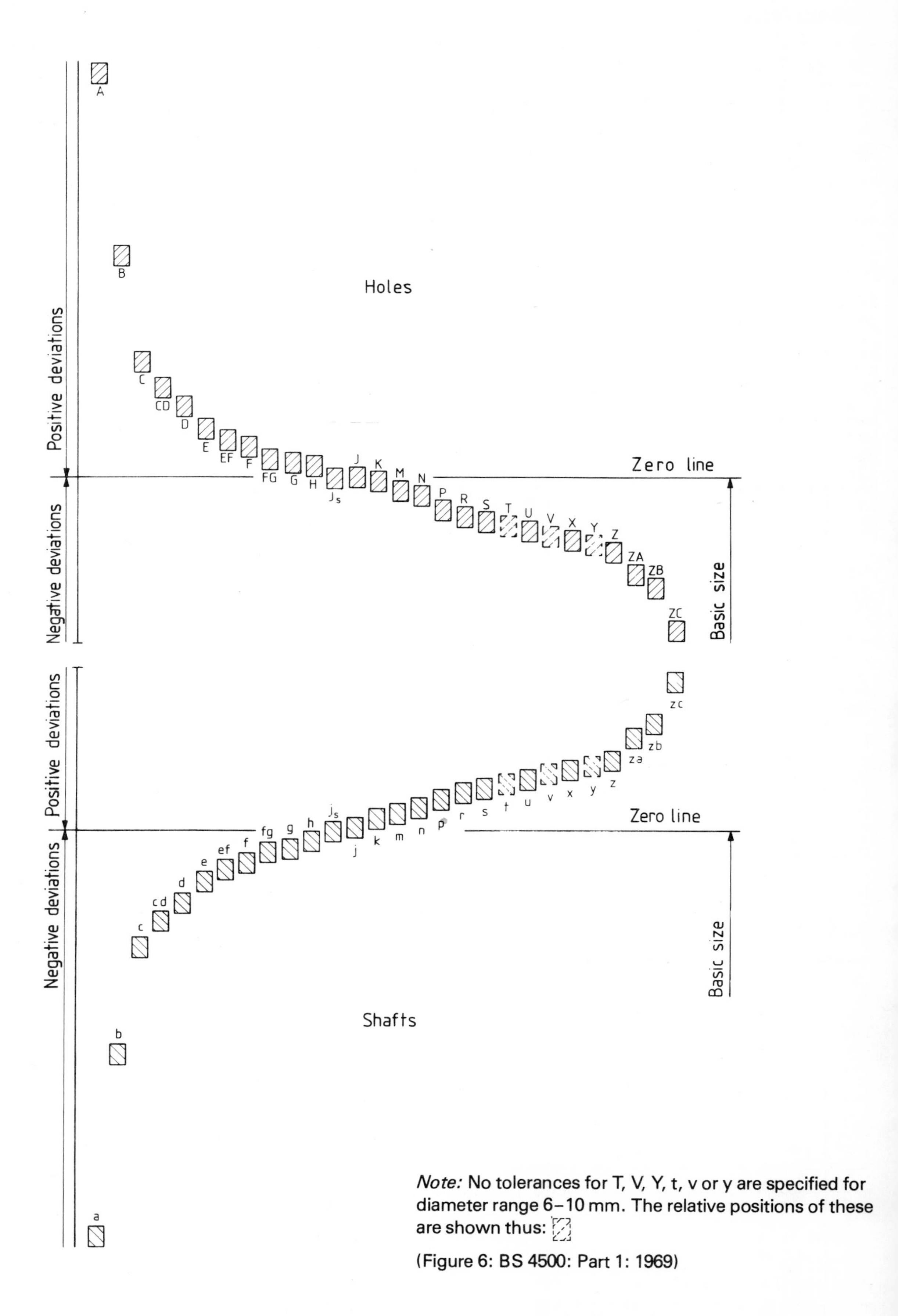

Note: No tolerances for T, V, Y, t, v or y are specified for diameter range 6–10 mm. The relative positions of these are shown thus:

(Figure 6: BS 4500: Part 1: 1969)

Figure 6.7 *Respective positions of grade 7 tolerance zones for the diameter range 6 to 10 mm*

BS 4500 provides for 18 tolerance grades and these are shown in Table 6.1. (IT stands for ISO series of tolerances.)

Thus, for any particular dimension, 18 different tolerances may be applied (note the exceptions for dimensions above 500 mm). For example, a hole 10 mm in diameter could have tolerances of 0.0004 mm; 0.0006 mm; 0.001 mm; 0.0015 mm . . . 0.900mm (see Table 6.2).

Fundamental deviations

The fundamental deviation positions the tolerance with relation to the basic size. BS 4500 provides 28 deviations each designated by a letter or letters; upper case (capital) letters being used for holes, and lower case (small) letters being used for shafts. Figure 6.7 illustrates the relative positions of the fundamental deviations for holes and shafts.

It should be noted that deviation j_s (J_s for holes) has no fundamental deviation and the tolerance zone (of whatever magnitude) is disposed equally about the zero line. This provides a symmetrical bilateral tolerance.

Thus, for any particular dimension with a particular tolerance grade applied, there are 28 positions for that tolerance grade (with some exceptions). For example (see Table 6.3), a shaft 10 mm in diameter with a tolerance of IT 9 (i.e. a tolerance of 0.036 mm) could have that tolerance positioned such that the fundamental deviation is −0.280 mm; −0.150 mm; −0.080 mm . . . +0.097 mm.

This would give toleranced dimensions of:

$10^{-0.280}_{-0.316}$ mm $\quad 10^{-0.150}_{-0.186}$ mm

$10^{-0.080}_{-0.116}$ mm . . . $\quad 10^{+0.133}_{+0.097}$ mm

Note: A table for fundamental deviations for holes (similar to Table 6.3) is given in BS 4500: Part 1, but is not reproduced in this manual.

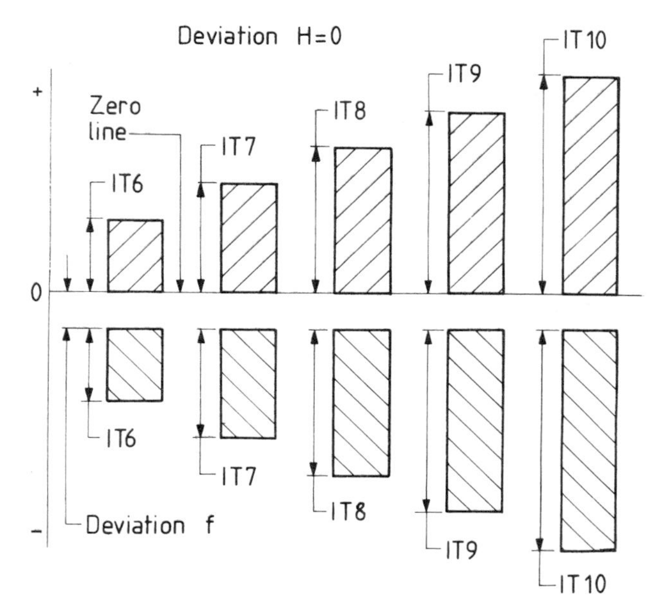

Figure 6.8 *Association of fundamental deviations and tolerances (for deviations H and f and various tolerances)*

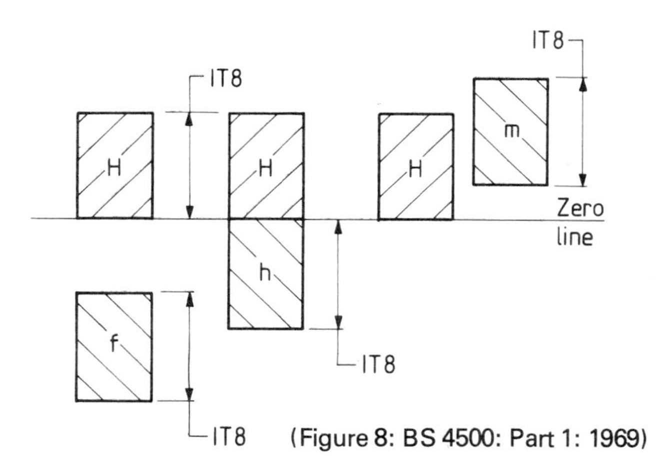

Figure 6.9 *Association of fundamental deviations and tolerances (for various deviations and IT 8 tolerances)*

Association of fundamental deviations and tolerances

The separate series of deviations and tolerances can be combined in any way that appears necessary to give a required fit condition. For example, the deviations H (for the hole) and f (for the shaft) could be combined, and with each of these deviations any one of the tolerance grades IT 01 to IT 16 could be used. Figure 6.8 shows, diagrammatically, the conditions which would be produced by combining the deviations H and f with the tolerance grades IT 6 to IT 10.

Similarly, the tolerance grade IT 8 could be used with any of the deviations. Figure 6.9 shows the conditions which would be produced by combining the tolerance grade IT 8 with the deviations f, h and m with deviation H.

Designation

The complete designation of the limits of tolerance for a shaft or a hole requires the use of the appropriate letter to indicate the fundamental deviation, followed by a suffix number denoting the tolerance grade. For this purpose the numerical part of the tolerance grade designation is used. For example, a hole tolerance with deviation H and tolerance grade IT 7 is designated 'H7'. Similarly, a shaft tolerance with deviation p and tolerance grade IT 6 is designated 'p6'.

The limits of size for a feature are defined by the basic size of the feature, say 45 mm, followed by the appropriate tolerance designation, for example, 45 H7 or 45 p6.

A fit is indicated by combining the basic size common

Table 6.3 *Fundamental deviations for shafts*

Unit = 0.001 mm

Fundamental deviation		*Upper deviation es*												*Lower deviation ei*				
Letter		*a**	*b**	*c*	*cd*	*d*	*e*	*ef*	*f*	*fg*	*g*	*h*	*j*$_s$†	*j*			*k*	
Grade		*01 to 16*												*5–6*	*7*	*8*	*4–7*	*≤3 >7*
Nominal sizes																		
Over	*To*																	
mm	mm																	
—	3	−270	−140	−60	−34	−20	−14	−10	−6	−4	−2	0	±IT/2	−2	−4	−6	0	0
3	6	−270	−140	−70	−46	−30	−20	−14	−10	−6	−4	0		−2	−4	—	+1	0
6	10	−280	−150	−80	−56	−40	−25	−18	−13	−8	−5	0		−2	−5	—	+1	0
10	14	−290	−150	−95	—	−50	−32	—	−16	—	−6	0		−3	−6	—	+1	0
14	18																	
18	24	−300	−160	−110	—	−65	−40	—	−20	—	−7	0		−4	−8	—	+2	0
24	30																	
30	40	−310	−170	−120	—	−80	−50	—	−25	—	−9	0		−5	−10	—	+2	0
40	50	−320	−180	−130														
50	65	−340	−190	−140	—	−100	−60	—	−30	—	−10	0		−7	−12	—	+2	0
65	80	−360	−200	−150														
80	100	−380	−220	−170	—	−120	−72	—	−36	—	−12	0		−9	−15	—	+3	0
100	120	−410	−240	−180														
120	140	−460	−260	−200	—	−145	−85	—	−43	—	−14	0		−11	−18	—	+3	0
140	160	−520	−280	−210														
160	180	−580	−310	−230														
180	200	−660	−340	−240	—	−170	−100	—	−50	—	−15	0		−13	−21	—	+4	0
200	225	−740	−380	−260														
225	250	−820	−420	−280														
250	280	−920	−480	−300	—	−190	−110	—	−56	—	−17	0		−16	−26	—	+4	0
280	315	−1050	−540	−330														
315	355	−1200	−600	−360	—	−210	−125	—	−62	—	−18	0		−18	−28	—	+4	0
355	400	−1350	−680	−400														
400	450	−1500	−760	−440	—	−230	−135	—	−68	—	−20	0		−20	−32	—	+5	0
450	500	−1650	−840	−480														
Grade		*6 to 16*																
500	630	—	—	—	—	−260	−145	—	−76	—	−22	0	±IT/2				0	
630	800	—	—	—	—	−290	−160	—	−80	—	−24	0					0	
800	1000	—	—	—	—	−320	−170	—	−86	—	−26	0					0	
1000	1250	—	—	—	—	−350	−195	—	−98	—	−28	0					0	
1250	1600	—	—	—	—	−390	−220	—	−110	—	−30	0					0	
1600	2000	—	—	—	—	−430	−240	—	−120	—	−32	0					0	
2000	2500	—	—	—	—	−480	−260	—	−130	—	−34	0					0	
2500	3150	—	—	—	—	−520	−290	—	−145	—	−38	0					0	

* Not applicable to sizes up to 1 mm.
† In grades 7 to 11, the two symmetrical deviations ±IT/2 should be rounded if the IT value in micrometres is an odd value by replacing it by the even value immediately below.

Fundamental deviation		*Lower deviation ei*													
Letter		*m*	*n*	*p*	*r*	*s*	*t*	*u*	*v*	*x*	*y*	*z*	*za*	*zb*	*zc*
Grade		*01 to 16*													
Nominal size															
Over	*To*														
mm	mm														
—	3	+2	+4	+6	+10	+14	—	+18	—	+20	—	+26	+32	+40	+60
3	6	+4	+8	+12	+15	+19	—	+23	—	+28	—	+35	+42	+50	+80
6	10	+6	+10	+15	+19	+23	—	+28	—	+34	—	+42	+52	+67	+97
10	14	+7	+12	+18	+23	+28	—	+33	—	+40	—	+50	+64	+90	+130
14	18								+39	+45	—	+60	+77	+108	+150
18	24	+8	+15	+22	+28	+35	—	+41	+47	+54	+63	+73	+98	+136	+188
24	30						+41	+48	+55	+64	+75	+88	+118	+160	+218
30	40	+9	+17	+26	+34	+43	+48	+60	+68	+80	+94	+112	+148	+200	+274
40	50						+54	+70	+81	+97	+114	+136	+180	+242	+325
50	65	+11	+20	+32	+41	+53	+66	+87	+102	+122	+144	+172	+226	+300	+405
65	80				+43	+59	+75	+102	+120	+146	+174	+210	+274	+360	+480
80	100	+13	+23	+37	+51	+71	+91	+124	+146	+178	+214	+258	+335	+445	+585
100	120				+54	+79	+104	+144	+172	+210	+254	+310	+400	+525	+690
120	140	+15	+27	+43	+63	+92	+122	+170	+202	+248	+300	+365	+470	+620	+800
140	160				+65	+100	+134	+190	+228	+280	+340	+415	+535	+700	+900
160	180				+68	+108	+146	+210	+252	+310	+380	+465	+600	+780	+1000
180	200	+17	+31	+50	+77	+122	+166	+236	+284	+350	+425	+520	+670	+880	+1150
200	225				+80	+130	+180	+258	+310	+385	+470	+575	+740	+960	+1250
225	250				+84	+140	+196	+284	+340	+425	+520	+640	+820	+1050	+1350
250	280	+20	+34	+56	+94	+158	+218	+315	+385	+475	+580	+710	+920	+1200	+1550
280	315				+98	+170	+240	+350	+425	+525	+650	+790	+1000	+1300	+1700
315	355	+21	+37	+62	+108	+190	+268	+390	+475	+590	+730	+900	+1150	+1500	+1900
355	400				+114	+208	+294	+435	+530	+660	+820	+1000	+1300	+1650	+2100
400	450	+23	+40	+68	+126	+232	+330	+490	+595	+740	+920	+1100	+1450	+1850	+2400
450	500				+132	+252	+360	+540	+660	+820	+1000	+1250	+1600	+2100	+2600
Grade		*6 to 16*													
500	560	+26	+44	+78	+150	+280	+400	+600							
560	630				+155	+310	+450	+660							
630	710	+30	+50	+88	+175	+340	+500	+740							
710	800				+185	+380	+560	+840							
800	900	+34	+56	+100	+210	+430	+620	+940							
900	1000				+220	+470	+680	+1050							
1000	1120	+40	+66	+120	+250	+520	+780	+1150							
1120	1250				+260	+580	+840	+1300							
1250	1400	+48	+78	+140	+300	+640	+960	+1450							
1400	1600				+330	+720	+1050	+1600							
1600	1800	+58	+92	+170	+370	+820	+1200	+1850							
1800	2000				+400	+920	+1350	+2000							
2000	2240	+68	+110	+195	+440	+1000	+1500	+2300							
2240	2500				+460	+1100	+1650	+2500							
2500	2800	+76	+135	+240	+550	+1250	+1900	+2900							
2800	3150				+580	+1400	+2100	+3200							

to both features with the designations appropriate to each, with the hole being quoted first.

Example 45 H7–p6 or H7/p6

Hole-basis and shaft-basis systems of fits

For most general applications using the ISO system it is usual to recommend hole-basis fits, i.e. a system of fits in which the different fits are obtained by combining shafts with varying fundamental deviations with a hole having a fundamental deviation of zero (H). This, in effect, means the hole is a constant and it is the shaft which is varied in order to obtain differing conditions of fit (see Figure 6.10).

The reason for a preference for the hole-basis system is that, except for very large sizes where the effects of temperature play a large part, it is usually considered easier to manufacture and measure the male member of a fit and it is thus desirable to be able to allocate the larger part of the total tolerance available to the hole and adjust the shaft to suit.

In some circumstances, however, it may be preferable to employ a shaft-basis system, i.e. a system of fits in which the different fits are obtained by combining holes with varying fundamental deviations with a shaft having a fundamental deviation of zero (h). This, in effect, means the shaft is a constant and the hole is varied in order to obtain the differing conditions of fit (see Figure 6.11).

The shaft-basis system can be used, for example, in the case of driving shafts where a single shaft may have to accommodate a variety of accessories such as couplings, bearings, collars and gears. In such applications it is preferable to maintain a constant diameter for the permanent member, the shaft, and vary the bores of the accessories.

Selected fits

The ISO system provides a great many hole and shaft tolerances, so as to cater for a very wide range of conditions. However, the fit conditions required for the majority of engineering products can be provided by a quite limited selection of tolerances and the following are most commonly applied:

Selected hole tolerances	H7, H8, H9, H11
Selected shaft tolerances	c11, d10, e9, f7, g6, h6, k6, n6, p6, s6

BS 4500A 'Selected hole-basis' and BS 4500B 'Selected shaft-basis' provide a selection which, for most applications, will be found adequate (see pages 129 and 130).

From BS 4500, manufacturers can make selections other than those in BS 4500A and BS 4500B, and careful selection can provide a wide range of fits to suit almost any application.

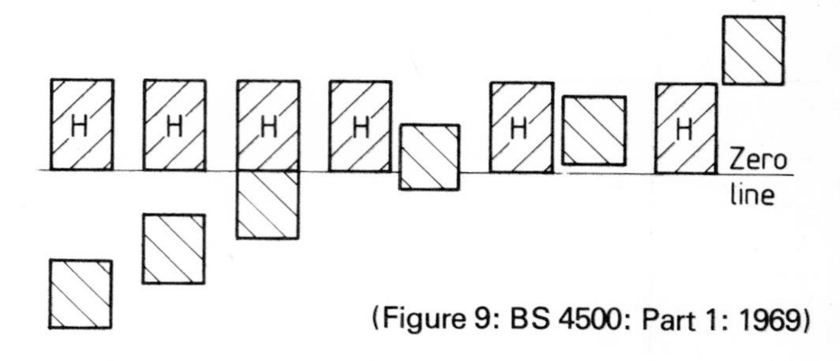

(Figure 9: BS 4500: Part 1: 1969)

Figure 6.10 *Hole and shaft relationships for a hole-basis system*

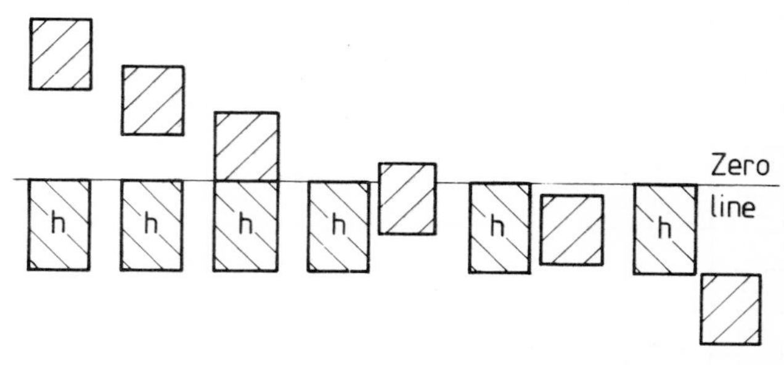

(Figure 10: BS 4500: Part 1: 1969)

Figure 6.11 *Hole and shaft relationships for a shaft-basis system*

Sizes above 500 mm

Although there is no fixed dividing line between what are regarded as 'small sizes' and 'large sizes' it is recognized that manufacturing and measuring problems which are not encountered for small sizes do occur in large sizes. The effects of temperature, for example, are more significant where very large sizes are concerned. Factors which should be taken into account when dealing with fits in the large size range are covered in BS 4500: Part 1: Appendix B. Appendix C, of the same publication, provides guidance where interchangeable manufacture is not practicable and the technique of matched fits may be used. (Appendices B and C are not included in this publication.)

Surface texture and errors of form

The recommendations in BS 4500 enable limits of size for mating features to be derived from a standard series of tolerances and deviations in order to provide various types of fit. However, the mating features of a fit are made up of surfaces and the surface texture and geometrical form of these surfaces frequently has a considerable bearing on satisfactory functioning.

Surface texture is particularly important in the case of precise fits involving relative movement (see Chapter 10).

Although ordinary limits of size, in themselves, only comprise limiting sizes between which the measured

Extracted from BS 4500

BRITISH STANDARD

SELECTED ISO FITS—HOLE BASIS

Data Sheet 4500A

Diagram to scale for 25 mm. diameter

Clearance fits: H11/c11, H9/d10, H9/e9, H8/f7, H7/g6, H7/h6 — Transition fits: H7/k6, H7/n6 — Interference fits: H7/p6, H7/s6 — Holes / Shafts

Nominal sizes		Clearance fits												Transition fits				Interference fits				Nominal sizes	
		Tolerance		Tolerance		Tolerance		Tolerance		Tolerance		Tolerance		Tolerance		Tolerance		Tolerance		Tolerance			
Over	To	H11	c11	H9	d10	H9	e9	H8	f7	H7	g6	H7	h6	H7	k6	H7	n6	H7	p6	H7	s6	Over	To
mm	mm	0·001 mm	0·001 mm	0·001 mm	0·001 mm	0·001 mm	0·001 mm	0·001 mm	0·001 mm	0·001 mm	0·001 mm	0·001 mm	0·001 mm	0·001 mm	0·001 mm	0·001 mm	0·001 mm	0·001 mm	0·001 mm	0·001 mm	0·001 mm	mm	mm
—	3	+ 60 0	− 60 − 120	+ 25 0	− 20 − 60	+ 25 0	− 14 − 39	+ 14 0	− 6 − 16	+ 10 0	− 2 − 8	+ 10 0	− 6 0	+ 10 0	+ 6 + 0	+ 10 0	+ 10 + 4	+ 10 0	+ 12 + 6	+ 10 0	+ 20 + 14	—	3
3	6	+ 75 0	− 70 − 145	+ 30 0	− 30 − 78	+ 30 0	− 20 − 50	+ 18 0	− 10 − 22	+ 12 0	− 4 − 12	+ 12 0	− 8 0	+ 12 0	+ 9 + 1	+ 12 0	+ 16 + 8	+ 12 0	+ 20 + 12	+ 12 0	+ 27 + 19	3	6
6	10	+ 90 0	− 80 − 170	+ 36 0	− 40 − 98	+ 36 0	− 25 − 61	+ 22 0	− 13 − 28	+ 15 0	− 5 − 14	+ 15 0	− 9 0	+ 15 0	+ 10 + 1	+ 15 0	+ 19 + 10	+ 15 0	+ 24 + 15	+ 15 0	+ 32 + 23	6	10
10	18	+ 110 0	− 95 − 205	+ 43 0	− 50 − 120	+ 43 0	− 32 − 75	+ 27 0	− 16 − 34	+ 18 0	− 6 − 17	+ 18 0	− 11 0	+ 18 0	+ 12 + 1	+ 18 0	+ 23 + 12	+ 18 0	+ 29 + 18	+ 18 0	+ 39 + 28	10	18
18	30	+ 130 0	− 110 − 240	+ 52 0	− 65 − 149	+ 52 0	− 40 − 92	+ 33 0	− 20 − 41	+ 21 0	− 7 − 20	+ 21 0	− 13 0	+ 21 0	+ 15 + 2	+ 21 0	+ 28 + 15	+ 21 0	+ 35 + 22	+ 21 0	+ 48 + 35	18	30
30	40	+ 160 0	− 120 − 280	+ 62 0	− 80 − 180	+ 62 0	− 50 − 112	+ 39 0	− 25 − 50	+ 25 0	− 9 − 25	+ 25 0	− 16 0	+ 25 0	+ 18 + 2	+ 25 0	+ 33 + 17	+ 25 0	+ 42 + 26	+ 25 0	+ 59 + 43	30	40
40	50	+ 160 0	− 130 − 290																			40	50
50	65	+ 190 0	− 140 − 330	+ 74 0	− 100 − 220	+ 74 0	− 60 − 134	+ 46 0	− 30 − 60	+ 30 0	− 10 − 29	+ 30 0	− 19 0	+ 30 0	+ 21 + 2	+ 30 0	+ 39 + 20	+ 30 0	+ 51 + 32	+ 30 0	+ 72 + 53	50	65
65	80	+ 190 0	− 150 − 340																	+ 30 0	+ 78 + 59	65	80
80	100	+ 220 0	− 170 − 390	+ 87 0	− 120 − 260	+ 87 0	− 72 − 159	+ 54 0	− 36 − 71	+ 35 0	− 12 − 34	+ 35 0	− 22 0	+ 35 0	+ 25 + 3	+ 35 0	+ 45 + 23	+ 35 0	+ 59 + 37	+ 35 0	+ 93 + 71	80	100
100	120	+ 220 0	− 180 − 400																	+ 35 0	+ 101 + 79	100	120
120	140	+ 250 0	− 200 − 450	+ 100 0	− 145 − 305	+ 100 0	− 84 − 185	+ 63 0	− 43 − 83	+ 40 0	− 14 − 39	+ 40 0	− 25 0	+ 40 0	+ 28 + 3	+ 40 0	+ 52 + 27	+ 40 0	+ 68 + 43	+ 40 0	+ 117 + 92	120	140
140	160	+ 250 0	− 210 − 460																	+ 40 0	+ 125 + 100	140	160
160	180	+ 250 0	− 230 − 480																	+ 40 0	+ 133 + 108	160	180
180	200	+ 290 0	− 240 − 530	+ 115 0	− 170 − 355	+ 115 0	− 100 − 215	+ 72 0	− 50 − 96	+ 46 0	− 15 − 44	+ 46 0	− 29 0	+ 46 0	+ 33 + 4	+ 46 0	+ 60 + 31	+ 46 0	+ 79 + 50	+ 46 0	+ 151 + 122	180	200
200	225	+ 290 0	− 260 − 550																	+ 46 0	+ 159 + 130	200	225
225	250	+ 290 0	− 280 − 570																	+ 46 0	+ 169 + 140	225	250
250	280	+ 320 0	− 300 − 620	+ 130 0	− 190 − 400	+ 130 0	− 110 − 240	+ 81 0	− 56 − 108	+ 52 0	− 17 − 49	+ 52 0	− 32 0	+ 52 0	+ 36 + 4	+ 52 0	+ 66 + 34	+ 52 0	+ 88 + 56	+ 52 0	+ 190 + 158	250	280
280	315	+ 320 0	− 330 − 650																	+ 52 0	+ 202 + 170	280	315
315	355	+ 360 0	− 360 − 720	+ 140 0	− 210 − 440	+ 440 0	− 125 − 265	+ 89 0	− 62 − 119	+ 57 0	− 18 − 54	+ 57 0	− 36 0	+ 57 0	+ 40 + 4	+ 57 0	+ 73 + 37	+ 57 0	+ 98 + 62	+ 57 0	+ 226 + 190	315	355
355	400	+ 360 0	− 400 − 760																	+ 57 0	+ 244 + 208	355	400
400	450	+ 400 0	− 440 − 840	+ 155 0	− 230 − 480	+ 155 0	− 135 − 290	+ 97 0	− 68 − 131	+ 63 0	− 20 − 60	+ 63 0	− 40 0	+ 63 0	+ 45 + 5	+ 63 0	+ 80 + 40	+ 63 0	+ 108 + 68	+ 63 0	+ 272 + 232	400	450
450	500	+ 400 0	− 480 − 880																	+ 63 0	+ 292 + 252	450	500

Extracted from BS 4500

BRITISH STANDARD

SELECTED ISO FITS—SHAFT BASIS

Data Sheet
4500B

Diagram to scale for 25 mm. diameter

Clearance fits: C11 / h11, D10 / h9, E9 / h9, F8 / h7, G7 / h6, H7 / h6

Transition fits: K7 / h6, N7 / h6

Interference fits: P7 / h6, S7 / h6

Holes — Shafts

Nominal sizes		Clearance fits: Tolerance		Tolerance		Tolerance		Tolerance		Tolerance		Tolerance		Transition fits: Tolerance		Tolerance		Interference fits: Tolerance		Tolerance		Nominal sizes	
Over	To	h11	C11	h9	D10	h9	E9	h7	F8	h6	G7	h6	H7	h6	K7	h6	N7	h6	P7	h6	S7	Over	To
mm	mm	0·001 mm	0·001 mm	0·001 mm	0·001 mm	0·001 mm	0·001 mm	0·001 mm	0·001 mm	0·001 mm	0·001 mm	0·001 mm	0·001 mm	0·001 mm	0·001 mm	0·001 mm	0·001 mm	0·001 mm	0·001 mm	0·001 mm	0·001 mm	mm	mm
—	3	0 − 60	+ 120 + 60	0 − 25	+ 60 + 20	0 − 25	+ 39 + 14	0 − 10	+ 20 + 6	0 − 6	+ 12 + 2	0 − 6	+ 10 0	0 − 6	0 − 10	0 − 6	− 4 − 14	0 − 6	− 6 − 16	0 − 6	− 14 − 24	—	3
3	6	0 − 75	+ 145 + 70	0 − 30	+ 78 + 30	0 − 30	+ 50 + 20	0 − 12	+ 28 + 10	0 − 8	+ 16 + 4	0 − 8	+ 12 0	0 − 8	+ 3 − 9	0 − 8	− 4 − 16	0 − 8	− 8 − 20	0 − 8	− 15 − 27	3	6
6	10	0 − 90	+ 170 + 80	0 − 36	+ 98 + 40	0 − 36	+ 61 + 25	0 − 15	+ 35 + 13	0 − 9	+ 20 + 5	0 − 9	+ 15 0	0 − 9	+ 5 − 10	0 − 9	− 4 − 19	0 − 9	− 9 − 24	0 − 9	− 17 − 32	6	10
10	18	0 − 110	+ 205 + 95	0 − 43	+ 120 + 50	0 − 43	+ 75 + 32	0 − 18	+ 43 + 16	0 − 11	+ 24 + 6	0 − 11	+ 18 0	0 − 11	+ 6 − 12	0 − 11	− 5 − 23	0 − 11	− 11 − 29	0 − 11	− 21 − 39	10	18
18	30	0 − 130	+ 240 + 110	0 − 52	+ 149 + 65	0 − 52	+ 92 + 40	0 − 21	+ 53 + 20	0 − 13	+ 28 + 7	0 − 13	+ 21 0	0 − 13	+ 6 − 15	0 − 13	− 7 − 28	0 − 13	− 14 − 35	0 − 13	− 27 − 48	18	30
30	40	0 − 160	+ 280 + 120	0 − 62	+ 180 + 80	0 − 62	+ 112 + 50	0 − 25	+ 64 + 25	0 − 16	+ 34 + 9	0 − 16	+ 25 0	0 − 16	+ 7 − 18	0 − 16	− 8 − 33	0 − 16	− 17 − 42	0 − 16	− 34 − 59	30	40
40	50	0 − 160	+ 290 + 130																			40	50
50	65	0 − 190	+ 330 + 140	0 − 74	+ 220 + 100	0 − 74	+ 134 + 60	0 − 30	+ 76 + 30	0 − 19	+ 40 + 10	0 − 19	+ 30 0	0 − 19	+ 9 − 21	0 − 19	− 9 − 39	0 − 19	− 21 − 51	0 − 19	− 42 − 72	50	65
65	80	0 − 190	+ 340 + 150																	0 − 19	− 48 − 78	65	80
80	100	0 − 220	+ 390 + 170	0 − 87	+ 260 + 120	0 − 87	+ 159 + 72	0 − 35	+ 90 + 36	0 − 22	+ 47 + 12	0 − 22	+ 35 0	0 − 22	+ 10 − 25	0 − 22	− 10 − 45	0 − 22	− 24 − 59	0 − 22	− 58 − 93	80	100
100	120	0 − 220	+ 400 + 180																	0 − 22	− 66 − 101	100	120
120	140	0 − 250	+ 450 + 200	0 − 100	+ 305 + 145	0 − 100	+ 185 + 85	0 − 40	+ 106 + 43	0 − 25	+ 54 + 14	0 − 25	+ 40 0	0 − 25	+ 12 − 28	0 − 25	− 12 − 52	0 − 25	− 28 − 68	0 − 25	− 77 − 117	120	140
140	160	0 − 250	+ 460 + 210																	0 − 25	− 85 − 125	140	160
160	180	0 − 250	+ 480 + 230																	0 − 25	− 93 − 133	160	180
180	200	0 − 290	+ 530 + 240	0 − 115	+ 355 + 170	0 − 115	+ 215 + 100	0 + 46	+ 122 + 50	0 − 29	+ 61 + 15	0 − 29	+ 46 0	0 − 29	+ 13 − 33	0 − 29	14 − 60	0 − 29	− 33 − 79	0 − 29	− 105 − 151	180	200
200	225	0 − 290	+ 550 + 260																	0 − 29	− 113 − 159	200	225
225	250	0 − 290	+ 570 + 280																	0 − 29	− 123 − 169	225	250
250	280	0 − 320	+ 620 + 300	0 − 130	+ 400 + 190	0 − 130	+ 240 + 110	0 − 52	+ 137 + 56	0 − 32	+ 62 + 17	0 − 32	+ 52 0	0 − 32	+ 16 − 36	0 − 32	− 14 − 66	0 − 32	− 36 − 88	0 − 32	− 138 − 190	250	280
280	315	0 − 320	+ 650 + 330																	0 − 32	− 150 − 202	280	315
315	355	0 − 360	+ 720 + 360	0 − 140	+ 440 + 210	0 − 140	+ 265 + 125	0 − 57	+ 151 + 62	0 − 36	+ 75 + 18	0 − 36	+ 57 0	0 − 36	+ 17 − 40	0 − 36	− 16 − 73	0 − 36	− 41 − 98	0 − 36	− 169 − 226	315	355
355	400	0 − 360	+ 760 + 400																	0 − 36	− 187 − 244	355	400
400	450	0 − 400	+ 840 + 440	0 − 155	+ 480 + 230	0 − 155	+ 290 + 135	0 − 63	+ 165 + 68	0 − 40	+ 83 + 20	0 − 40	+ 63 0	0 − 40	+ 18 − 45	0 − 40	− 17 − 80	0 − 40	− 45 − 108	0 − 40	− 209 − 272	400	450
450	500	0 − 400	+ 880 + 480																	0 − 40	− 229 − 292	450	500

sizes of the features at any cross-section must fall, conventions have been established by which such limits exert some control on geometric form. In selecting limits of size for mating features it is important to consider whether this degree of control is adequate or whether a specific form tolerance should be applied. BS 308 'Engineering drawing practice' contains information on the interpretation of limits of size and on methods by which, if necessary, limits can be specified to control the form of a feature separately from its size.

Introduction to plain limit gauges

The use of a system of limits and fits leads, logically, to the use of limit gauges. When inspection is undertaken using limit gauges no attempt is made to determine the actual size of the feature, only to find whether the feature is within the specified limits.

In 1905 William Taylor put forward concepts relating to the design of limit gauges and these are known as the 'Taylor Principle'. The design of limit gauges is still very much influenced by the Taylor Principle although certain deviations are accepted (see Appendix B of BS 969, on page 137, for both the Taylor Principle and departures from its strict application.)

Gauge blanks

BS 1044 'Gauge blanks', Part 1 'Plug, ring and calliper gauges' relates to the blanks for commonly used types of gauges. (A blank is a gauge which is complete except for the final finishing to size.) Those features of design necessary to ensure interchangeability are specified in some detail and general information on other features is included for guidance in the manufacture of these gauges. No material from BS 1044: Part 1 has been included in this manual although the examples of limit gauges (Figures 6.12 to 6.31) are based on the provisions of BS 1044: Part 1.

Gauges for holes

While plug gauges in which the gauging portion was integral with the handle, the whole gauge being machined from one piece of metal were common, such gauges are now becoming obsolete. They have been replaced by the renewable-end type which has a number of advantages. In such a gauge the gauging portion and the handle are manufactured separately, and for use, are engaged together to form a rigid assembly by means of a locking device. Thus, when the gauging member becomes damaged or worn it may be replaced without discarding the whole gauge. A further advantage is that the handle need not be made of metal but may be of a suitable plastics material, thus reducing both the weight and cost of the gauge and the risk of heat transference.

The GO and NOT GO gauges may be in the form of separate 'single-ended' gauges, or may be combined on

(1) = GO gauging member
(2) = NOT GO gauging member
(3) = Progressive gauging member

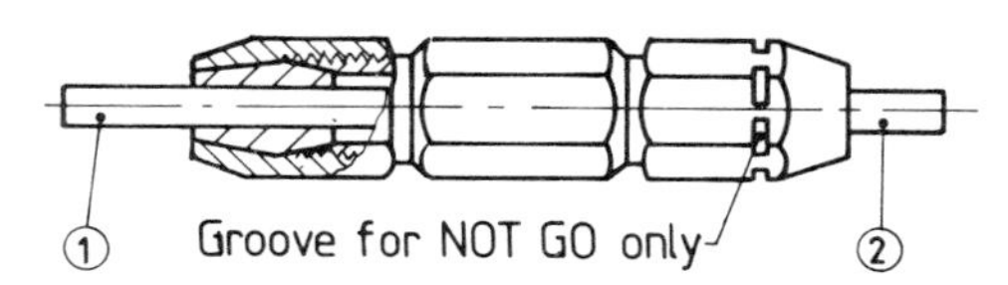

Figure 6.12 *Double-ended collet type plug gauge (range from 0.4 to 20 mm)*

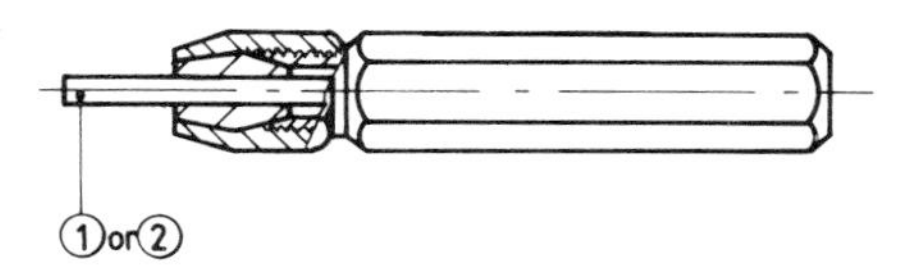

Figure 6.13 *Single-ended collet type plug gauge (range from 0.4 to 20 mm)*

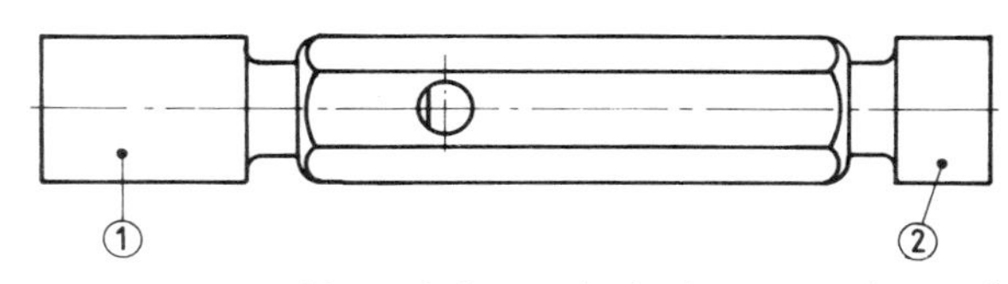

Figure 6.14 *Double-ended taper lock plug gauge (range from 1.5 to 64 mm)*

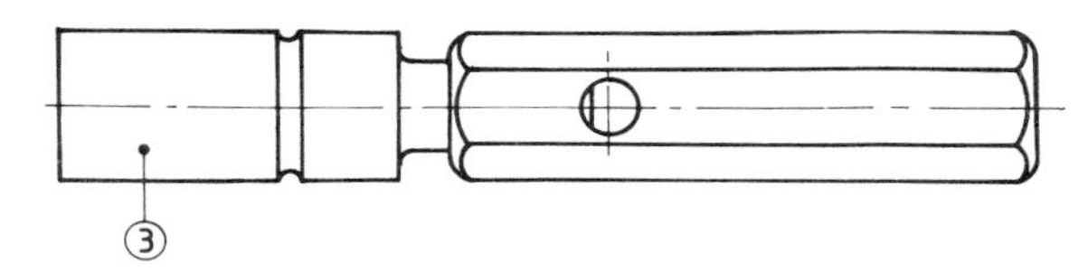

Figure 6.15 *Single-ended progressive taper lock type plug gauge (range from 1.5 to 64 mm)*

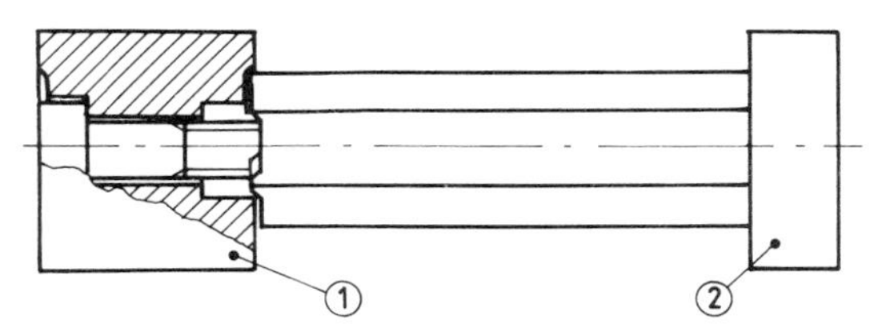

Figure 6.16 *Double-ended trilock type plug gauge (range 40 to 64 mm)*

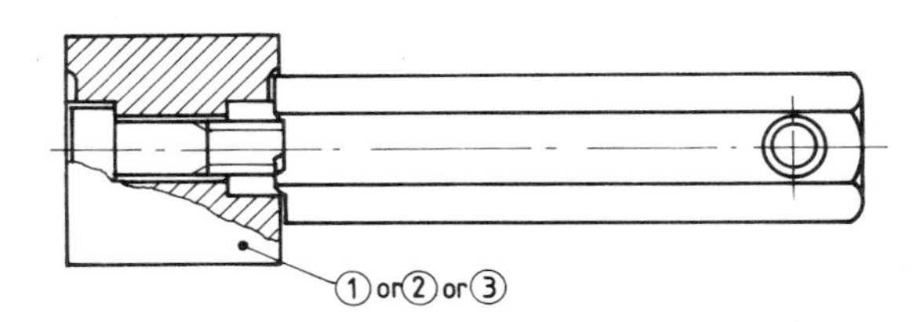

Figure 6.17 *Single-ended trilock type plug gauge (range 40 to 64 mm)*

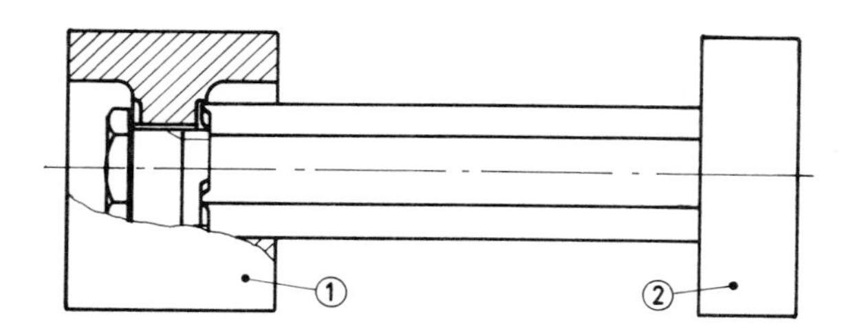

Figure 6.18 *Double-ended trilock type plug gauge (range 64 to 200 mm)*

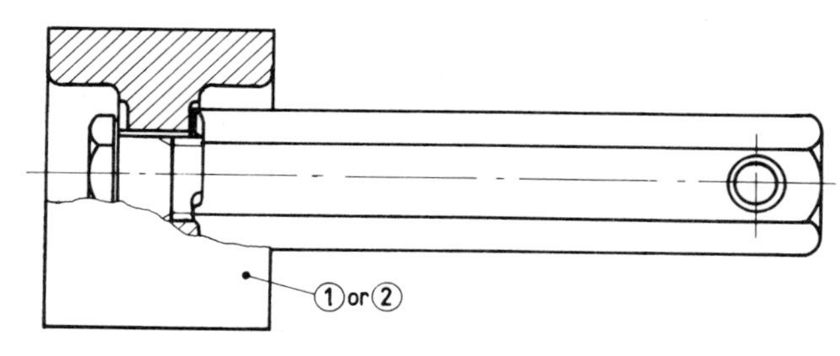

Figure 6.19 *Single-ended trilock type plug gauge (range 64 to 200 mm)*

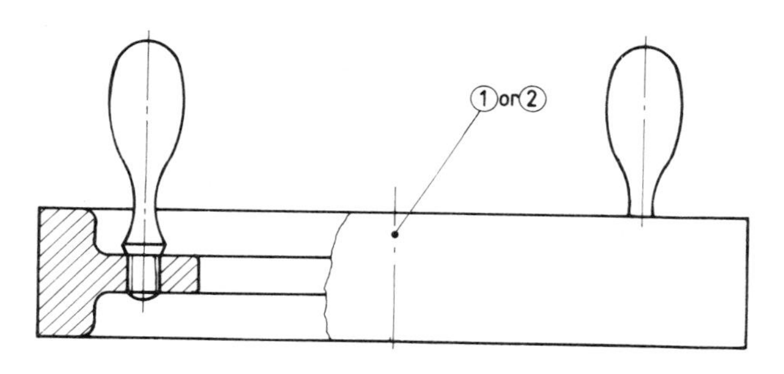

Figure 6.20 *Annular design, ball handle type plug gauge (range 200 to 300 mm). Not recommended for NOT GO gauging*

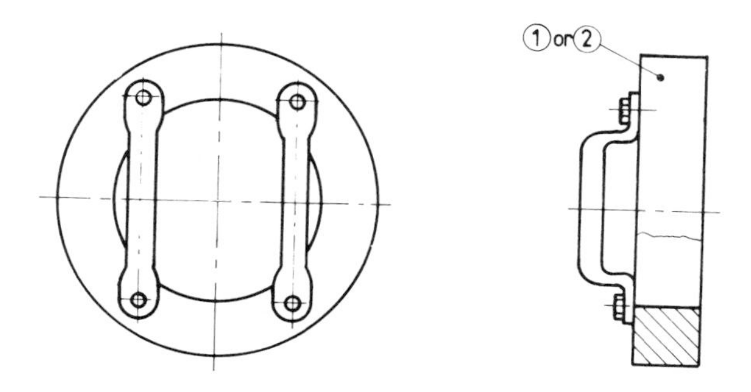

Figure 6.21 *Annular design, bar handle type plug gauge (range 200 to 250 mm). Not recommended for NOT GO gauging*

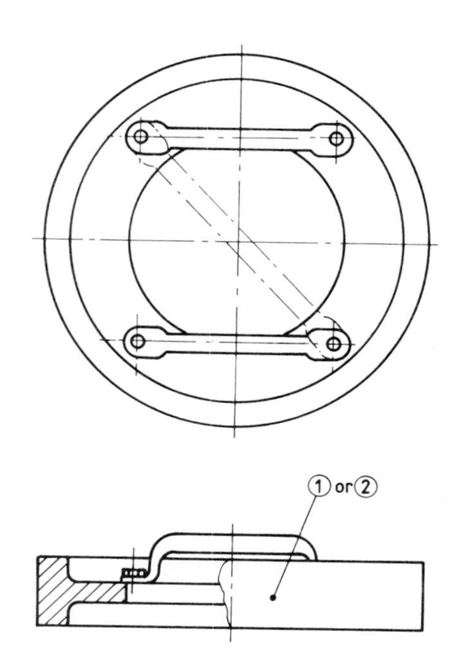

Figure 6.22 *Annular design, bar handle type cylindrical plug gauge (range 250 to 300 mm). Not recommended for NOT GO gauging*

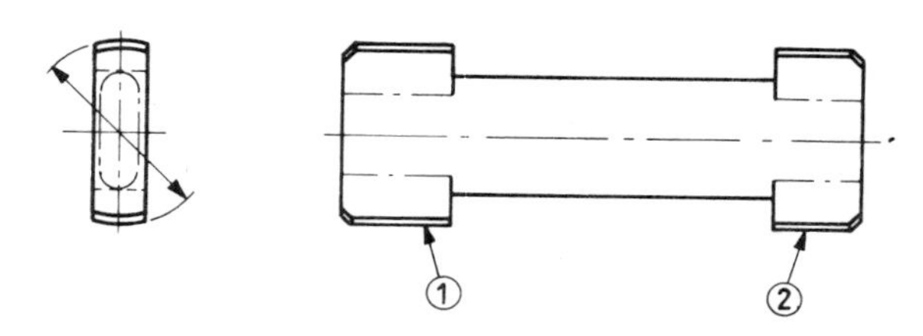

Figure 6.23 *Segmented cylindrical gauges (range 32 to 64 mm)*

one handle to form a 'double-ended' gauge. It is also possible for plain plug gauges to have a combined gauging portion on which both the GO and NOT GO elements are provided to give a 'progressive' gauge. Collet type gauges, in which the gauging portions are held in a collet, are particularly suitable for small diameters At the other end of the scale, provision is made for plug gauges suitable for large diameters. These may be of annular design or of the bar type with segmental cylindrical or spherical ends.

Gauges for shafts

Gauges for shafts are customarily in the form of either a ring gauge or a gap gauge.

Gap gauges may be of the solid type or of the adjustable type. Separate gap gauges may be used for GO and NOT GO gauging or, alternatively, a combined GO and NOT GO gauge may be used; adjustable gap gauges are invariably of this latter type.

Note: BS 1044: Part 1 is in imperial units. The ranges of the gauges shown in Figures 6.12 to 6.30 are approximate metric conversions. See also the size ranges given in BS 969 on page 136.

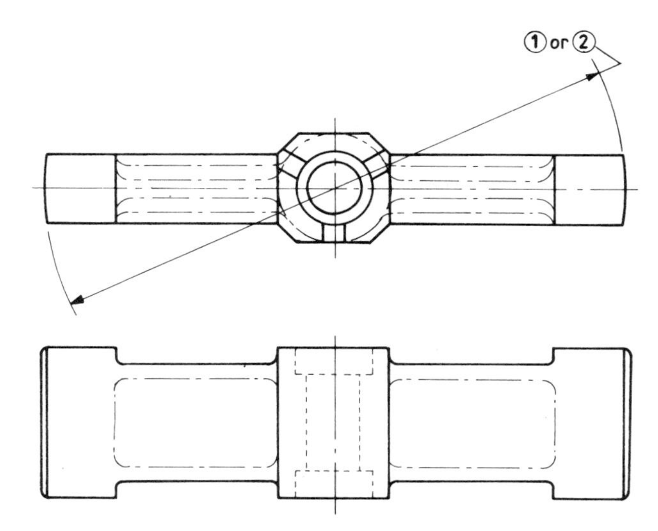

Figure 6.24 *Segmented cylindrical gauge of trilock design for use with a handle (range 64 to 200 mm)*

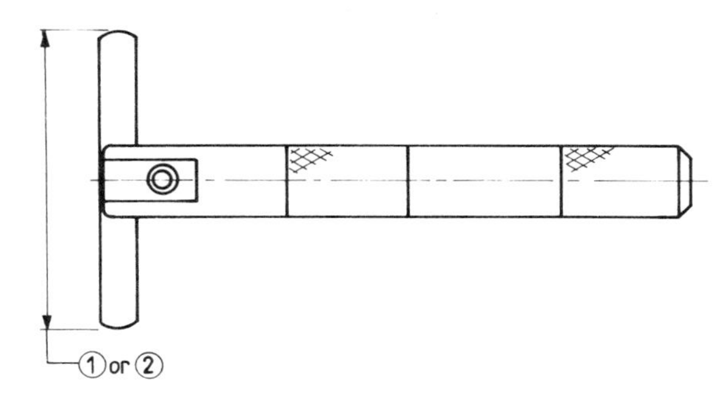

Figure 6.25 *Spherical-ended rod gauges (range 13 to 150 mm)*

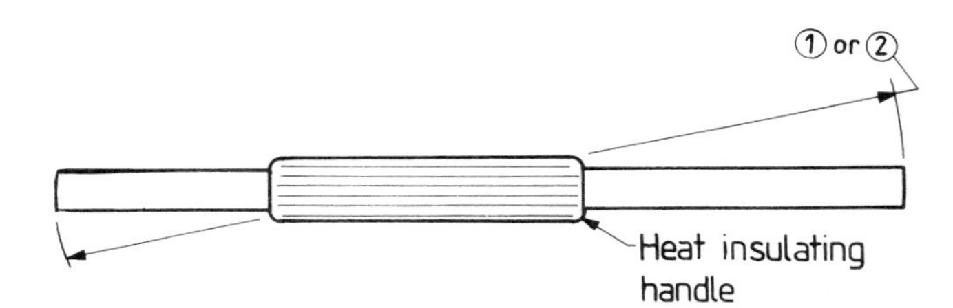

Figure 6.26 *Spherical-ended rod gauge with insulating handle (range 150 to 300 mm)*

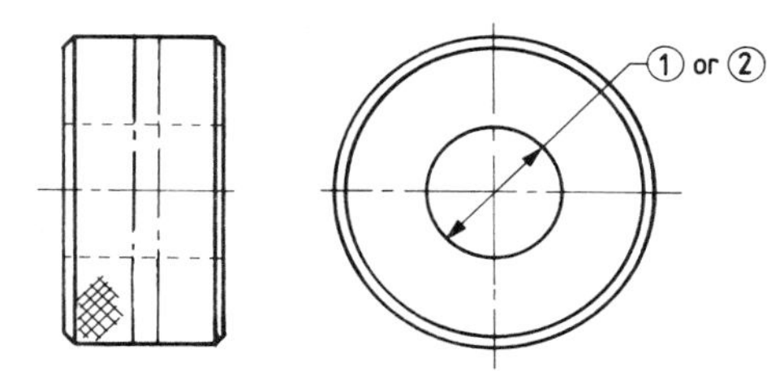

Figure 6.27 *Plain ring gauges, solid type (range up to 90 mm)*

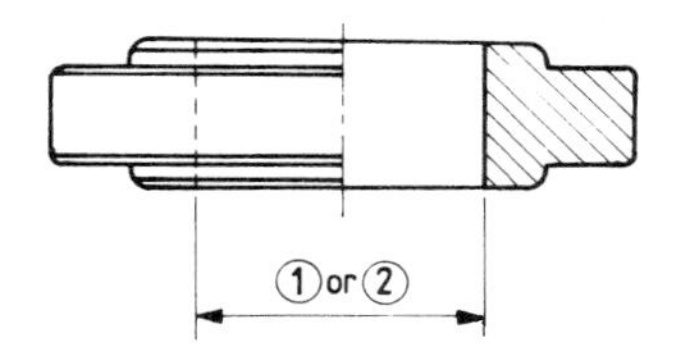

Figure 6.28 *Plain ring gauge, solid type (range 90 to 300 mm)*

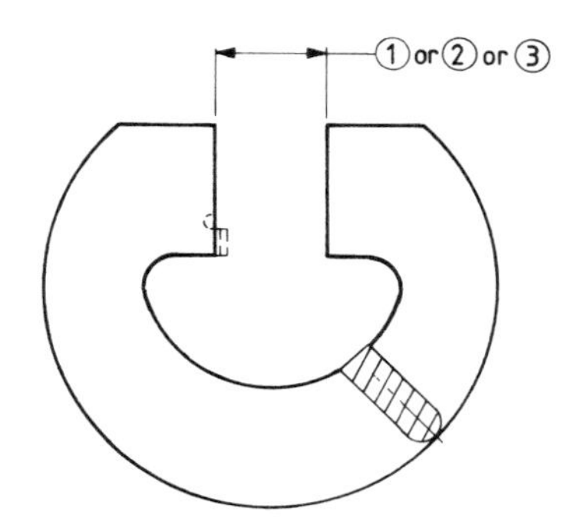

Figure 6.29 *Plain gap gauge, solid type. May be single function or dual GO/NOT GO combined (range 6 to 270 mm)*

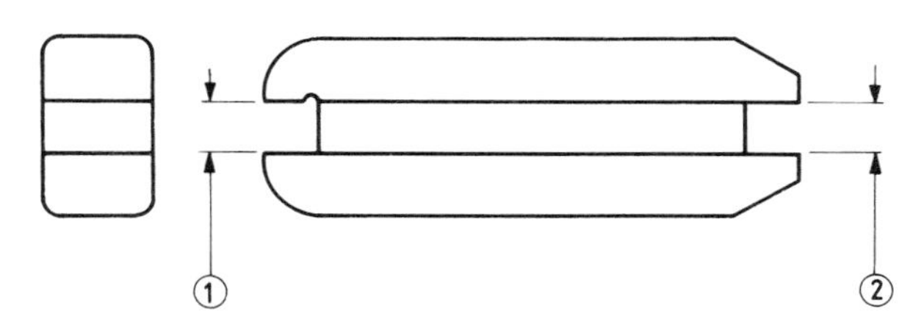

Figure 6.30 *Built up gap gauge (range up to 6.5 mm)*

Limits for limit gauges

It has already been stated in this chapter that nothing can be made to an exact size, and limit gauges are no exception. Gauge purchasers should state the gauge limits they require. However, when workpiece limits only (i.e. without instructions) are supplied to a gauge maker, reference will be to the appropriate British Standard. BS 969: 1941, BS 969: 1953 and BS 4500: Part 2: 1974 have each, in turn, prescribed tolerances for plain limit gauges. These standards differed in their way of specifying the placing of the gauge limits, i.e. within, outside or straddling the workpiece limits.

BS 4500: Part 2: 1974 was aligned with an ISO recommendation in which the amount of gauge tolerance was coupled with the various tolerances for fits of interchangeable workpieces and also according to ranges of workpiece size, resulting in the gauge tolerance impinging on the workpiece tolerance.

In the current standard, BS 969: 1982, the workpiece limits are considered to be boundaries in that

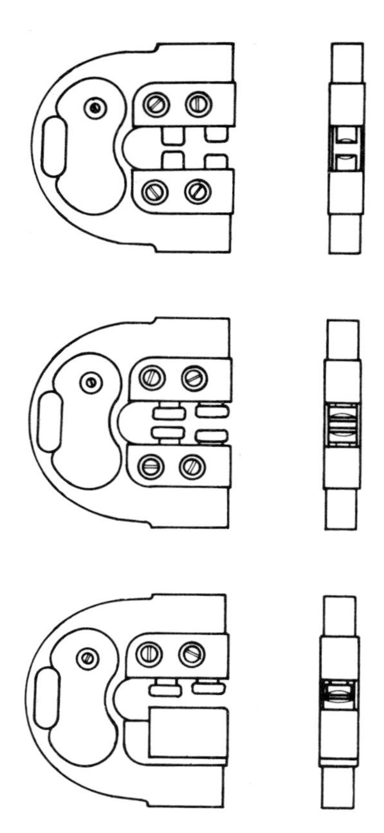

Figure 6.31 *Adjustable calliper gauges*

Note: It is possible to set well-made adjustable gauges to within about 0.003 mm of a desired size. Their use thus enables fuller advantage to be taken of the manufacturing tolerance on the work than when solid gauges, with an appreciable manufacturing tolerance of their own, are employed.

The adjustability also enables wear of the GO anvils to be taken up at any time. Should the anvil faces lose their flatness with use, they can be reground quite readily.

When setting a gauge for testing cylindrical work to fine tolerances, it is preferable for it to be adjusted to fit setting disks of the correct size rather than combinations of slip gauges which would not offer the same delicacy of 'feel'.

Several different types of adjustable callipers are available. Those shown are typical examples and are not necessarily representative of the whole range.

gauge tolerances will be such that the size of any workpiece accepted by a gauge has to lie within the workpiece limits. Furthermore, the magnitude of the gauge tolerance is consequent upon the amount of the workpiece tolerance only. A wear allowance is provided on all 'GO' gauges, i.e. those that control the maximum material condition of the workpiece.

BS 969 covers gauge tolerances for workpiece tolerances in ranges between 0.009 mm and 3.2 mm. When the workpiece tolerance is outside this range, or if it is not possible to comply with BS 969 regarding the type of gauge, gauge tolerance and workpiece size, direct measurement should be used. The standard specifies the tolerances and limits for simple forms of limit gauges, such as cylindrical plain plug, ring and gap gauges, and is intended for reference by component producers and gauge makers when gauge limits have not been stipulated.

BS 969

Limits and tolerances on plain limit gauges

. . . .

3. Definitions

For the purposes of this British Standard the following definitions apply.

3.1 maximum material limit of size. The dimension defining the maximum material condition of a feature, and is at the maximum limit of size for an external feature, e.g. a shaft, and at the minimum limit of size for an internal feature, e.g. a hole.

3.2 least material limit of size. The dimension defining the minimum limit of local size for an external feature, e.g. a shaft, or the maximum limit of local size for an internal feature, e.g. a hole.

3.3 GO gauge. A gauge that controls the maximum material limit of size of the workpiece.

3.4 NOT GO gauge. A gauge that controls the least material limit of size of the workpiece.

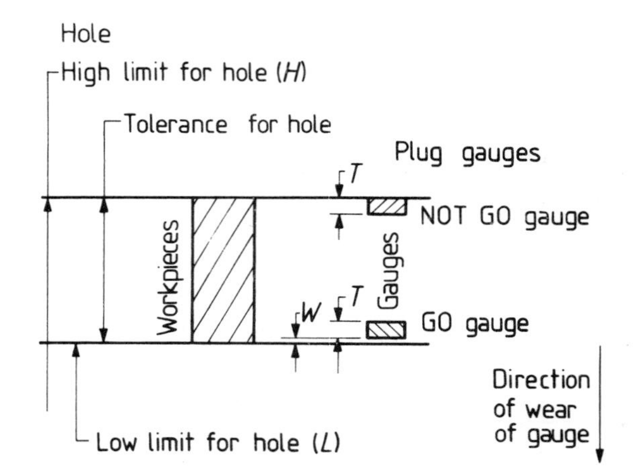

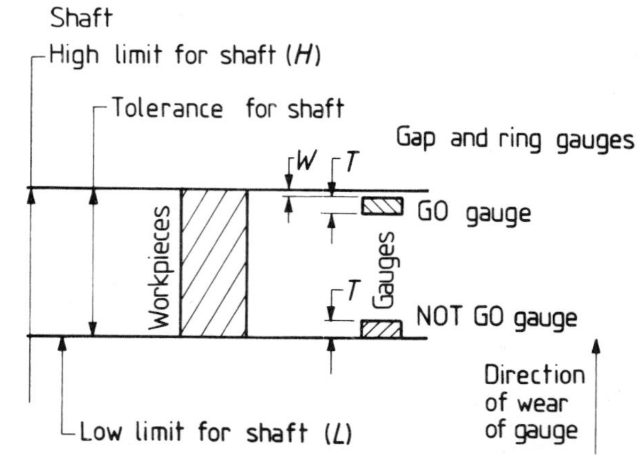

Figure 1. Relationships of workpiece and gauge tolerance zones

3.5 workpiece tolerance. The difference between the high and low limits of size of the workpiece.

4. Gauge limits and tolerances

4.1 Gauge limits. Gauge purchasers should state the gauge limits they require. However, when workpiece limits only (i.e. without instructions) are supplied to the gauge maker, gauge limits for workpiece tolerances in ranges between 0.009 mm and 3.2 mm shall be as given in table 1. (See appendix A for examples of application of the table.)

NOTE. It will be seen that gauge tolerance increases with workpiece tolerance only. To appreciate this apparent disregard of workpiece size, it should be remembered that the size factor plays its proper part when the workpiece tolerance is fixed initially.

4.2 Tolerances on gauge types. The minimum tolerance on various types and sizes of gauge shall be as given in table 2.

NOTE 1. In most cases it is uneconomical to manufacture, and impractical to use, limit gauges with tolerances on sizes smaller than those given in table 2.
NOTE 2. The GO gauges should be of full form; the NOT GO gauges should be designed to make contact at only the two ends of a diameter. (The principle and reservations are described in appendix B.)
NOTE 3. Recommended gauge designs are given in BS 1044 : Part 1.

. . . .

Table 1. Gauge size limits at 20 °C for ranges of workpiece tolerance (see figure 1)

1	2	3	4	5	6	7
			Plug type gauges		**Ring or gap gauges**	
			*Limits expressed with respect to *H* minus *L* for the workpiece (hole)		†Limits expressed with respect to *H* minus *L* for the workpiece (shaft)	
Workpiece tolerance Difference between high (*H*) limit and low (*L*) limit shaft or hole	**Tolerance, *T* for GO and for NOT GO gauges**	**Wear allowance, *W*, for GO gauges only**	**GO**	**NOT GO**	**GO**	**NOT GO**
mm	mm	mm	mm	mm	mm	mm
0.009† up to and including 0.018	0.001	0.001	$L\,^{+0.002}_{+0.001}$	$H\,^{+0}_{-0.001}$	$H\,^{-0.001}_{-0.002}$	$L\,^{+0.001}_{-0}$
Above 0.018 up to and including 0.032	0.002	0.001	$L\,^{+0.003}_{+0.001}$	$H\,^{+0}_{-0.002}$	$H\,^{-0.001}_{-0.003}$	$L\,^{+0.002}_{-0}$
Above 0.032 up to and including 0.058	0.003	0.002	$L\,^{+0.005}_{+0.002}$	$H\,^{+0}_{-0.003}$	$H\,^{-0.002}_{-0.005}$	$L\,^{+0.003}_{-0}$
Above 0.058 up to and including 0.100	0.004	0.004	$L\,^{+0.008}_{+0.004}$	$H\,^{+0}_{-0.004}$	$H\,^{-0.004}_{-0.008}$	$L\,^{+0.004}_{-0}$
Above 0.100 up to and including 0.180	0.006	0.007	$L\,^{+0.013}_{+0.007}$	$H\,^{+0}_{-0.006}$	$H\,^{-0.007}_{-0.013}$	$L\,^{+0.006}_{-0}$
Above 0.180 up to and including 0.320	0.009	0.012	$L\,^{+0.021}_{+0.012}$	$H\,^{+0}_{-0.009}$	$H\,^{-0.012}_{-0.021}$	$L\,^{+0.009}_{-0}$
Above 0.320 up to and including 0.580	0.014	0.025	$L\,^{+0.039}_{+0.025}$	$H\,^{+0}_{-0.014}$	$H\,^{-0.025}_{-0.039}$	$L\,^{+0.014}_{-0}$
Above 0.580 up to and including 1.000	0.025	0.048	$L\,^{+0.073}_{+0.048}$	$H\,^{+0}_{-0.025}$	$H\,^{-0.048}_{-0.073}$	$L\,^{+0.025}_{-0}$
Above 1.000 up to and including 1.800	0.040	0.080	$L\,^{+0.120}_{+0.080}$	$H\,^{+0}_{-0.040}$	$H\,^{-0.080}_{-0.120}$	$L\,^{+0.040}_{-0}$
Above 1.800 up to and including 3.200	0.050	0.155	$L\,^{+0.205}_{+0.155}$	$H\,^{+0}_{-0.050}$	$H\,^{-0.155}_{-0.205}$	$L\,^{+0.050}_{-0}$

* Errors of size and form of the gauge are to be contained within these limits.
† Workpieces with a tolerance less than 0.009 mm should be measured directly, or by means other than the gauges described in this standard.

NOTE. *Gauge wear and wear allowance*. Provision is made for the wear of GO gauges by the introduction of a wear allowance (*W*) between the tolerance zone for the gauge and the maximum material limit for the workpiece. Wear allowance is not applied to the NOT GO gauge.

Gauge users have to watch for the effect of wear upon sizes of their gauges. Regular examination and measurement of gauges in use is essential so that a gauge, and in particular a GO gauge, which has worn outside its limit is detected and withdrawn from service to avoid accepting workpieces exceeding the maximum material limit.

6. Marking

Gauges shall be marked with:

(a) the limiting size of the workpiece which the gauge controls;
(b) GO or NOT GO, as appropriate.

NOTE. Gauges may also be marked with:

(a) the manufacturer's name or trademark;
(b) a serial number.

7. Gauges in disagreement

It is possible that similar, satisfactory gauges, using a different region of the permitted gauge tolerance, may respectively accept and reject a workpiece. Such cases are likely to be rare and a dispute may be resolved by measuring the workpiece directly.

Table 2. Minimum gauge tolerance appropriate to type and size of gauge

Gauge size		Type	Minimum tolerance appropriate to the type and size of gauge
Above	Up to and including		
mm	mm		mm
—	25	Cylindrical plug	0.001
25	50		0.002
50	100		0.003
100	150		0.006
150	200		0.009
100	250	Cylindrically ended bar	0.009
250	500		0.016
375	750	Spherically ended rod and pin gauges	0.016
750	1000		0.030
—	13	Ring and gap gauges	0.001
13	25		0.002
25	50		0.003
50	75		0.006
75	175		0.009
175	400		0.016
400	750		0.030

Appendix A

Examples of the use of table 1

Example 1. The limits for GO and NOT GO gauges for an internal diameter component (see figure 2) are found as follows:

The workpiece tolerance is 0.200 mm.

From column 4 of table 1, the limits for the GO gauge are:

+0.021 mm
+0.012 mm.

Therefore the size of the GO gauge is:

$$L\begin{matrix}+0.021 \text{ mm}\\+0.012 \text{ mm}\end{matrix} = \begin{matrix}75.021 \text{ mm}\\75.012 \text{ mm.}\end{matrix}$$

From column 5 of table 1, the limits for the NOT GO gauge are:

+0
−0.009 mm.

Therefore the size of the NOT GO gauge is:

$$H\begin{matrix}+0\\-0.009 \text{ mm}\end{matrix} = \begin{matrix}75.200 \text{ mm}\\75.191 \text{ mm.}\end{matrix}$$

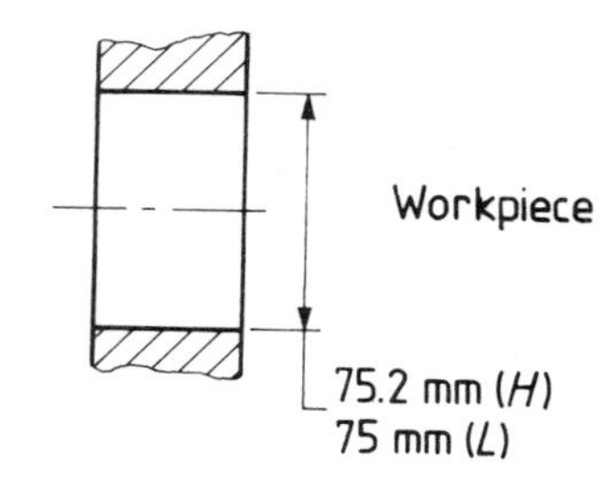

Figure 2. Limits for hole

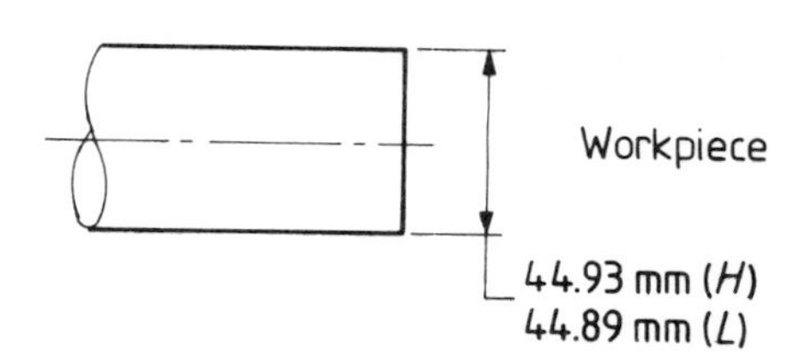

Figure 3. Limits for shaft

Example 2. The limits for GO and NOT GO gauges for a shaft (see figure 3) are found as follows:

The workpiece tolerance is 0.040 mm.

From column 6 of table 1, the limits for the GO gauge are:

−0.002 mm
−0.005 mm.

Therefore the size of the GO gauge is:

$$H\begin{matrix}-0.002 \text{ mm}\\-0.005 \text{ mm}\end{matrix} \quad \begin{matrix}44.928 \text{ mm}\\44.925 \text{ mm.}\end{matrix}$$

From column 7 of table 1, the limits for the NOT GO gauge are:

+0.003 mm
−0.

Therefore the size of the NOT GO gauge is:

$$L\begin{matrix}+0.003 \text{ mm}\\-0\end{matrix} = \begin{matrix}44.893 \text{ mm}\\44.890 \text{ mm.}\end{matrix}$$

Appendix B

Principles of inspection using plain limit gauges

B.1 Limits. The workpiece limits of size within the prescribed length of assembly are considered to be as follows.

(a) *Holes*. The diameter of a perfect cylinder just contacting the high points of the workpiece is to be not less than the workpiece lower limit. The maximum diameter at any position in the hole does not exceed the workpiece upper limit.
(b) *Shafts*. The diameter of a perfect cylinder just circumscribing the shaft is to be not more than the workpiece upper limit. The minimum diameter at any position on the shaft is to be not less than the workpiece lower limit.

The above considerations mean that if the workpiece is everywhere at its maximum material limit, that workpiece should be perfectly round and straight.

Unless specific tolerances on circularity of section and straightness are specified, departures of the workpiece from cylindrical form may reach the diametral tolerance only if the workpiece size is at the least material limit. Examples of extreme errors in form are shown in figures 4 and 5.

B.2 Application of limit gauges to workpieces

NOTE. The relationship of the limits for gauges to the workpiece limits is given in figure 1.

B.2.1 The maximum material limit of a workpiece, i.e. the upper limit for shaft or the lower limit for hole, should be checked with a full form GO gauge that should be of the same length as the intended assembly of workpieces, shafts and holes concerned. The workpiece has to pass into or over the gauge.

B.2.2 The least material limit of a workpiece, i.e. the lower limit for shaft or the upper limit for hole, should be checked with a NOT GO gauge designed to contact the workpiece at two diametrically opposite points. The workpiece is not to pass into or over the gauge at any diametral position around and along the workpiece length.

B.2.3 The system of checking described in **B.2.1** and **B.2.2** is known as the Taylor Principle. For practical reasons the following departures from the strict application of the Taylor Principle are recognized.

(a) *Gauging at the maximum material limit*
(1) The length of a GO cylindrical plug or ring gauge may be less than the length of engagement of the mating workpieces if it is known that with the manufacturing process used, the error of straightness of the hole or shaft is so small that it does not affect the character of fit of the assembled workpieces.

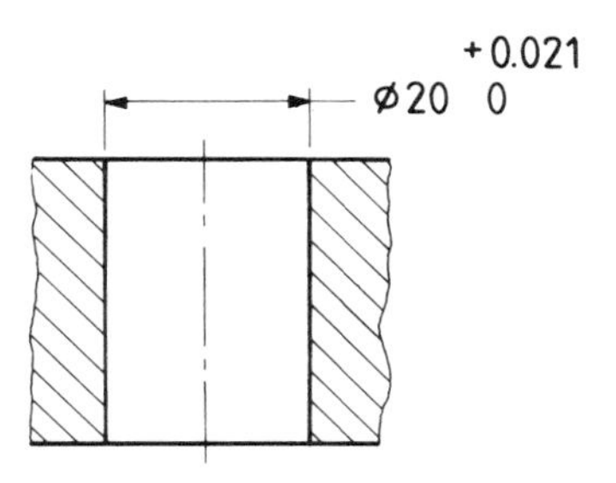

This presentation on drawing allows any of the following deviations.

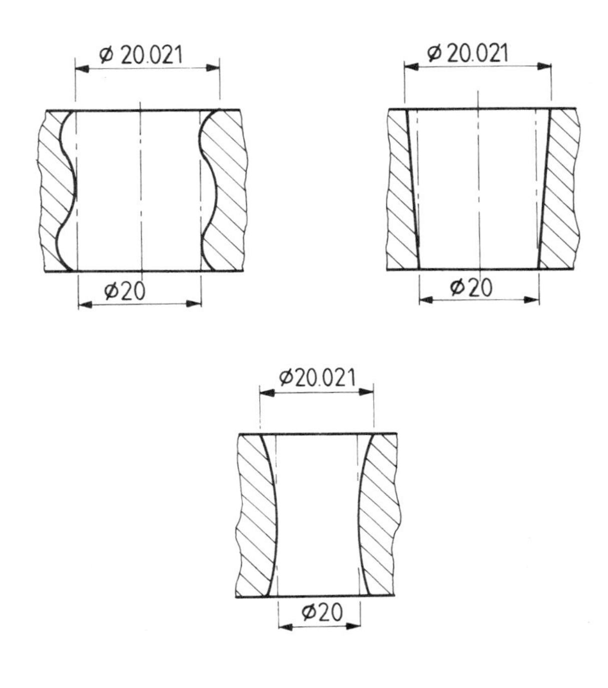

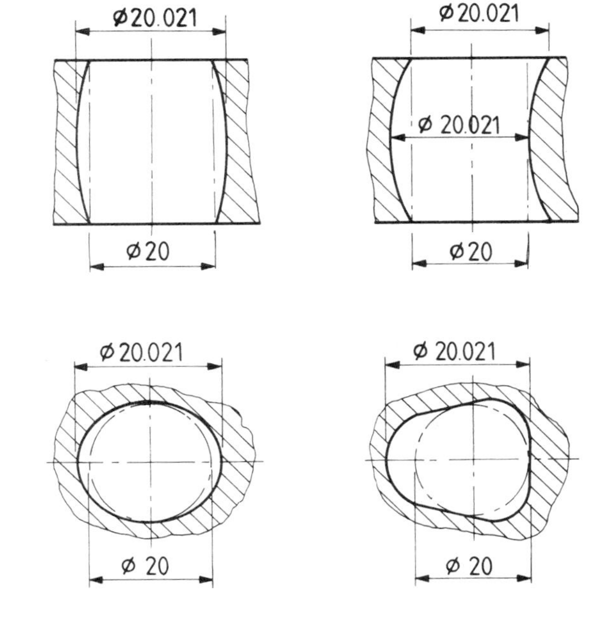

Figure 4. Possible extreme errors of form allowed by the effects of limits of workpiece size: holes

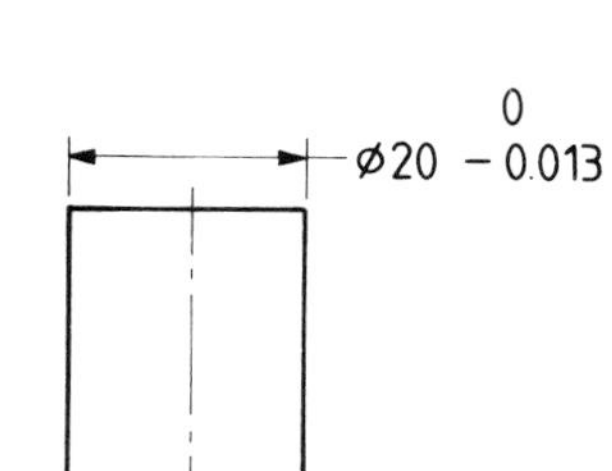

This presentation on drawing allows any of the following deviations.

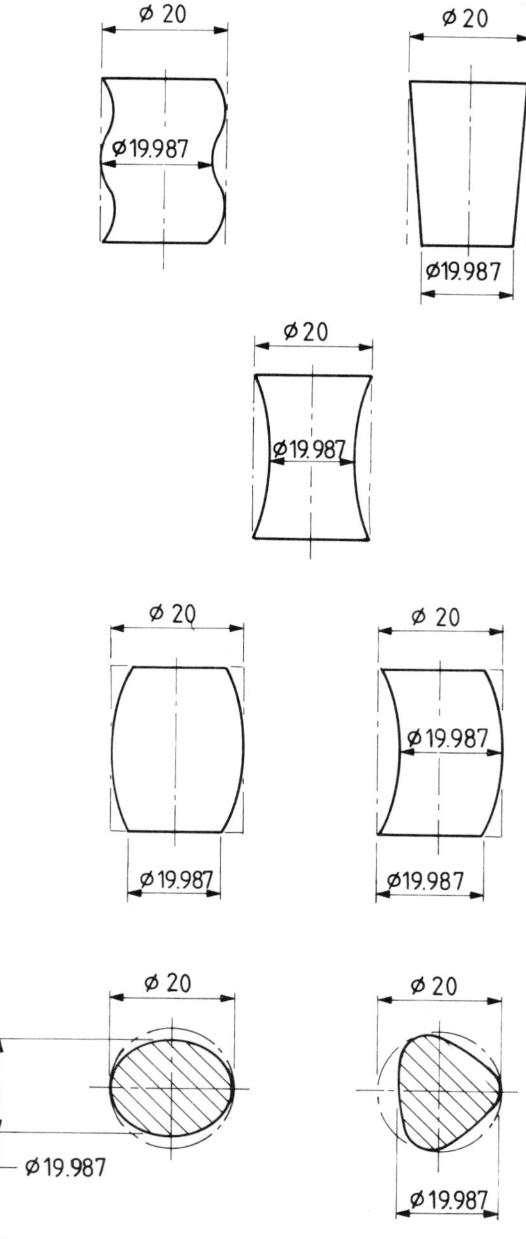

Figure 5. Possible extreme errors of form allowed by the effects of the limits of workpiece size: shafts

This departure from the ideal facilitates the use of standard gauge blanks.

(2) For gauging a large hole, a GO cylindrical plug gauge may be too heavy for convenient use. It is permissible to use a segmental cylindrical bar or spherical gauge if it is known that with the manufacturing process used, the error of roundness or straightness of the hole is so small that it does not affect the character of fit of the assembled workpieces.

(3) A GO cylindrical ring gauge is often inconvenient for gauging shafts and may be replaced by a gap gauge if it is known that with the manufacturing process used, the errors of roundness (especially lobing) and of straightness of the shaft are so small that they do not affect the character of fit of the assembled workpieces. The straightness of long shafts that have a small diameter should be checked separately.

(b) *Gauging at the least material limit*

(1) Point contacts are subject to rapid wear and in most cases may be replaced, where appropriate, by small plain cylindrical or spherical surfaces.

(2) For gauging very small holes, a two-point checking device is difficult to design and manufacture. NOT GO plug gauges of full cylindrical form may have to be used but the user has to be aware that there is a possibility of accepting workpieces having a diameter outside the NOT GO limit.

(3) Non-rigid workpieces may be deformed to an oval by a two-point mechanical contact device operating under a finite contact force. If it is not possible to reduce the contact force almost to zero, then it is necessary to use NOT GO ring or plug gauges of full cylindrical form.

Thin-walled workpieces may be out-of-round due to internal stresses or heat treatment. In these cases the NOT GO limit means that the circumference of the cylinder corresponding to that limit is not to be transgressed. Therefore NOT GO gauges of full cylindrical form have to be applied with a force that is sufficient to convert the elastic deformation into circularity but does not expand or compress the wall of the workpiece.

B.2.4 The sizes of gauges cannot be made exactly to the appropriate workpiece limit; they have to be made to specified tolerances.

. . . .

Standards publications referred to in this chapter

BS 308	Engineering drawing practice (in 3 parts)
BS 969	Limits and tolerances on plain limit gauges
BS 1044	Gauge blanks Part 1 Plug, ring and calliper gauges
BS 4500	ISO limits and fits (in 3 parts)
BS 4500A	Data sheet: Selected ISO fits – hole basis
BS 4500B	Data sheet: Selected ISO fits – shaft basis

7 Screw threads

Introduction

Definitions of terms relating to screw threads are given in BS 2517 'Definitions for use in mechanical engineering'.

Note 1: All references to taper threads have been removed from the following extract.
Note 2: Figure 11 from BS 2517 shows only Unified and Whitworth forms of thread. An additional figure (Figure 1), taken from BS 3643 'ISO metric screw threads', Part 1 'Principles and basic data', showing the ISO metric thread, is given at the end of the extract from BS 2517.
Note 3: The February 1983 edition of *BSI News* listed BS 2517 as *proposed for withdrawal*. In the event of its being withdrawn no British Standard will supersede it at present. There is, however, work being undertaken within the International Organization for Standardization to provide internationally standardized terms for screw threads. When published, the international standard will, subject to committee approval, form the basis for a British Standard. (The latest draft of the work of the international committee is DIS 5408.2.)

BS 2517

Definitions for use in mechanical engineering

Section six. Screw threads

A. General

screw thread. The ridge produced by forming, on the surface of a cylinder, a continuous helical or spiral groove of uniform section such that the distance measured parallel to the axis between two corresponding points on its contour is proportional to their relative angular displacement about the axis.

NOTE. This definition describes a *perfect* screw thread.

external (male) screw thread. A thread formed on the external surface of a cylinder. See figure 8.

NOTE. The thread on a bolt is a typical example of an external screw thread.

internal (female) screw thread. A thread formed on the internal surface of a hollow cylinder. See figure 9.

NOTE. The threads in nuts, tapped holes, or screwed sockets are typical examples of internal screw threads.

right-hand screw thread. A thread which, if assembled with a stationary mating thread, recedes from the observer when rotated in a clockwise direction. See figure 7*a*.

left-hand screw thread. A thread which, if assembled with a stationary mating thread, recedes from the observer when rotated in an anti-clockwise direction. See figure 7*b*.

parallel screw thread. A thread formed on the surface of a cylinder. See figures 8 and 9.

single-start screw thread. A thread formed by a single continuous helical groove. See figures 7*a* and 7*b*.

multi-start screw thread. A thread formed by a combination of two or more helical grooves equally spaced along the axis. See figure 7*c*.

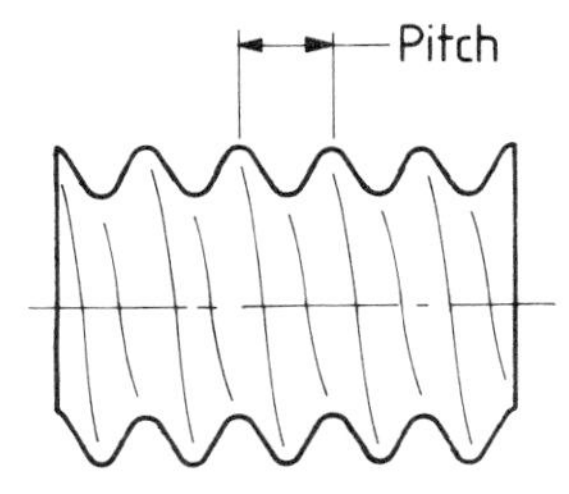

Figure 7*a*. Single-start screw thread (right hand)

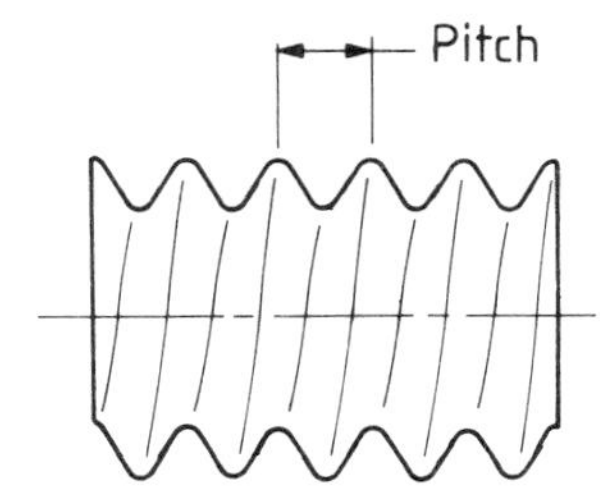

Figure 7*b*. Single-start screw thread (left hand)

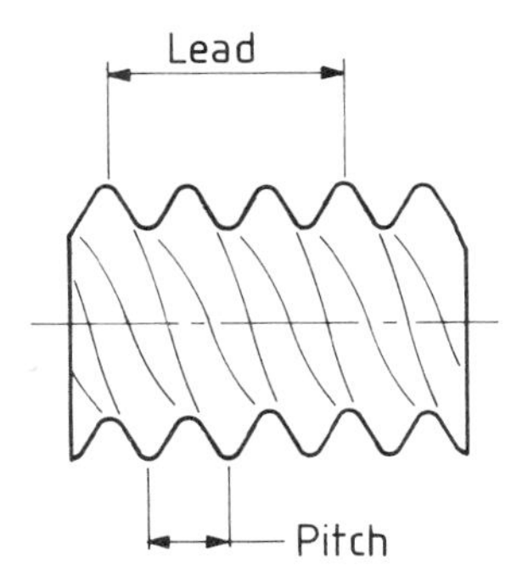

Figure 7*c*. Multi-start screw thread (triple-start right hand)

B. Geometry of screw threads

form. The shape of one complete profile of the thread between corresponding points, at the bottom of adjacent grooves, as shown in axial plane section.

basic form. The theoretical form on which the design forms for both the external and internal threads are based. See figures 11, 12*a* and 17*a*.

design forms. The forms of the external and internal threads in relation to which the limits of tolerances are assigned. See figure 11.

NOTE 1. The two design forms normally represent the maximum metal forms for the respective threads.

NOTE 2. Screw threads may have different design forms, derived from the same basic form, for the internal and external members respectively.

NOTE 3. The design forms for screw threads are not necessarily shown in detail on the drawing; both they and the associated limits of tolerance may be defined by an appropriate reference to a standard specification.

NOTE 4. The general form of the screw thread is allowed to vary from the design form within the zone defined by the associated limits of tolerance.

flanks. Those parts of the surface, on either side of the thread the inter-sections of which with an axial plane are theoretically straight lines. See figure 12*a*.

crest. That part of the surface of a thread which connects adjacent flanks at the top of the ridge. See figures 8 and 9.

root. That part of the surface of a thread which connects adjacent flanks at the bottom of the groove. See figures 8 and 9.

included angle (angle of thread). The angle between the flanks of the thread, measured in an axial plane section. See figure 12*a*.

flank angles. The angles between the individual flanks and the perpendicular to the axis of the thread measured in an axial plane section. See figure 12*a*.

fundamental triangle. A triangle of which two sides represent the form of a theoretical thread with sharp crest and roots, having the same pitch and flank angles as the basic thread form and whose third side, or *base* is parallel to a generator of the cylinder on which the thread is formed. See figure 12*a*.

NOTE. The fundamental triangle provides the framework on which the basic and design forms of the thread are set out.

apex. The sharp corner of the fundamental triangle opposite to its base. See figure 12*a*.

height (or depth) of the fundamental triangle. The distance, measured perpendicular to the axis from its apex to its base. See figure 12*a*.

basic truncation. The distance, measured perpendicular to the axis, between the basic major or minor cylinder and the adjacent apex of the fundamental triangle. See figure 12*a*.

NOTE 1. The terms 'basic major truncation' and 'basic minor truncation' may also be used, and are self-explanatory. The basic major and minor truncations are not necessarily equal.

NOTE 2. For definitions of major and minor cylinders see below.

C. Pitch of screw threads

axis. The axis of the pitch cylinder of a screw thread. See figures 8, 9, and 16. (See also definitions for pitch cylinder below.)

NOTE. References to the 'axis' of a screw thread are

usually in general terms. This definition is, however, required for precision of statement.

pitch. The distance, measured parallel to the axis, between corresponding points on adjacent thread forms in the same axial plane section and on the same side of the axis. See figures 7*a*, 7*b*, 7*c*, 17*b* and 17*c*.

NOTE. The pitch (in inches) is the reciprocal of the number of threads per inch.

lead. The distance, measured parallel to the axis, between corresponding points on consecutive contours of the same thread helix in the same axial plane section and on the same side of the axis. See figure 7*c*.

NOTE 1. The lead is the distance the thread advances axially in one revolution.
NOTE 2. For a single-start thread the lead is identical with the pitch. The use of the term lead is normally confined to multi-start threads.
NOTE 3. The lead (in inches) is the reciprocal of the number of turns per inch.

cumulative pitch. The distance, measured parallel to the axis of the thread between corresponding points on any two thread forms whether in the same axial plane or not.

pitch cylinder. An imaginary cylinder, co-axial with the thread, which intersects the surface of a parallel thread in such a manner that the intercept on a generator of the cylinder between the points where it meets the opposite flanks of the thread groove is equal to half the basic pitch of the thread. See figures 8 and 9.

pitch line. The generator of the pitch cylinder. See figures 8, 9, 17*b* and 17*c*.

pitch point. The point where the pitch line intersects the flank of the thread. See figures 8 and 9.

lead angle. On a parallel thread the angle made by the helix of the thread at the pitch point with a plane perpendicular to the axis.

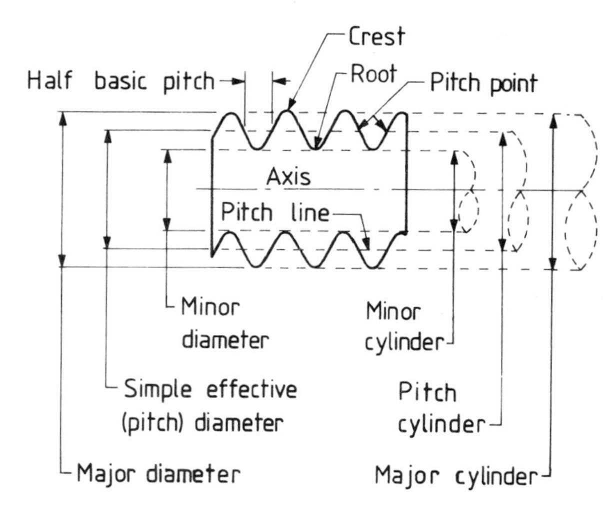

Figure 8. External parallel screw thread

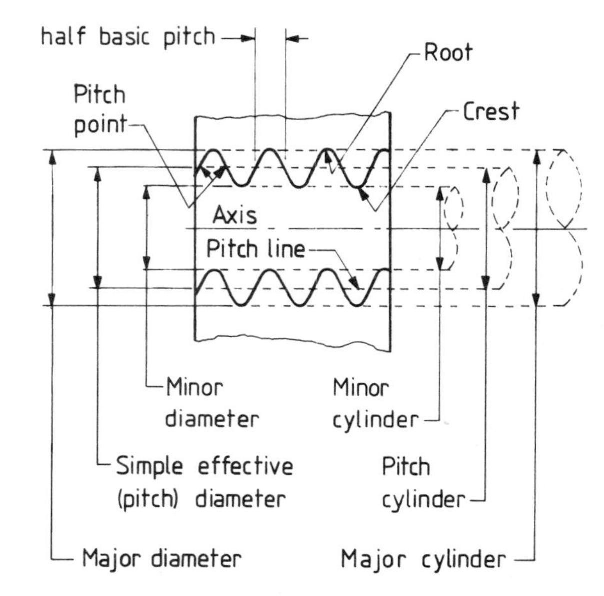

Figure 9. Internal parallel screw thread

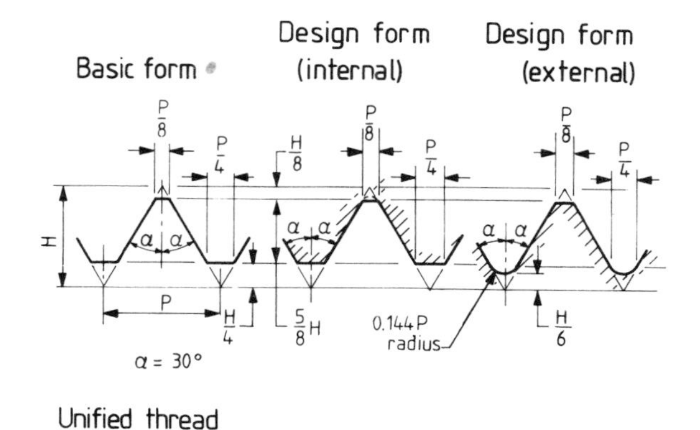

Figure 11. Basic and design forms of threads

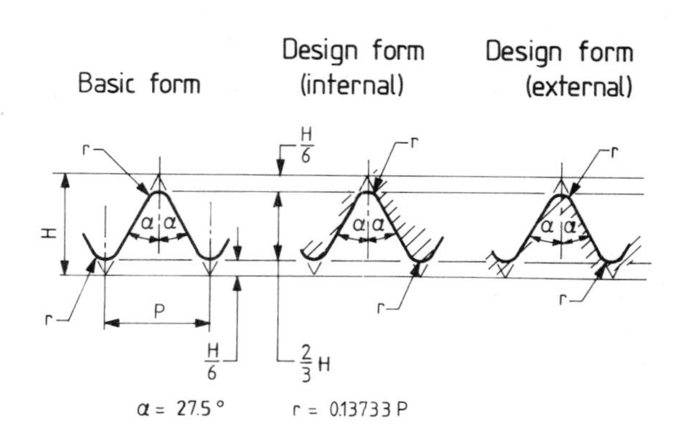

D. Diameter of screw threads

major cylinder. An imaginary cylindrical surface which just touches the crests of an external thread or the roots of an internal thread. See figures 8 and 9.

minor cylinder. An imaginary cylindrical surface which just touches the roots of an external thread or the crests of an internal thread. See figures 8 and 9.

major diameter. The diameter of the major cylinder of a parallel thread, in a specified plane normal to the axis. See figures 8 and 9.

minor diameter. The diameter of the minor cylinder of a parallel thread, in a specified normal to the axis. See figures 8 and 9.

effective (or pitch) diameter. The diameter of the pitch cylinder of a parallel thread, in a specified plane normal to the axis. See figures 8 and 9.

NOTE 1. USA practice uses the term 'pitch diameter' for this measurement.
NOTE 2. It is necessary to draw a distinction between the 'simple' effective diameter, as defined here, and the 'virtual' effective diameter. (See definition for virtual effective diameter below.)

virtual effective diameter. The effective diameter of an imaginary thread of perfect pitch and flank angle, having the full depth of flanks, but clear at the crests and roots, which would just assemble with the actual thread over the prescribed length of engagement.

NOTE 1. The 'virtual' effective diameter exceeds the simple effective diameter in the case of an external thread, but is less than the simple effective diameter in the case of an internal thread, by an amount corresponding to the combined diametral effects due to any errors in the pitch and/or the flank angles of the thread.
NOTE 2. USA practice uses the term 'effective size' for this measurement.

. . . .

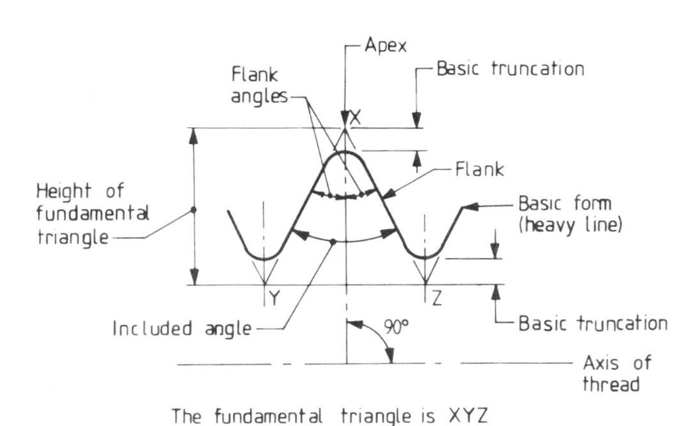

Figure 12*a*. Parallel screw thread

F. Assembly of screw threads

length to the end of full thread (length of full thread). The distance from the plane defining the end of the thread to the parallel plane, normal to the axis, which passes through the point on the root diameter helix at which the thread ceases to be fully formed at the root. See figures 14*a* and 14*b*.

NOTE 1. The term 'full thread' applies only to parallel threads, and should not be confused with 'complete thread' which applies only to taper threads.
NOTE 2. The full thread excludes that part of the thread over which, owing to the form and mode of operation of the threading tool, the root ceases to be fully formed. It includes the length of any permissible chamfer at the free end of the thread.
NOTE 3. The root diameter is the minor diameter of an external thread, or the major diameter of an internal thread.

depth of engagement. The radial distance by which the thread forms of two mating threads overlap each other. The radial distance between the basic major cylinder of the external thread and the minimum minor cylinder of the internal thread represents 100 per cent depth of engagement. See figure 15.

length of engagement. The axial distance over which two mating threads are designed to make contact. See figure 16.

leading flank. The flank which, when the thread is about to be assembled with a mating thread, faces the mating thread. See figure 16.

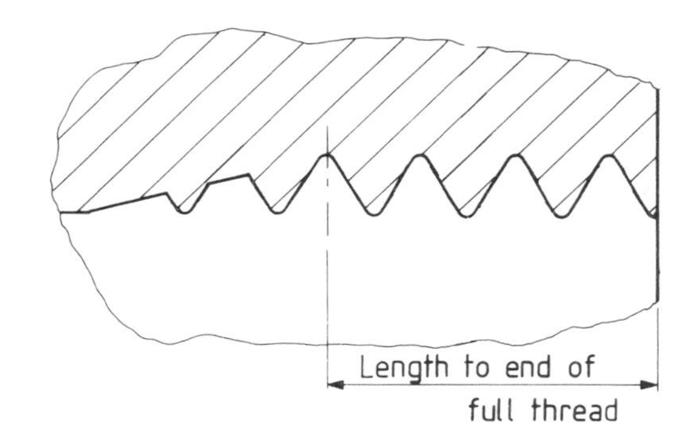

Figure 14*a*.

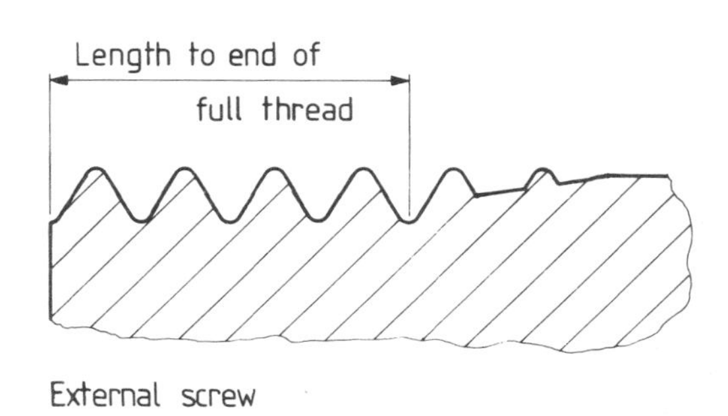

Figure 14*b*.

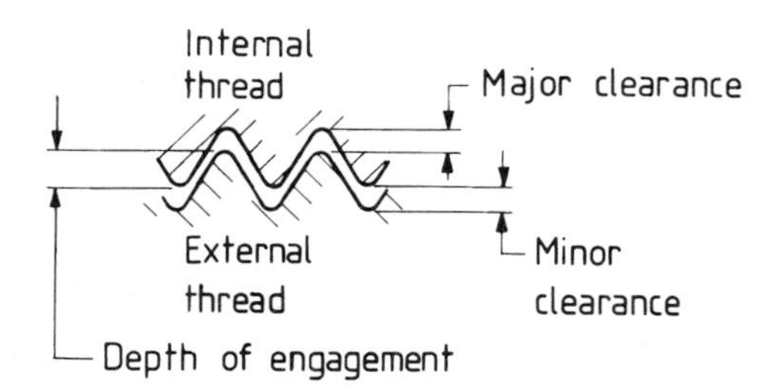

Figure 15.

following flank. The flank of a thread which is opposite to the leading flank. See figure 16.

pressure flank. The flank that takes the thrust or load in an assembly. See figure 16.

NOTE. The terms 'pressure' and 'clearing' flanks are used particularly in relation to buttress and similar threads.

clearing flank. The flank that does not take the thrust or load in an assembly. See figure 16.

grade (or class) of a thread. The characteristic of a thread which is determined by the relationship between the tolerance, and the associated allowance, if any, on the thread, and its basic size.

NOTE. The grade of a thread is not related solely to the accuracy of workmanship, but also to the fit which is to be expected when mating threads are assembled.

grade of fit of a threaded pair. That characteristic of a fit which is determined by the associated grades of the two mating threads.

blunt start. The condition resulting from the removal of the partial thread at the entering end.

NOTE. Blunt start is commonly provided on threaded parts which are repeatedly assembled by hand, such as loose couplings and thread gauges, to facilitate the entry of the threads without damage and to prevent cutting of the hands.

major crest truncation. The distance, if any, measured perpendicular to the axis, between the generators of the major cylinders for the basic and design forms of the external thread, assuming no allowances. See figure 17*b*.

NOTE. The major crest truncation is *additional* to the basic truncation.

minor crest truncation. The distance, if any, measured perpendicular to the axis, between the generators of the minor cylinders for the basic and design forms of the internal thread. See figure 17*c*.

NOTE. The minor crest truncation is *additional* to the basic truncation.

major clearance. The distance, measured perpendicular to the axis, between the design forms at the root of the internal thread and the crest of the external thread. See figure 15.

minor clearance. The distance, measured perpendicular to the axis, between the design forms at the crest of the internal thread and the root of the external thread. See figure 15.

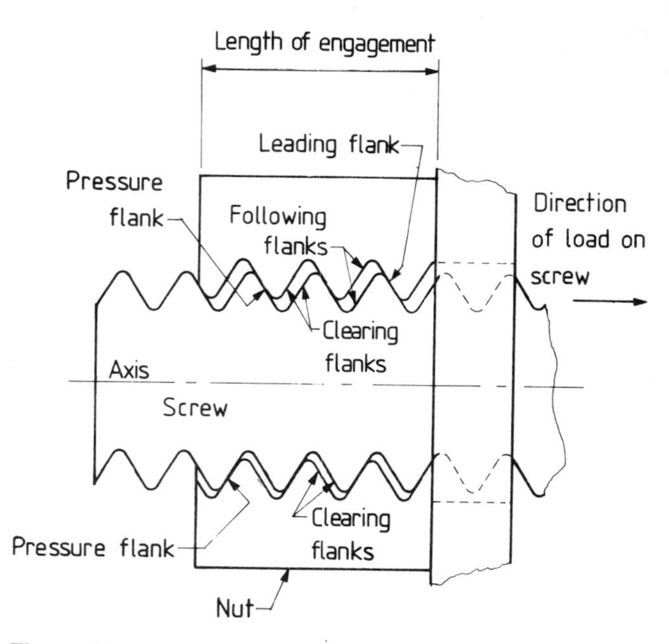

Figure 16.

Figure 17*a*. Basic form

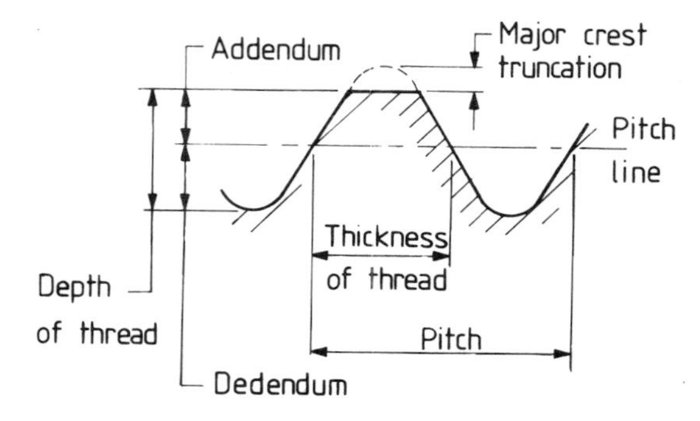

Figure 17*b*. Design form (external thread)

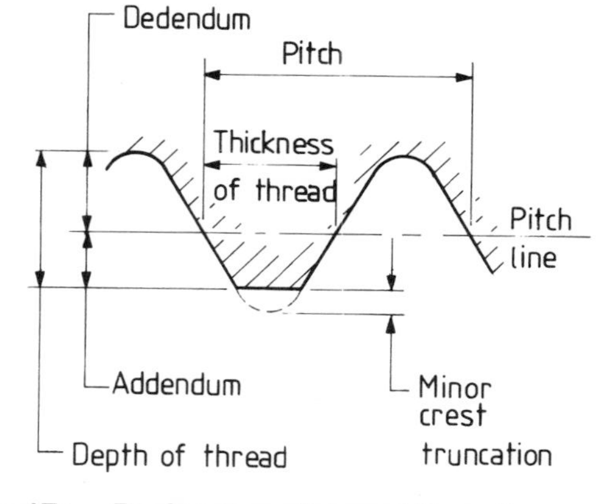

Figure 17*c*. Design form (internal thread)

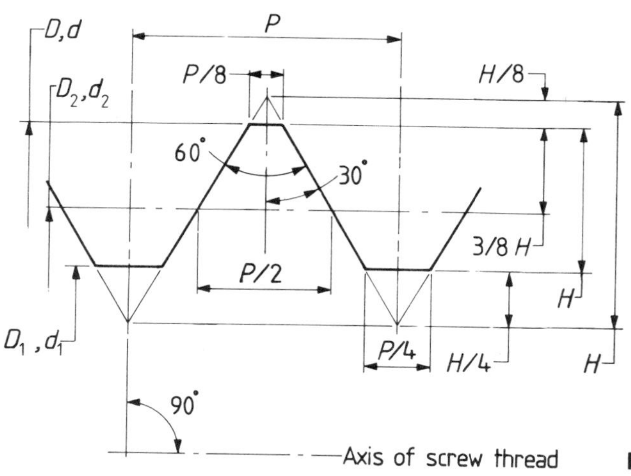

D = major diameter of internal thread
d = major diameter of external thread
D_2 = pitch diameter of internal thread
d_2 = pitch diameter of external thread
D_1 = minor diameter of internal thread
d_1 = minor diameter of external thread
P = pitch
H = height of fundamental triangle

Figure 1. Basic profile of ISO metric thread

addendum. The radial distance between the major and pitch cylinders of an external thread; the radial distance between the pitch and minor cylinders of an internal thread. See figures 17*b* and 17*c*.

dedendum. The radial distance between the pitch and minor cylinders of an external thread; the radial distance between the major and pitch cylinders of an internal thread. See figures 17*b* and 17*c*.

depth of thread. The radial distance between its major and minor cylinders. See figures 17*b* and 17*c*.

NOTE. The depth of thread is equal to the sum of the addendum and the dedendum.

thickness of thread. The distance between the flanks of a thread measured parallel to the axis at the design pitch line. See figures 17*b* and 17*c*.

NOTE. This definition applies only to parallel threads.

The inspection of screw threads may be by gauging or measuring.

Inspection by gauging

BS 919 'Screw gauge limits and tolerances', Part 3 'Gauges for ISO metric screw threads' contains the recommended gauging system for checking threads of nominal diameters 1 mm and larger which have been made in accordance with BS 3643 'ISO metric screw threads'.

Provision is made for the following types of gauges:

(a) Screw gauges
- (1) GO and NOT GO screw plug gauges
- (2) GO and NOT GO screw ring gauges, solid type
- (3) GO and NOT GO screw ring gauges, adjustable type
- (4) GO and NOT GO screw calliper gauges, adjustable type

(b) Plain gauges (for the crest diameters)
- (1) GO and NOT GO plug gauges
- (2) GO and NOT GO calliper gauges
- (3) GO and NOT GO ring gauges

(c) Setting plugs, in both single and double length
- (1) for GO and NOT GO adjustable screw ring gauges
- (2) for GO and NOT GO screw calliper gauges

(d) Check plugs
GO and NOT GO check plug gauges for solid type
GO and NOT GO screw ring gauges

(e) Wear check plugs
NOT GO wear check plug gauges for solid type
GO and NOT GO screw ring gauges

Information is given about the function and methods of use of the various types of gauges and a recommended procedure is given for the settlement of disputes which may arise when borderline products are inspected.

BS 919

Screw gauge limits and tolerances

Part 3. Gauges for ISO metric screw threads

....

Specification

....

3. Function and method of use of the various types of gauges

3.1 Gauges for external screw threads

3.1.1 *Solid or adjustable GO screw ring gauge.* This gauge checks that the virtual pitch (effective) diameter of the product threat is not too large. The major and minor diameters of the product thread are not checked by this gauge. It must be possible to screw the gauge by hand, without using excessive force, over the complete length of the product thread.

3.1.2 *Solid or adjustable NOT GO screw ring gauge.* This gauge checks that the simple pitch (effective) diameter of the product thread is not too small. Ideally it should not be possible for the product thread to enter the gauge, but as external threads are often slightly tapered at the leading end it is permissible to allow entry provided that, on withdrawal, disengagement takes place within two full turns of thread. If the product has a length of thread of three turns or less the gauge should not screw completely on to the product thread. The gauge should be applied by hand without using excessive force.

3.1.3 *Adjustable GO screw calliper gauge.* This gauge is intended to check that the virtual pitch (effective) diameter of the product thread is not too large. The major and minor diameters of the product thread are not checked by this gauge. The gauge is generally applied to the product thread under its own weight or in accordance with a fixed working load at three positions at least evenly spaced around the circumference of the thread. The gauge must pass completely over the product thread at any of the positions at which it is applied.

3.1.4 *Adjustable NOT GO screw calliper gauge.* This gauge checks that the simple pitch (effective) diameter of the product thread is not too small. It is generally applied to the product thread under its own weight or under a fixed working load at three positions at least evenly spaced around the circumference of the thread. The gauge must not pass over the product thread except possibly for the first two turns of thread.

3.1.5 *GO and NOT GO plain ring or calliper gauges.* These gauges check that the major diameter of the product thread is between the specified limits. The gauges are applied to the product thread under the same conditions as the corresponding screw gauges.

3.1.6 *Setting plugs (double length) for adjustable GO and NOT GO screw ring gauges.* These setting plugs are used to set the adjustable screw ring gauges to the specified pitch (effective) diameters. Each setting plug has a length of thread approximately equal to twice the length of the screw ring gauge to be controlled. The pitch (effective) diameter of the setting plug is constant throughout, but half the length of the setting plug has a full form thread and the remaining half has a truncated form of thread. The screw ring gauge is adjusted to be a snug fit on the full form portion of the setting plug. The setting plug is then unscrewed by hand, without using excessive force, through the screw ring gauge until the truncated portion of the setting plug completely engages the screw ring gauge. In the latter condition there shall be no perceptible shake or play between the setting plug and the screw ring gauge: shake or play is an indication of an unacceptable error of thread form of the adjustable screw ring gauge.

3.1.7 *Setting plugs (single length) for adjustable GO and NOT GO screw calliper gauges.* These setting plugs are used to set the adjustable screw calliper gauges. The setting plugs for the GO and NOT GO screw calliper gauges have full form threads. The screw calliper is adjusted so that it just passes over the appropriate setting plug under its own weight or under a fixed working load.

3.1.8 *Screw check plugs for new solid GO and NOT GO screw ring gauges.* These GO and NOT GO screw check plugs are used to check that the pitch (effective) diameters of the screw ring gauges are within the specified limits. It should be possible to screw the GO check plug by hand completely through the appropriate screw ring gauge. The NOT GO check plug when screwed by hand, without using excessive force, may be allowed to enter both ends of the screw ring gauge provided that on withdrawal, disengagement takes place within one full turn of thread.

When setting a new calliper gauge with the setting plugs recommended in **3.1.7**, an additional check should be made with the NOT GO check plug for the corresponding new screw ring gauge; this latter check may also be applied to a use calliper gauge to ensure that the thread form is not worn. The wear check plug for the screw ring gauge should not be used as a wear check plug for the calliper gauge.

3.1.9 *Plain check plugs for new solid GO and NOT GO screw ring gauges.* These GO and NOT GO check plugs are used to verify that the minor diameter of the screw ring gauge is between the specified limits. The GO check plug should assemble completely with the screw ring gauge. The NOT GO check plug may enter both ends of the screw ring gauge by not more than a distance of one turn.

3.1.10 *NOT GO screw wear check plugs for used solid GO and NOT GO of screw ring gauges.* These wear check plug gauges are used to check that the appropriate solid screw ring gauges have not worn beyond the specified limits of wear.

The wear check plug gauge, when screwed by hand, without using excessive force, may be allowed to enter both ends of the screw ring gauge provided that on withdrawal, disengagement takes place within one full turn of thread.

3.2 Gauges for internal screw threads

3.2.1 *GO screw plug gauge.* This gauge checks that the virtual pitch (effective) diameter and the major diameter of the product thread are not smaller than the minimum limits specified. It does not check the minimum minor diameter of the product thread. It must be possible to screw the GO screw plug gauge by hand, without using excessive force, into the complete length of the product thread.

3.2.2 *NOT GO screw plug gauge.* This gauge checks that the maximum simple pitch (effective) diameter of the product thread is not too large. Ideally it should not be possible for the gauge to enter the product thread but, as internal threads are often slightly bell-mouthed, it is permissible to allow entry provided that, on withdrawal, disengagement takes place within two full turns of thread. If the product has a length of thread of three turns or less, the gauge should not screw completely through the product thread. The gauge should be applied by hand without using excessive force.

3.2.3 *GO and NOT GO plain plug gauges.* These gauges check that the minor diameter of the product thread is between its specified limits. The GO plain plug gauge should assemble completely with the product thread. The NOT GO plain plug gauge may enter the product thread by not more than a distance of two full turns of the thread. If the product has a length of thread of three turns or less, the NOT GO gauge should not pass completely through the product thread.

3.3 Notes on gauging product threads. It is not necessary for all the different types of screw gauges described in **3.1** to be used for checking external threads on products, but it is essential for one of the types of GO screw gauges and one of the types of NOT GO screw gauges to be used.

For preference a solid or adjustable GO screw ring gauge should always be used for gauging the maximum effective diameter of an external thread, but to save time in checking, a GO screw calliper gauge may be employed. It is recommended that gauging with GO screw calliper gauges should be supplemented by random sampling with a GO screw ring gauge, thus giving greater assurance of not accepting parts outside the limits. In cases of dispute, gauging with the GO screw ring gauge is decisive. The GO screw calliper gauge should *not* be used if the manufacturing process is likely to introduce errors in the product thread which this gauge is not certain to detect, e.g. lobing, local pitch errors in milled threads and burrs at the start of the thread. Furthermore, the GO screw calliper gauge is not suitable for checking non-rigid, e.g. thin-walled, parts which would deform when the gauge is applied; in such cases checking must be carried out with a GO screw ring gauge.

For large screw threads (over about 100 mm diameter) the size to which the GO screw calliper gauge is set may be reduced to compensate for the effect of form errors that may occur in these threads; this procedure reduces the possibility of accepting screw threads that are outside the specified limits.

A NOT GO screw calliper gauge is generally recommended for checking the minimum pitch (effective) diameter of an external thread. The solid or adjustable NOT GO screw ring gauge should only be used to check non-rigid product threads, e.g. those on thin-walled parts which would deform if a NOT GO screw calliper were applied.

It is recommended that all solid screw ring gauges be inspected periodically for wear by means of the appropriate NOT GO wear check plugs. GO and NOT GO screw plug gauges should similarly be regularly checked by measurement. The thread profiles of all screw gauges should be inspected regularly for wear.

The above remarks concerning the precautions to be observed when using screw calliper gauges also apply to the use of plain calliper gauges for checking the major diameter of the product external thread.

4. Inspection procedures and the settlement of disputes

4.1 Inspection by the manufacturer. Generally speaking, where product threads are made and also checked at the manufacturer's works, his inspection department can use the same kind of gauges as the workshop. In order to avoid differences between the results obtained by the workshop and the inspection department, it is recommended that the workshop use new or only slightly worn GO gauges while the inspection department is provided with GO gauges the pitch (effective) diameter of which lies nearer the permissible wear limit. The reverse procedure should be adopted for NOT GO gauges.

In case of dispute, the product threads are regarded as satisfactory when the gauges with which they are checked and passed conform to the sizes specified for those gauges, including the permissible wear limits. This can be proved by measuring the gauges or by checking them with the check plugs specified.

NOTE. In some special cases where it is essential to ensure that threads which may be outside the limits are not accepted and in the case of GO gauges used for sample inspection, gauges which have worn beyond the maximum metal limit of the product thread should not be used.

4.2 Inspection by the purchaser. There are three possible procedures for inspection on behalf of the purchaser by an inspector who is independent of the manufacturing plant concerned:

(a) The inspector may gauge the product threads with the manufacturer's own gauges. In this case, he should check the accuracy of the ring or calliper gauges employed by means of check plugs and setting plugs which belong either to the manufacturer or to the inspector (purchaser). Screw plugs should be checked by direct measurement.

(b) The inspector may use his own gauges made in accordance with this standard for gauging product threads. In this case, the remarks in **4.1** concerning checking during manufacture and the procedure for settling disputes apply.

(c) The inspector may use his own inspection gauges for checking the product threads. The disposition of the tolerance zones for these gauges should be such as to ensure that the purchaser does not reject product threads the actual size (e.g. pitch diameter) of which lies within the limits specified for the product.

4.3 This standard does not specify which gauges are to be used but recommends that the purchaser inform the manufacturer when ordering what procedure will be employed for the inspection of the product threads.

5. Reference temperature

In accordance with ISO Recommendation R1*, the sizes of both the gauge and the workpiece are related to the standard temperature of 20 °C.

If the workpieces and the gauges have the same thermal coefficient of expansion (e.g. steel workpieces and steel gauges), the checking temperature may deviate from 20 °C without detriment to the result, always provided that the temperatures of both gauges and workpieces are the same at the time of gauging.

If the workpieces and gauges have different thermal coefficients of expansion (e.g. steel workpieces and carbide gauges or brass workpieces and gauges of steel or carbide), the temperatures of both gauges and workpieces should, in principle, be close to 20 °C at the time of gauging.

*ISO/R1, 'Standard reference temperature for industrial length measurements'.

. . . .

7. Marking

Each gauge shall be plainly and permanently marked with the minimum marking essential for positive identification. In the case of plug gauges of the renewable end type, in addition to marking the handle, the marking shall also appear on the face of the gauging member where practicable.

Unless otherwise specified by the purchaser, the following particulars shall be included in the marking:

(a) The designation of the corresponding product thread in accordance with the recommendations in BS 3643 'ISO metric screw threads'.

NOTE. In the case of left-hand screw gauges, the symbol LH follows the designation.

(b) The size of the gauge, i.e. the limiting size of the product thread which the gauge is intended to control.

(c) GO or NOT GO.

(d) SET in the case of setting plugs for adjustable screw ring or calliper gauges.

(e) CHECK in the case of check plugs for solid screw rings or calliper gauges.

(f) The manufacturer's name or trademark.

(g) A serial number if required for recording purposes.

Examples

Type of gauge	*Marking*
GO screw plug	M8 × 1–6H, GO 7.350,X Co.(35)
NOT GO screw plug	M8 × 1–6H, NOT GO 7.500,X Co.(3)
Plain plugs for minor diameter	M8 × 1–6H, GO 6.917,X Co.(9) M8 × 1–6H, NOT GO 7.153,X Co.(15)
GO screw ring or calliper	M10 × 1.5–6g, GO 8.994,X Co.(56)
NOT GO screw ring or calliper	M10 × 1.5–6g, NOT GO 8.862,X Co.(75)

Plain callipers for major diameter	M10 × 1.5–6g, GO 9.968,X Co.(43) M10 × 1.5–6g, NOT GO 9.732,X Co.(64)
GO setting plug	M10 × 1.5–6g, GO SET 8.994,X Co.(24)
NOT GO check plug	M10 × 1.5–6g, NOT GO CHECK 8.862,X Co.(97)

Inspection by measurement

Measurement of a screw thread can be very complex; there being a number of elements to be measured, some of which are interrelated.

Figure 7.1 shows the elements of a parallel ISO metric screw thread.

Of the elements shown in Figure 7.1, it is the following which will normally be measured:

major diameter
minor diameter
form, particularly flank angles }
pitch } virtual effective diameter
pitch diameter }

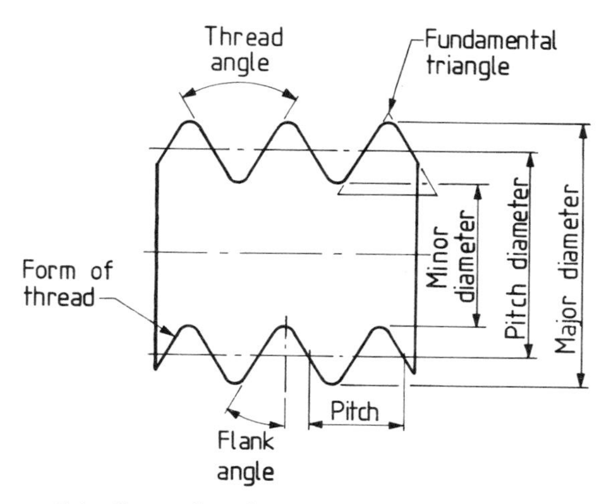

Figure 7.1 *Screw thread terms*

Measurement of the major diameter

An ordinary hand or, preferably, a bench micrometer, serves for measuring the major diameter. A bench micrometer designed by the National Physical Laboratory is shown in Figure 7.2.

Instead of a fixed anvil the NPL designed bench micrometer has a fiducial indicator so that all measurements are made at the same pressure. Since the fiducial indicator has positional adjustment, the range of the micrometer is increased but requires it to be used as a comparator and not for direct readings. The setting standard may be a gauge block or preferably a calibrated setting cylinder which gives greater similarity of contact at the anvils. Provided a setting standard which is similar in size to the dimension to be measured is used, the micrometer thread is used over a short length of travel and any pitch errors it contains will be virtually eliminated.

From the calibrated size of the setting cylinder, the micrometer reading over the setting cylinder and the micrometer reading over the major diameter of the thread, the major diameter of the thread may be obtained as follows. If:

calibrated size of the setting cylinder = B
micrometer reading on setting cylinder = R_c
micrometer reading on thread = R_t

then, major diameter = $B + R_t - R_c$

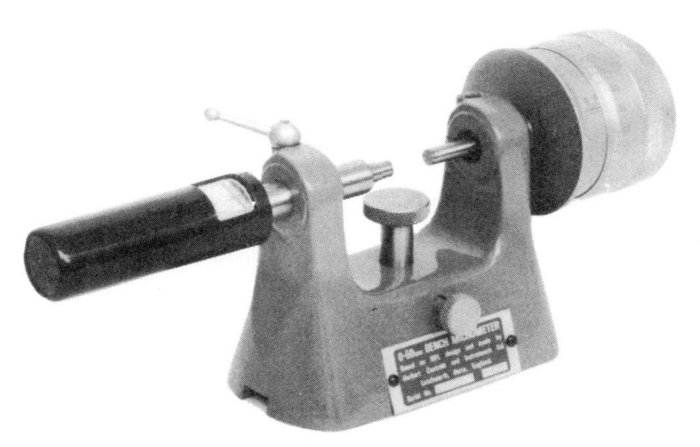

Figure 7.2 *Bench micrometer of NPL design*
Reproduced by permission of Sigma Ltd, Letchworth

Measurement should be made in different diametral planes along the axis and around the helix of the gauge so as to determine any variation in the major diameter. Care should be taken to make a check on the setting cylinder after measurement to ensure that the setting of the micrometer has not altered.

Measurement of the minor diameter

The minor diameter is measured using a similar process to that used for the major diameter except that hardened and ground steel prisms are used to obtain contact with the root of the screw thread. A reading is taken on the setting cylinder with the prisms in position and on the thread, as shown in Figure 7.3.

The minor diameter will thus be found if:

calibrated size of setting cylinder = B
micrometer reading on setting cylinder with prisms = R_c
micrometer reading on thread with prisms = R_t

then, minor diameter = $B + R_t - R_c$

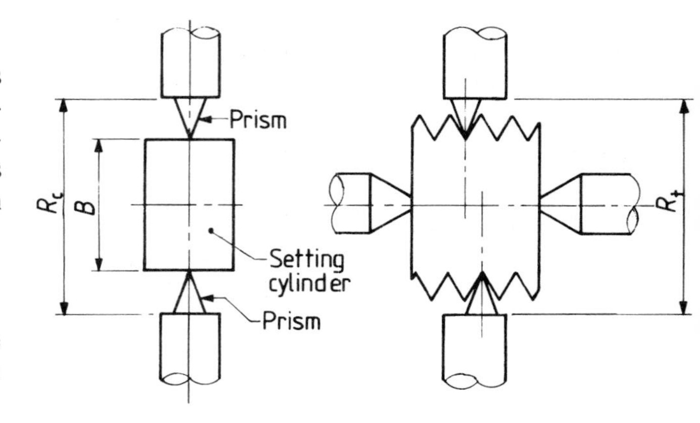

Figure 7.3 *Measurement of the minor diameter*

With this method it is essential that the micrometer is held at right angles to the axis of the screw thread being measured and this is ensured by using a floating carriage diameter measuring machine (see Figure 7.4).

The floating carriage diameter measuring machine employs centres to hold the thread under inspection. A bench micrometer is mounted on vee grooves so that it is constrained to move only in such a manner that a right angle is maintained between the direction of measurement and the axis of the screw thread under inspection.

Measurements of minor diameter should be made in different diametral planes along the axis and around the helix of the screw thread to determine any variations. Care should be taken to make a check reading on the setting cylinder after measurement to ensure that the setting of the machine has not altered.

Figure 7.4 *Floating carriage diameter measuring machine*
Reproduced by permission of Sigma Ltd, Letchworth

Measurement of the pitch diameter

The pitch diameter is measured using a similar method to that used for the minor diameter except that two small cylinders are used instead of vee pieces. The cylinders should be such that when placed between the threads, contact is made about half way down the flanks. (BS 5590 'Screw thread metric series measuring cylinders' relates to standard measuring cylinders for standard pitches of ISO metric threads given in BS 3643 'ISO metric screw threads'.)

As for the minor diameter, the use of the floating carriage diameter measuring machine is necessary to ensure the correct relationship between the direction of measurement and the axis of the screw thread under inspection.

A reading is taken on the setting cylinder with the cylinders in position and on the thread, as shown in Figure 7.5.

Thus, the pitch diameter will be found if:

E = measured pitch diameter
T = measured diameter under the cylinders
p = nominal pitch of thread
w = mean measured diameter of cylinders used
α = flank angle of thread
then, $E = T + w - w \operatorname{cosec} \alpha + p/2 \cot \alpha$

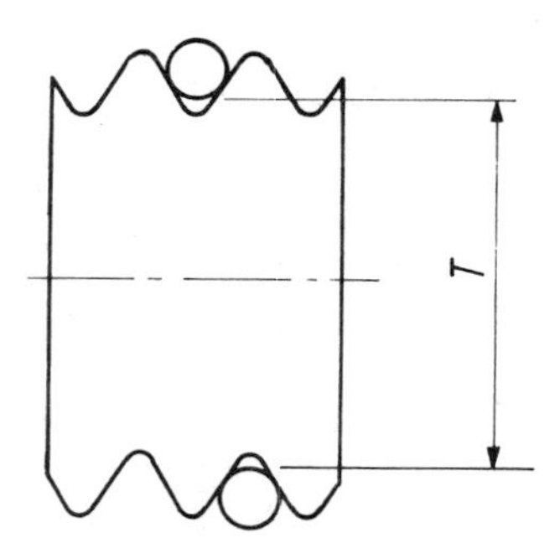

Figure 7.5 *Measurement of pitch diameter*

Since the nominal values of p, cosec α and cot α, together with the value of w are known before measurement, they can be gathered into a single value called the P value. That is, for a given diameter of cylinder, w, in a given thread form of flank angle α and pitch p.

P value = $\frac{1}{2}p \cot \alpha - (\operatorname{cosec} \alpha - 1)w$

and

$E = T + P$ value

By substituting the appropriate values of cosec α and cot α in the equation for P value, a general expression can be obtained for the P value of each thread form as follows:

ISO metric	P value $= 0.866025\,p - w$
Unified	P value $= 0.866025\,p - w$
Whitworth	P value $= 0.960491\,p - 1.165681\,w$
BA	P value $= 1.136336\,p - 1.482950\,w$

Measurements of pitch diameter should be made in different diametral planes along the axis and around the helix of the screw thread to determine any variations. Care should be taken to make a check reading on the setting cylinder after measurement to ensure that the reading of the machine has not altered.

Pitch measurement

Any error in the pitch of a screw thread is reflected in a pitch diameter error of approximately twice the pitch error. Two types of pitch error are known as:

periodic pitch error
progressive pitch error

Periodic pitch error produces a cyclic graph of the type shown in Figure 7.6.

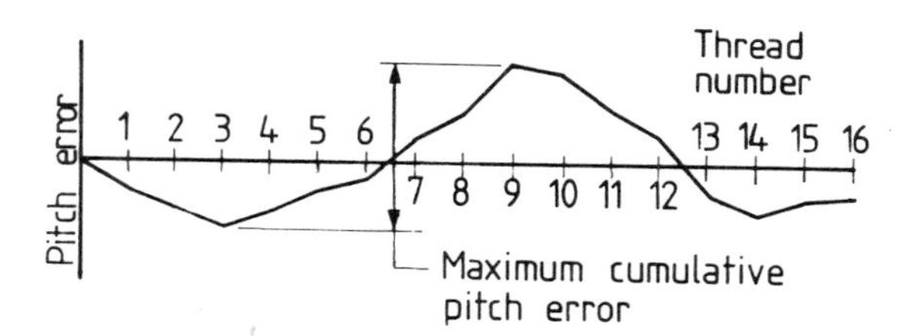

Figure 7.6 *Periodic pitch error*

Progressive pitch error produces a straight line graph of the type shown in Figure 7.7.

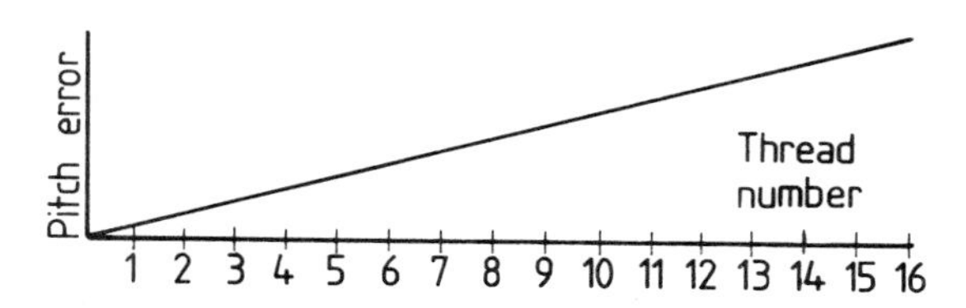

Figure 7.7 *Progressive pitch error*

The measurement of a type of periodic pitch error, known as thread drunkenness, requires a special instrument, since the error occurs during each revolution of the thread and results from the thread not being cut to a true helix. Such errors, except on very large threads, will not have any great effect.

With the exception of drunken threads, pitch errors may be measured using the NPL-designed pitch measuring machine (see Figure 7.8).

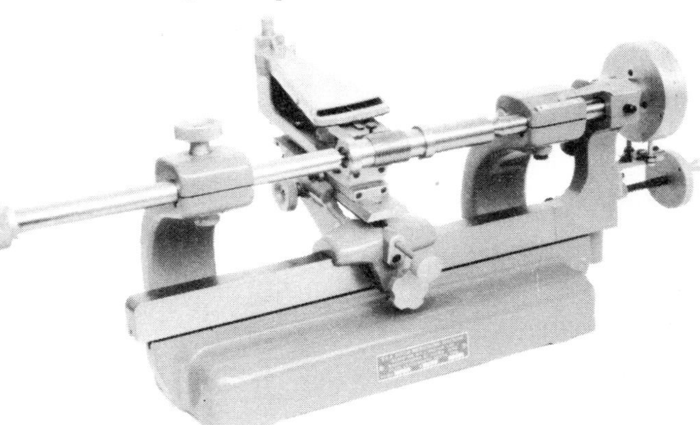

Figure 7.8 *Pitch measuring machine of NPL design*
Reproduced by permission of Sigma Ltd, Letchworth

A male thread is held between centres and a round nose stylus is engaged with the thread as shown in Figure 7.9 such that the pressure at P_1 and P_2 brings a fiducial indicator to the same position in each thread (see Figure 7.9). Micrometer readings are taken at each point as indicated by the fiducial indicator and from these the pitch of the screw thread can be obtained.

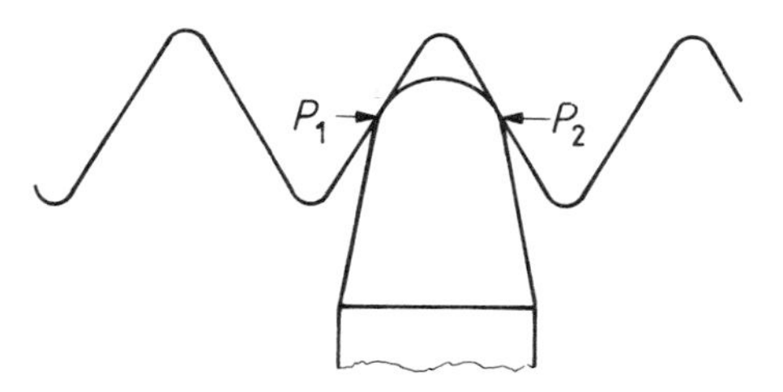

Figure 7.9 *Pressure on the round nose stylus*

The test is repeated along another line at 180 ° round the screw, and the mean of the two sets of measurements is taken to eliminate the effects any misalignment of the measuring axis to the axis of the screw thread under inspection.

It is important to ensure that the pitch measuring machine is calibrated and the measurements of a screw thread adjusted accordingly (see Figure 7.10).

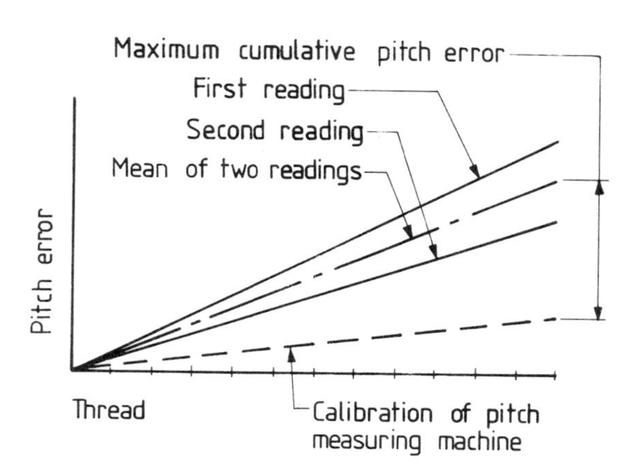

Figure 7.10 *Graphical representation of pitch measuring*

Measurement of thread form

The measurement of the flank angles is the most important measurement of thread form which, on very large screw threads, may be made by contact methods. The more common, and only practical method for smaller

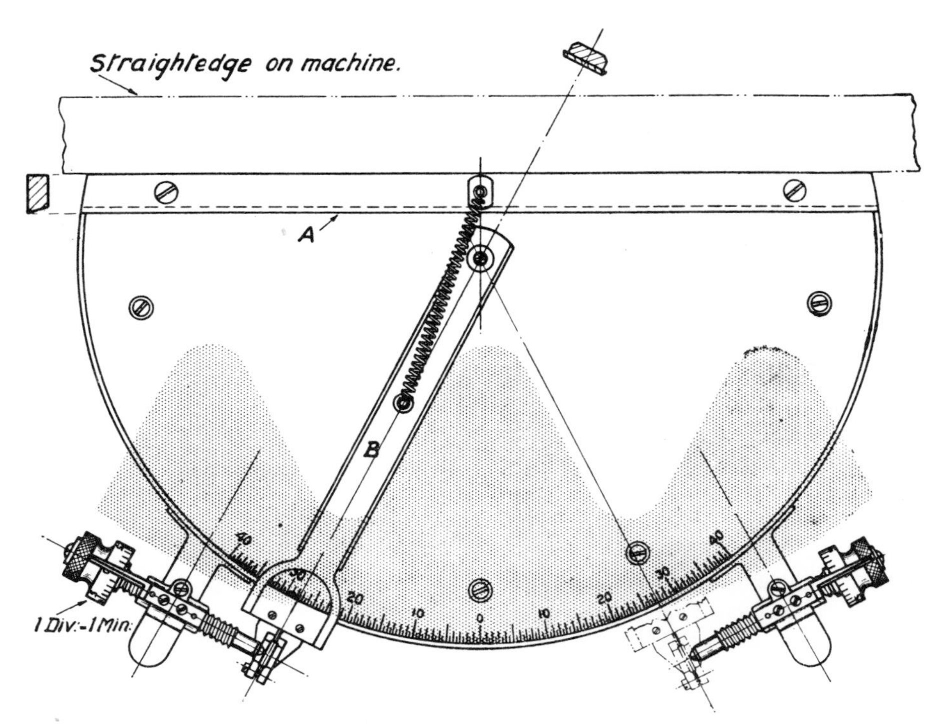

Figure 7.11 *Shadow protractor for measuring flank angles*
Taken from *Notes on Applied Science No. 1* (HMSO, 1958) and reproduced by permission of the NPL

threads, is to use optical equipment, such as the NPL projector.

The image of the screw thread is magnified and projected on to a screen so that measurement of the flank angles may be made with a shadow protractor, as shown in Figure 7.11.

The basic procedure for measuring flank angles, known as the 'throw over' method is as follows:

'The pivoted arm of the protractor is rotated so as to bring the edge of its shadow on the white background parallel to the sloping flank of which the angle is required. The position of the pivoted arm is controlled by the tangent screws. The screw plug is then moved across the field of the lens until the thread form of the other side appears on the white background of the protractor. Without moving the position of the protractor on the table of the machine, the pivoted arm is swung across to measure the angle of the *same* flank as before. The mean of the two readings gives the inclination of the flank with respect to the normal to the axis of the screw.'

National Physical Laboratory, *Notes on Applied Science, No. 1: Gauging and measuring screw threads* (HMSO 1958)

Virtual effective (or pitch) diameter

The effect of pitch and/or flank angle errors on an external screw thread is always to cause a virtual *increase* in the effective (pitch) diameter. (For internal screw threads such errors always cause a virtual *decrease* in the effective (pitch) diameter.)

The equivalent difference in pitch diameter necessary to compensate for a pitch error is given for metric threads by $+ 1.732\ \delta p$, where δp is the pitch error over the length of engagement.

The virtual increase in pitch diameter to compensate for flank angle errors (where δa_1 and δa_2 represents the errors in the two flank angles) is given for metric threads by $0.0115 \times p \times (\delta a_1 + \delta a_2)$, where p = nominal pitch of thread.

Thus, virtual effective diameter equals pitch diameter $+0.0115\,p\,(\delta a_1 + \delta a_2) + 1.732\,\delta p$.

Measurement of internal screw threads

The measurement of internal screw threads is similar to that of external threads with the added problems caused by the inaccessibility of the thread.

For the measurement of the minor diameter of small diameter threads, a pair of accurately made sliding wedges (known as taper parallels) may be used, as shown in Figure 7.12.

Alternatively, the minor diameter may be sized by a range of cylindrical plugs, differing in size by small increments.

For larger threads, measurement can be made by the use of gauge blocks set between two rollers of known size placed diametrically opposite in the screw thread.

The measurement of the major and pitch diameter of internal screw threads is only practical using measuring machines of special design.

The pitch of an internal screw thread may be measured on a screw pitch measuring machine using an attachment to allow the stylus to enter the thread.

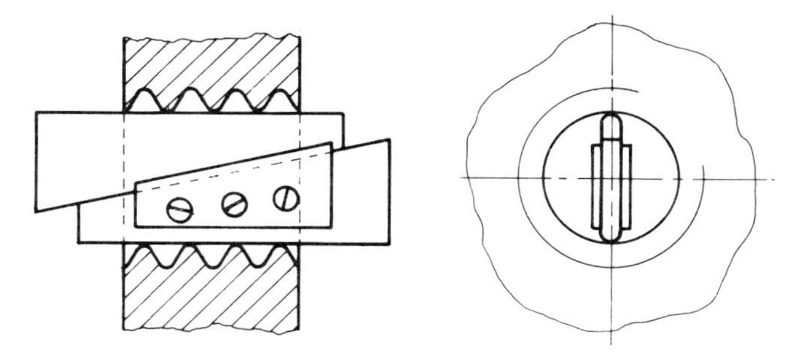

Figure 7.12 *Measurement of minor diameter by means of taper parallels*

The flank angle of an internal screw thread may only be measured by making a cast of the thread, either in plaster or dental wax, which is then projected as for external threads. When making the cast, a portion equal to less than half the diameter is made to enable the cast to be lifted out and not screwed out to avoid distortion.

Note: Full information on the measuring of screw threads will be found in the NPL booklet *Notes on Applied Science, No. 1: Gauging and measuring screw threads*, published by HMSO. It should be noted, however, that the second edition was published in 1958 and is, understandably, rather out of date. Although it contains a lot of useful information, it is now out of print.

Standards publications referred to in this chapter

BS 919	Screw gauge limits and tolerances Part 3 Gauges for ISO metric screw threads
BS 2517	Definitions for use in mechanical engineering
BS 3643	ISO metric screw threads (in 3 parts)
BS 5590	Screw thread metric series measuring cylinders

8 Gear measurement

Introduction

This chapter provides practical advice and information on gear measurement as specified in BS 436 'Spur and helical gears', Part 2 'Basic rack form, modules and accuracy (1 to 50 metric module)'.

Why measure gears?

Everyone concerned in gear production, be it manufacturing personnel, or designers, must be quite clear why it is necessary to measure gears. So often it is said we do not need gear inspection, all our gears are accurately ground, or they do not transmit load, etc.

It cannot be over-emphasized that gear inspection is an aid to production. It is not an unnecessary refinement. It can, and should, be a means of smoothing out manufacturing problems and ensuring ease of assembly and reliable service.

The paramount reason for gear measurement is to ensure that gears have been manufactured to the specifications laid down by the designer.

Basic gear geometry

The majority of gears are manufactured by the generating method. Hobbing being one example of this process. Gear hobbing is a continuous process in which the tooth flanks of the constantly moving workpiece are formed by equally spaced cutting edges of the hob.

The profile produced on the gear is a curve comprising a number of flats varying with the number of flutes in the hob which passes a given tooth during the generating movement. If the hob selected has a normal pitch of p_n and the gear to be cut has T teeth, then the gear will have a reference diameter equal to $Tp_n/\pi \sec \delta$, where δ is the helix angle of the teeth at the reference diameter.

Although the reference diameter is often known as the pitch diameter a gear does not strictly have a pitch diameter until it is meshed with another gear. If the hob is further defined as having a flank angle ψ_n then the gear generated will have a base diameter equal to $Tp_n/\pi \sec \delta \cos \psi_t$, where ψ_t is the transverse pressure angle which equals $\tan^{-1} (\tan \psi_n \sec \delta)$. The hob will generate a series of involute curves from the base diameter with an angular pitch of $360/T$ degrees.

The depth of insertion of the hob into the gear blank then determines the thickness of the teeth and the part of the involute curve which appears on the teeth. If the hob is inserted so that the gear reference diameter touches a line where the hob tooth thickness is equal to the hob tooth gap thickness, then a so-called standard proportion gear is generated.

There is no necessity to generate standard proportion gears, however, it is often preferable to apply 'correction' or 'addendum modification', in which case the blank outside diameter is non-standard. In the case of positive addendum modification, the outside diameter of the gear blank is increased. The contrary is the case with negative correction.

Elemental errors

It is now possible to list the elemental measurements which are required to completely define the errors in a gear.

(a) Gear blank or reference surface meaurement
(b) Pitch measurement
(c) Tooth profile measurement
(d) Tooth alignment measurement
(e) Gear tooth size measurement
(f) Composite error measurement

The choice of which measurement to make for a particular gear depends upon many factors. The following notes are intended as a guide. It is certainly not necessary to employ all these measurement methods on one particular gear.

Gear blank or reference surface measurements

Gears are intended to transmit rotary motion so the first step in the measurement process is to define a reference axis of the gear. There is no physical line which can be measured so the reference axis must be defined as the centre of one or more datum diameters.

In the case of a shaft gear, the reference diameter would be the journal diameter or the centre holes. In the case of a bored gear, the datum diameter will be the hole.

The reference surfaces will themselves have errors of roundness and cylindricity. This is sometimes forgotten and as a result quite unrealistic gear tolerances are demanded for gears which have a poor reference surface.

In addition to the tolerances of the reference diameter, the type of fit of these diameters with their mating parts is sometimes given insufficient thought. It is clearly absurd to require the gear teeth to have accumulative tolerance with the reference diameter of, say, 0.010 mm, and to have a 0.040 mm clearance of the reference surface on its mating part. The gear may be tested from the bore and found to be within tolerance but it is when it is finally assembled that the run out of the face has caused the gear to be outside tolerance. The tolerances of gear blanks are adequately specified in BS 436: Part 2.

BS 436

Spur and helical gears

Part 2. Basic rack form modules and accuracy (1 to 50 metric module)

. . . .

Table 2. Tolerances on gear blanks

(All values in micrometres except tip diameter in millimetres)

<table>
<tr><th colspan="2">Grade</th><th>1</th><th>2</th><th>3</th><th>4</th><th>5</th><th>6</th><th>7</th><th>8</th><th>9</th><th>10</th><th>11</th><th>12</th></tr>
<tr><td rowspan="2">Bore</td><td>Size tolerance*</td><td>IT4</td><td>IT4</td><td>IT4†</td><td>IT4</td><td>IT5</td><td>IT6</td><td>IT7</td><td>IT7</td><td>IT8</td><td>IT8</td><td>IT8</td><td>IT8</td></tr>
<tr><td>Form tolerance‡</td><td>IT1</td><td>IT2</td><td>IT3</td><td></td><td></td><td></td><td></td><td></td><td></td><td></td><td></td><td></td></tr>
<tr><td rowspan="2">Shaft</td><td>Size tolerance*</td><td>IT4</td><td>IT4</td><td>IT4</td><td>IT4</td><td>IT5</td><td>IT5</td><td>IT6</td><td>IT6</td><td>IT7</td><td>IT7</td><td>IT8</td><td>IT8</td></tr>
<tr><td>Form tolerance‡</td><td>IT1</td><td>IT2</td><td>IT3</td><td></td><td></td><td></td><td></td><td></td><td></td><td></td><td></td><td></td></tr>
<tr><td colspan="2">Radial run-out tip cylinder* (see figure 3)</td><td>IT6</td><td>IT6</td><td colspan="2" rowspan="3">$0.01d_a + 5$</td><td colspan="2" rowspan="3">$0.016d_a + 10$</td><td colspan="2" rowspan="3">$0.025d_a + 15$</td><td colspan="2" rowspan="3">$0.04d_a + 25$</td><td colspan="2" rowspan="3">$0.04d_a + 25$</td></tr>
<tr><td colspan="2">Radial run-out of reference</td><td colspan="2" rowspan="2">$0.004d_a + 2.5$</td></tr>
<tr><td colspan="2">Axial run-out of reference surface (see figure 3)</td></tr>
<tr><td colspan="2">Blank diameter</td><td>IT6</td><td>IT6</td><td>IT7</td><td>IT7</td><td>IT7</td><td>IT8</td><td>IT8</td><td>IT8</td><td>IT9</td><td>IT9</td><td>IT11</td><td>IT11</td></tr>
</table>

* Where the tip cylinder (blank diameter) is used as a datum for a checking instrument.
† The IT values are the internationally agreed fundamental tolerances of BS 4500 'ISO limits and fits', Part 1 'General, tolerances and deviations' to which reference should be made.
‡ Applies to grades 1, 2 and 3 only.

. . . .

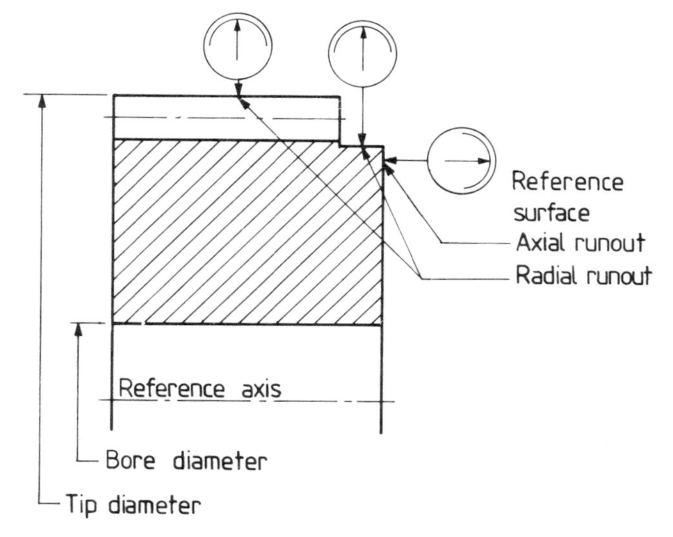

Figure 3. Radial and axial run-out (full indicator movement)

NOTE. In the case of double helical gears it may be desirable to have reference bands on each helix

Pitch measurement

Dual flank testing is sometimes used for pitch measurement but in fact it only measures eccentricity, and not true pitch error, since it is possible that some pitch errors do not produce a variation in centre distance. Moreover, the dual flank test inevitably measures the combined effect of both flanks without any means of determining the contribution from each set. The pitch error of the two sets of flanks may be quite different, so this is a serious drawback. Pitch errors can be measured either by the absolute indexing method or by the successive adjacent pitch method.

Indexing method
In this method the gear is indexed through a successive angular interval of $360/T$ degrees and the positional error of each tooth is measured relative to some arbitrary datum flank.

Adjacent pitch method
In this method adjacent pitch errors are measured relative to an arbitrary datum pitch. In general, the datum pitch setting will not be exactly equal to the theoretically correct figure, and when the errors are summed, a closing error is obtained.

The usual method of correcting for this closing error is to tabulate the errors and subtract $1/T$ times the closing error from each reading

This method suffers from the danger of rounding errors in determining the value of $1/T$ times the closing error and is very time-consuming. A simple graphical approach is preferable. In the standard, tolerances of pitch are given from the formulas, or from the graph it is possible to determine the tolerances of adjacent pitch and accumulative pitch.

Tooth profile measurement

An attempt to measure involute profile errors is sometimes made by successive cordal measurement at different tooth heights. This method is unsatisfactory because again it is a measure of both profiles of the teeth and again it is not measuring relative to the reference diameter.

The only satisfactory method is to use a specifically designed gear-measuring instrument in which the gear is made to rotate, and simultaneously a small displace-

....

Table 3. Limits of tolerance on pitch

Grade	Limits of tolerance (in micrometres)
1	$0.25\sqrt{l} + 0.63$
2	$0.4\sqrt{l} + 1.0$
3	$0.63\sqrt{l} + 1.6$
4	$1.0\sqrt{l} + 2.5$
5	$1.6\sqrt{l} + 4.0$
6	$2.5\sqrt{l} + 6.3$
7	$3.55\sqrt{l} + 9.0$
8	$5.0\sqrt{l} + 12.5$
9	$7.1\sqrt{l} + 18.0$
10	$10.0\sqrt{l} + 25.0$
11	$14.0\sqrt{l} + 35.5$
12	$20.0\sqrt{l} + 50.0$

Where l = any selected length of arc (in mm) less than $\pi d/2$.

NOTE. The limits of tolerance for grades 3 to 12 are shown graphically in figure 4 in order to assist specification of the appropriate grade.

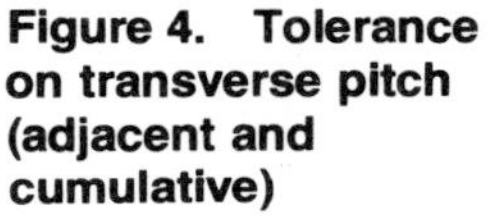
Figure 4. Tolerance on transverse pitch (adjacent and cumulative)

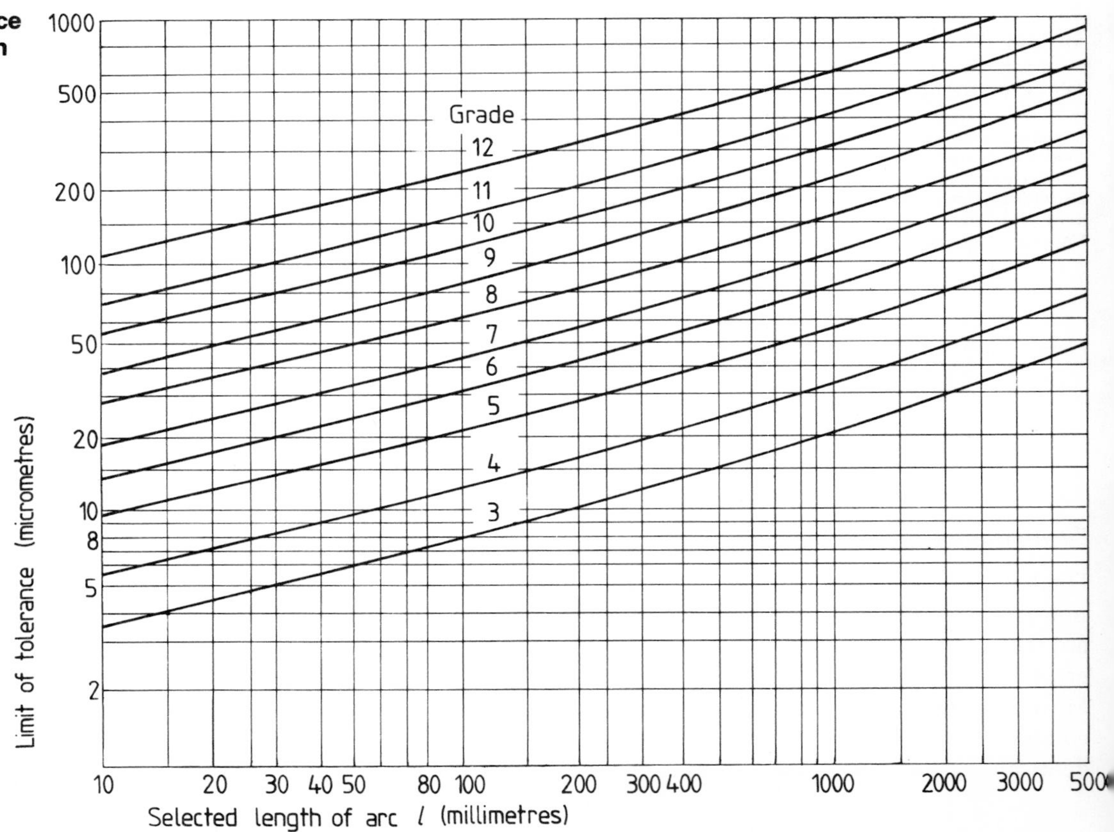

ment transducer is made to move along the base tangent. The relative movements are arranged by various means so that a perfect involute profile would produce a constant reading from the transducer. A profile error is sometimes represented as a departure from a straight line. Tolerances on tooth profile are given in the standard.

. . . .

4.4 Tolerance on tooth profile. The limits of tolerance on tooth profile shall be in accordance with the requirements of table 4. The values are based on a tolerance factor ϕ_f (see figures 6 and 7):

$$\phi_f = m_n + 0.1\sqrt{d}$$

where m_n = normal module
d = reference circle diameter expressed in millimetres.

Table 4. Limits of tolerance on tooth profile

Grade	Profile tolerance (in micrometres)
1	$0.063\ \phi_f + 2$
2	$0.10\ \phi_f + 2.5$
3	$0.16\ \phi_f + 3.15$
4	$0.25\ \phi_f + 4.0$
5	$0.40\ \phi_f + 5.0$
6	$0.63\ \phi_f + 6.3$
7	$1.0\ \phi_f + 8.0$
8	$1.6\ \phi_f + 10.0$
9	$2.5\ \phi_f + 16.0$
10	$4.0\ \phi_f + 25.0$
11	$6.3\ \phi_f + 40.0$
12	$10.0\ \phi_f + 63.0$

NOTE 1. Tolerance factors and limits of tolerance for grades 3 to 12 are shown graphically in figures 6 and 7.
NOTE 2. The disposition of the tolerance is indicated in figure 5*a*, the amount being determined by the particular functional requirement.

Figure 6. Tolerance factor ϕ_f (clause **4.4**)

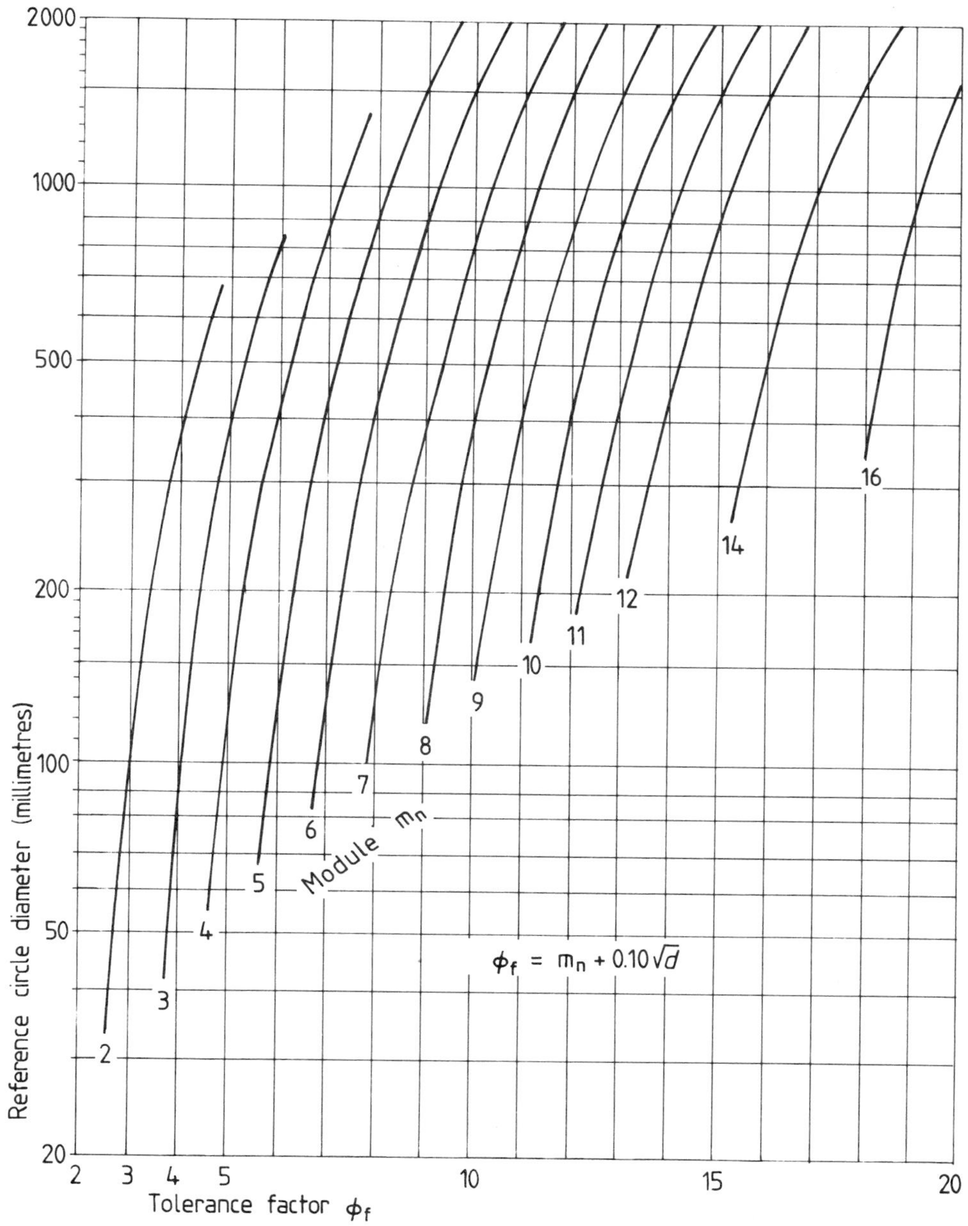

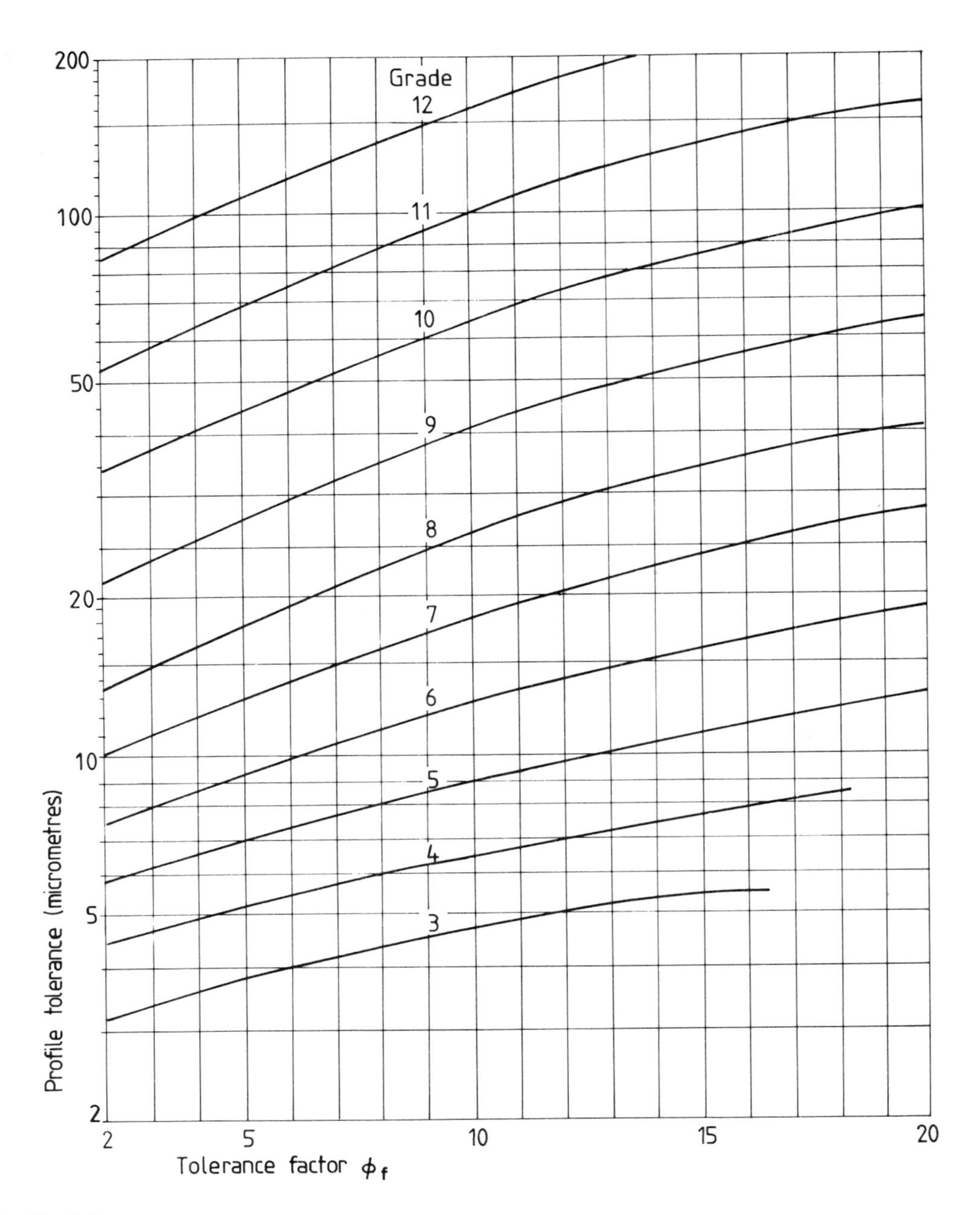

Figure 7. Profile tolerance

The limits of tolerance are given in relation to the design profile as shown in figure 5*a* which illustrates the tolerance zone bounded by the positive limit A and the negative limit B.

The surface irregularities of geometrical form of the actual profile, which constitute departures from the design profile of any one flank, shall be contained within the tolerance zone described by the parallel curves A and B.

In most gear applications positive departures from the design profile should not occur outside the central third of the working depth. The departures shall be controlled as indicated in figure 5*b*.

. . . .

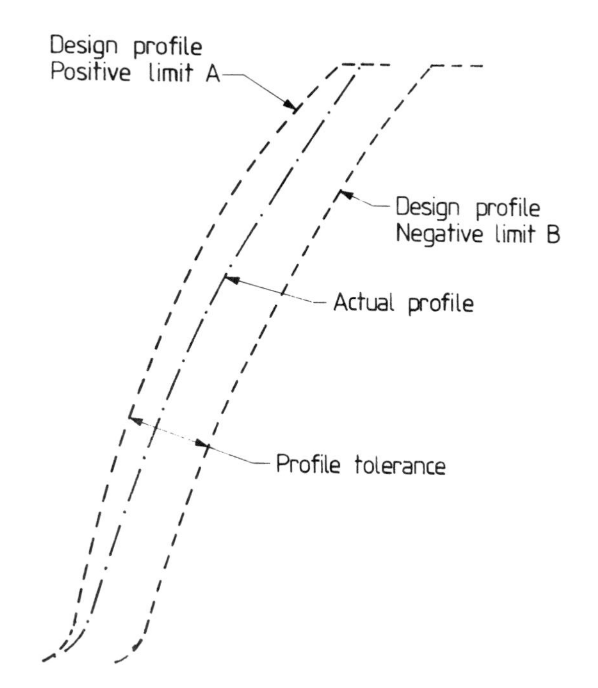

Figure 5a. Tolerance zone of tooth profile error

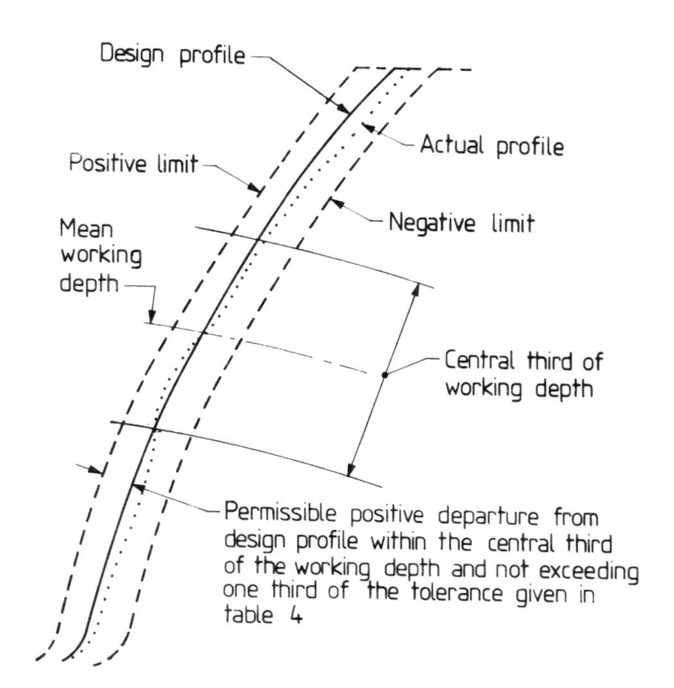

Figure 5b. Control of positive departures from design profile

Tooth alignment measurement (lead error)

The only convenient method of measuring alignment, or lead error, is to use a specifically designed lead-measuring instrument. The gear is made to rotate and a small displacement transducer measures the deviation from the correct lead as it moves accurately along the gear. The rotational and axial movements must be related by a generating mechanism set to give a theoretically correct relationship.

Sometimes facilities for both involute and lead measurement are provided on a combined measuring instrument. It is preferable to specify the lead of a gear rather than the helix angle since lead is independent of diameter. A note of caution should be made regarding the plane in which the lead error is measured.

Normally it is measured normal to the helix, but sometimes it is measured in the transverse plane, and sometimes in the axial plane. The conversion from one system to another is simple but can easily be overlooked.

Tolerances on tooth alignment or lead error are given in the standard. The tolerances are related to a proportion or to the whole of the face width of the gear.

Table 5. Limits of tolerance on tooth alignment

Grade	Tolerance on alignment (in micrometres)
1	$0.315\sqrt{b} + 1.6$
2	$0.40\sqrt{b} + 2$
3	$0.5\sqrt{b} + 2.5$
4	$0.63\sqrt{b} + 3.15$
5	$0.80\sqrt{b} + 4.0$
6	$1.0\sqrt{b} + 5.0$
7	$1.25\sqrt{b} + 6.3$
8	$2.0\sqrt{b} + 10.0$
9	$3.15\sqrt{b} + 16.0$
10	$5.0\sqrt{b} + 25.0$
11	$8.0\sqrt{b} + 40.0$
12	$12.5\sqrt{b} + 63.0$

NOTE 1. The limits of tolerance for grades 3 to 12 are shown graphically in figure 8.
NOTE 2. In both spur and helical gears, errors are measured as departures from nominal position in the transverse plane.
NOTE 3. In the case of double helical gears, the face width b applies to one helix.
NOTE 4. There are only limited data available on which to base tolerances for gears of greater face width than 150 mm, for which (as with all gears) the ultimate test is accuracy of meshing.

. . . .

4.5 Tolerance on tooth alignment. The limits of tolerance on tooth alignment shall be in accordance with the requirements given in table 5. Tolerances are related to a proportion or to the whole of the face width b (in mm) of the gear and apply up to a maximum of b = 150 mm. When it is necessary to specify accuracy of meshing, refer to **4.11** (below).

4.11 Accuracy of meshing. Accuracy of meshing shall be determined by contact marking under the following conditions (except in the case of master gears, which are subject to special agreement between the purchaser and the manufacturer).

The teeth of either the pinion or wheel, as appropriate, shall be coated with a thin film of toolmaker's blue. The gear pair shall then be accurately and suitably mounted (under just sufficient pressure to ensure contact between the teeth) and rotated slowly.

The accuracy of meshing shall be deemed satisfactory if the contact marking is not less than the following total percentages of tooth flank area:

Grades 3, 4 and 5. At least 40 % of the working depth for 50 % of the length and at least 20 % of the working depth for a further 40 % of the length.

Grades 6, 7 and 8. At least 40 % of the working depth for 35 % of the length and at least 20 % of the working depth for a further 35 % of the length.

Grades 9, 10, 11 and 12. At least 40 % of the working depth for 25 % of the length and at least 20 % of the working depth for a further 25 % of the length.

....

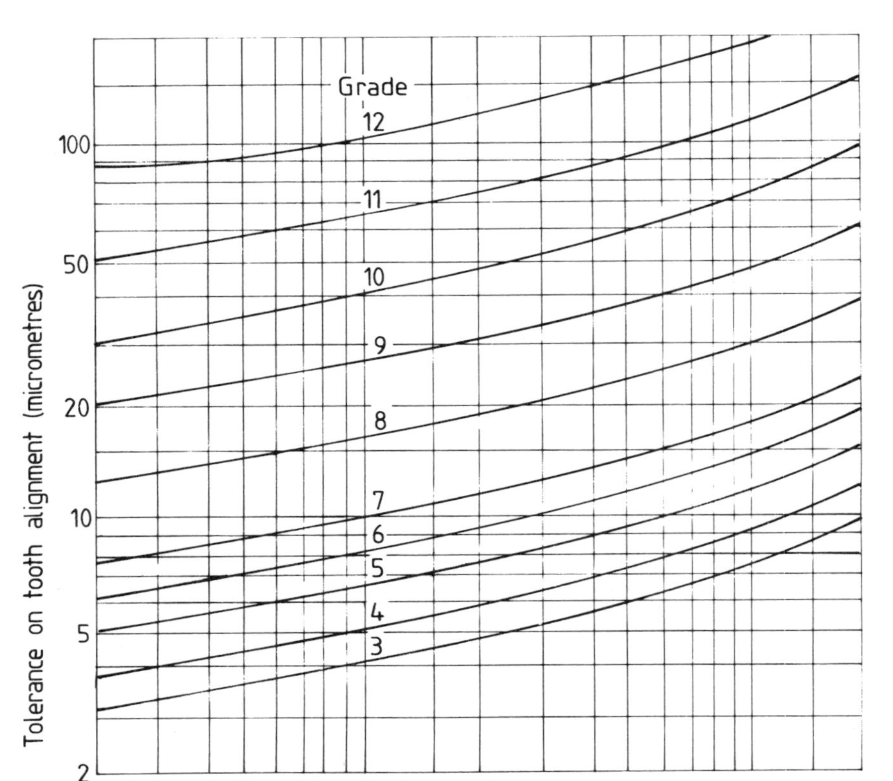

Figure 8. Limits of tolerance on tooth alignment (clause **4.5** and table 5)

Gear tooth size measurement

The method to be adopted for measuring the size of the gear depends upon the degree of backlash control required. If the gear is designed to run with a large backlash tolerance then it is adequate to use a simple hand-held tool. If minimum backlash is required, however, more precise techniques are required

Gears with large backlash tolerance
In this case there is no need to measure size relative to the reference axis as it is sufficient to make direct measurement on one or more teeth by means of a hand-measuring instrument.

There are three possible methods:

(a) Tooth cord thickness measurement at a known depth from the tip using a gear tooth calliper.
(b) Span gauge measurement across the opposite faces of two or more teeth using an ordinary micrometer or vernier calliper.
(c) Measuring the distance over gauging cylinders placed in tooth spaces which are as diametrically opposite as possible. If the gear has an even number of teeth then the gauging cylinders can be placed exactly diametrically opposite.

Gears with minimum backlash

In this case it is essential to measure gear tooth size relative to the reference axis otherwise gear errors such as eccentricity cannot be taken into account and these errors have a significant influence upon backlash.

It is possible to measure the distance between a gauging cylinder placed in successive teeth and the reference axis. However, this is a slow process and it is usually preferable to mesh the gear to be measured with a master gear, i.e. a gear with very small errors.

On fine pitch gears the master gear is sometimes replaced with a master worm. One of the gears is mounted on a floating saddle which is free to move along the line of centres, and the two gears are made to mesh tightly by a small force pressing the floating carriage towards the other gear.

The gear with the fixed axis is then made to rotate and the centre distance variation is noted by measuring and usually recording the movement of the floating carriage. This is known as the dual flank test. A typical graphical result is shown in Figure 8.1.

The straight line represents the position the floating carriage would occupy with zero backlash, for a perfect gear, obtained by initial calibration. The distance between this line and the centre distance variation curve is a measure of the backlash at any angular position.

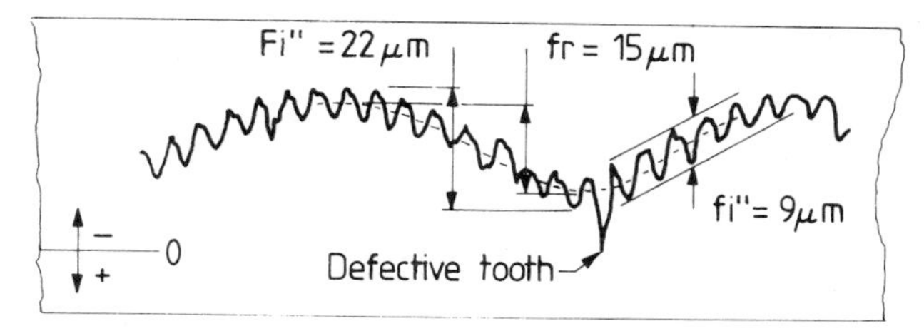

Figure 8.1 *Graphical result of tooth measurement*

If the minimum distance between the line and the curve is x, then the minimum backlash is $2x \tan \psi_n$, and similarly for the backlash at any angular position of the gear. The tolerances on tooth thickness are given in the standard. The standard also gives the dual flank composite error tolerances both for the tooth to tooth and the total composite.

. . . .

4.7 Tolerance on tooth thickness. The tooth thickness tolerance shall be determined as multiples of the adjacent pitch error tolerance, as shown relative to each grade in the curves of figure 4.

The magnitude of the values selected relative to designed tooth thickness will depend on functional considerations, due regard being give to the magnitude of the pitch error, tooth profile error and radial run-out error, since these features will have a direct effect on variations in tooth thickness around the gear.

4.8 Dual flank composite tolerance. When dual flank testing is applied, the limits of tolerance shall be in accordance with the requirements given in tables 7 and 8. The values derived from these tables represent variations in centre-distance when a product gear is rotated in close mesh with a master gear conforming to the requirements of BS 3696 in Part 1*.

4.8.1 *Tooth-to-tooth composite tolerance.* The limits of tolerance on tooth-to-tooth composite error shall be in accordance with the requirements given in table 7. Tolerances are based on the factor ϕ_p (see **4.6** on page 163).

*BS 3696 'Specification for master gears', Part 1 'Spur and helical gears (metric module)'.

Table 7. Limits of tolerance on tooth-to-tooth composite error

Grade	Tooth-to-tooth composite tolerance (in micrometres)
1	$0.16\ \phi_p + 2.0$
2	$0.224\ \phi_p + 2.8$
3	$0.32\ \phi_p + 4.0$
4	$0.45\ \phi_p + 5.6$
5	$0.63\ \phi_p + 8.0$
6	$0.9\ \phi_p + 11.2$
7	$1.25\ \phi_p + 16.0$
8	$1.8\ \phi_p + 22.4$
9	$2.24\ \phi_p + 28.0$
10	$2.8\ \phi_p + 35.5$
11	$3.55\ \phi_p + 45.0$
12	$4.5\ \phi_p + 56.0$

NOTE. The tolerance factor (ϕ_p) and limits of tolerance for grades 3 to 12 are shown graphically in figure 11.

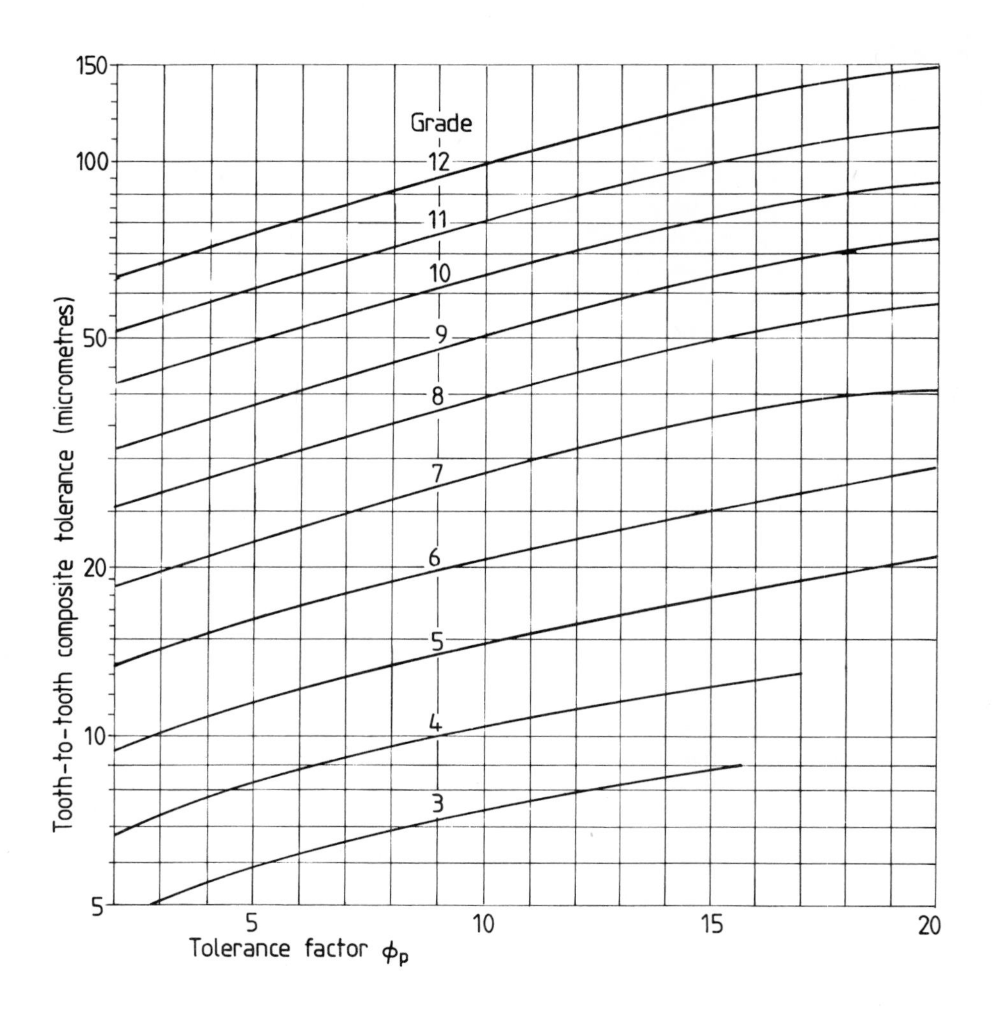

Figure 11. Tooth-to-tooth composite tolerance

4.8.2 *Total composite tolerance.* The limits of tolerance on total composite error shall be in accordance with the requirements given in table 8. Tolerances are based on the factor ϕ_p (see **4.6** on page 163).

Table 8. Limits of tolerance on total composite error

Grade	Total composite tolerance (in micrometres)
1	$0.315\ \phi_p + 4.0$
2	$0.5\ \phi_p + 6.3$
3	$0.8\ \phi_p + 10$
4	$1.25\ \phi_p + 16.0$
5	$2.0\ \phi_p + 25.0$
6	$3.15\ \phi_p + 40.0$
7	$4.5\ \phi_p + 56.0$
8	$5.6\ \phi_p + 71.0$
9	$7.1\ \phi_p + 90.0$
10	$9.0\ \phi_p + 112.0$
11	$11.2\ \phi_p + 140.0$
12	$14.0\ \phi_p + 180.0$

NOTE. The tolerance factor (ϕ_p) and limits of tolerance for grades 3 to 12 are shown graphically in figure 12.

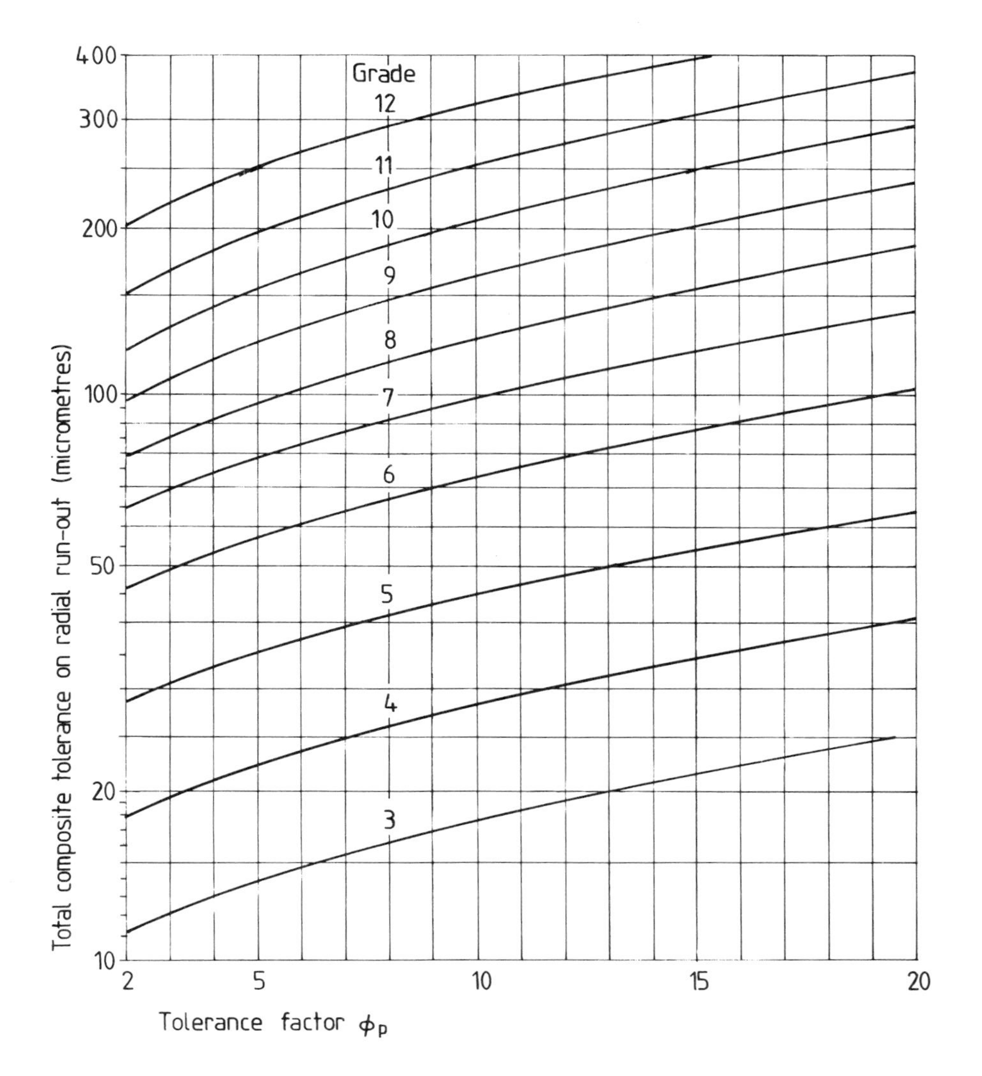

Figure 12. Total composite tolerance on radial run-out

. . . .

4.6 Tolerance on radial run-out of teeth. The limits of tolerance on radial run-out of teeth shall be in accordance with the requirements given in table 6. Tolerances are based on the factor ϕ_p:

$$\phi_p = m_n + 0.25\sqrt{d}$$

where m_n = normal module,
d = reference circle diameter in millimetres.

. . . .

Table 6. Limits of tolerance on radial run-out of teeth

Grade	Tolerance on radial run-out (in micrometres)
1	$0.224\ \phi_p + 2.8$
2	$0.355\ \phi_p + 4.5$
3	$0.56\ \phi_p + 7.1$
4	$0.90\ \phi_p + 11.2$
5	$1.40\ \phi_p + 18.0$
6	$2.24\ \phi_p + 28.0$
7	$3.15\ \phi_p + 40.0$
8	$4.0\ \phi_p + 50.0$
9	$5.0\ \phi_p + 63.0$
10	$6.3\ \phi_p + 80.0$
11	$8.0\ \phi_p + 100.0$
12	$10.0\ \phi_p + 125.0$

NOTE. Tolerance factors and limits of tolerances for grades 3 to 12 are shown graphically in figures 9 and 10 (not included in this manual).

Composite error measurement

It will be appreciated from the foregoing notes that to check a gear thoroughly in this way is very time-consuming. This is particularly true if it is necessary to measure the profile and lead errors of several teeth and on several parts of the tooth.

For example, to obtain a complete picture of the tooth surface a complete grid of profile and lead checks must be made (see Figure 8.2). In addition, it is usually considered that at least four teeth at approximately 90 ° spacing should be checked.

Because of the time needed to check in this way it is worthwhile to consider the alternative composite method of checking, in which the gear to be tested is meshed with a master gear.

In this way a gear can be checked at a rate of several teeth per second and the whole surface of the tooth has the opportunity to make contact with the master gear. The dual flank composite method has already been mentioned for the measurement of size and eccentricity.

It is natural that the method should be extended to attempt to measure other types of errors such as profile and pitch.

It is therefore common to give tolerance of tooth-to-tooth and total composite error in an attempt to control profile, pitch and run-out. This method is often quite adequate for low or average quality gears, but there are difficulties with this method which are not always appreciated.

Firstly, the tooth-to-tooth error is a crude measurement of the profile since it is impossible to separate the effect of the two sets of profiles in contact. Secondly, while this method undoubtedly measures eccentricity and backlash it is not a reliable method for measuring pitch errors as previously stated. Not only because it is impossible to separate the effects of the two flanks, but also because some types of pitch errors may not produce a change in centre-distance.

Because of these difficulties, attention has turned during the last few years to single flank composite methods of checking. It retains the advantages of speed of checking, and covering the whole meshing surface of the gear, but does not suffer from the difficulties of interpretation of the dual flank method.

In the single flank method it is possible to directly measure all pitch and eccentricity errors and also to measure the profile errors of the active part of the profile. When tests are made on the single flank tester, the gear is meshed with a master gear, at fixed centre-distance, with backlash, so that the contact is made on one set of flanks only (see Figure 8.3). A continuous measurement is then made of the angular position of the driven gear relative to where it should be with perfect gear pair, as the driving gear rotates at uniform rate. This is often called the transmission error of the gear pair (see Figure 8.4).

This method is becoming increasingly important, not

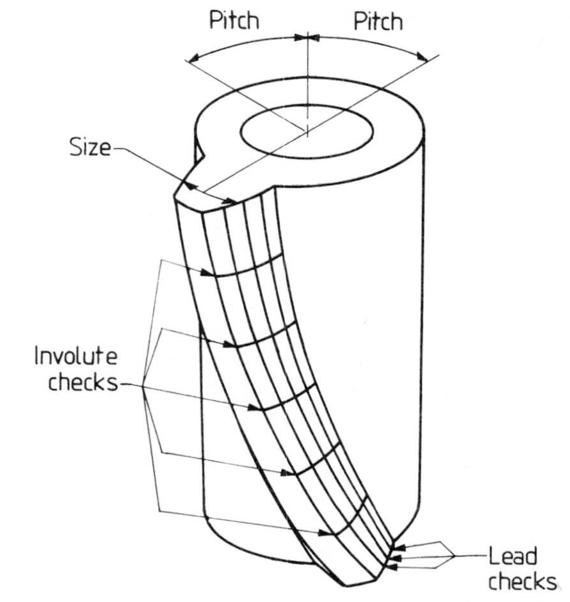

Figure 8.2 *Complete grid measurement*

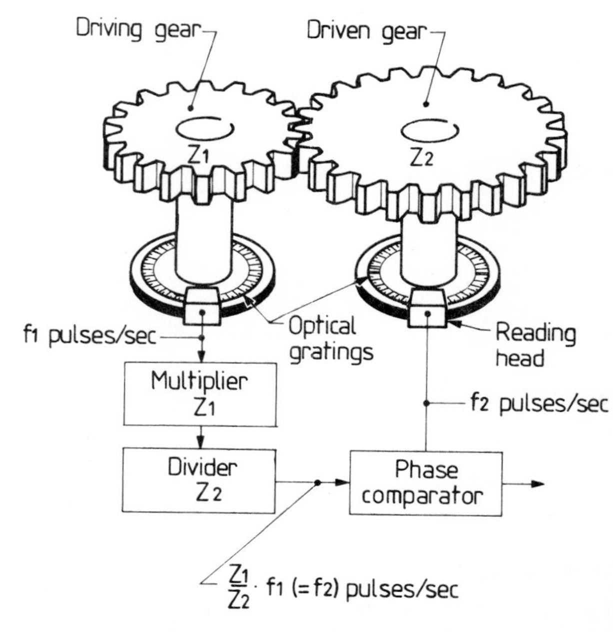

Figure 8.3 *Principle of operation of the single flank tester*

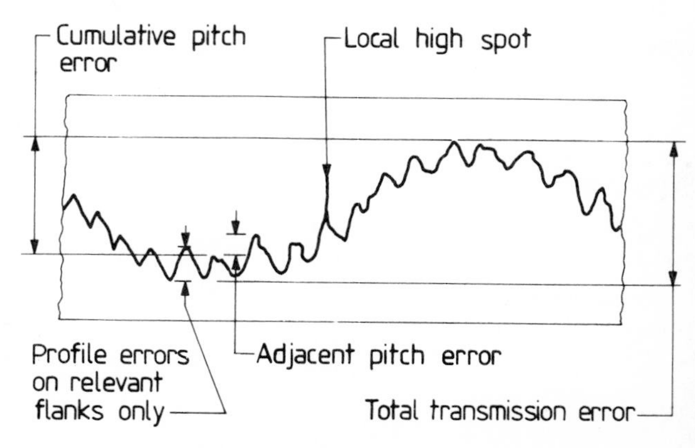

Figure 8.4 *Typical single flank error graph*

only for the reasons outlined above, but also because it offers quantitative measurement of errors in gear types which have hitherto been inspected only by marking tests. These include bevel gears, worm and wheels, and hypoid gears. Single flank composite tolerances are specified in the standard.

. . . .

4.9 Single flank composite tolerances. When single flanking testing is applied, the limits of tolerance shall be in accordance with the requirements given in **4.9.1** and **4.9.2**. The values derived represent errors in angular transmission when a product gear is rotated in mesh with a master gear conforming to BS 3696 : Part 1*.

*BS 3696 'Specification for master gears', Part 1 'Spur and helical gears'.

4.9.1 *Tooth-to-tooth composite tolerance.* For all grades, the tolerance is derived from: single pitch tolerance + tooth profile tolerance.

4.9.2 *Total composite tolerances.* For all grades, the tolerance is derived from: maximum cumulative pitch tolerance† + tooth profile tolerance.

. . . .

†Where the maximum cumulative pitch tolerance is derived from the value of $l = \pi d/2$.

Other tests

Surface texture control

When surface texture is to be controlled this should be applied in metric R_a values in accordance with BS 1134 'Assessment of surface texture' (see Chapter 10).

Accuracy of meshing

Accuracy of meshing should be determined by contact markings under the following conditions. The tooth of either the pinion or the wheel, as appropriate, should be coated with a thin film of toolmakers blue. The gear pair should then be accurately and suitably mounted, under just sufficient pressure to ensure contact between the teeth, and rotated slowly. The accuracy of meshing shall be deemed satisfactory if the contact marking is not less than the total percentage of the tooth flank area, as specified in the standard.

Information to be given on the drawing

Component drawings for spur and helical gears should be in accordance with BS 308 'Engineering drawing practice' and should contain data which is relevant to the finished tooth form dimensions, and accuracy of the gear.

This information shall be given in tables on the drawing.

Section one

Manufacturing and auxiliary data – all gears
Number of teeth
Normal module
Basic rack form
Axial pitch
Tooth profile modification
Addendum modification
Reference circle diameter
Helix angle reference circle
Tooth thickness reference circle
Grade of gear
Drawing number of mating gear
Working centre distance
Backlash

Single helical gears
Hand lead of helix angle

Double helical gears
Hand in relation to specific part of face width
Lead of tooth helix

Section two

Checking data
A table containing inspection data should also be included with the drawing. When inserting information in this table, care should be exercised to avoid conflict between requirements of accuracy for individual elements, dual flank testing, and single flank testing requirements.

Section three

Supplementary data
It will sometimes be necessary to add data to meet particular design, manufacturing and inspection requirements, or limitations. This should be added at the discretion of the purchaser and with the agreement of the manufacturer.

Section four

Other information
Dimensions, limits of tolerance on such dimensions, material, heat treatment, hardness, case depth, surface texture, protective finish, scale, etc., should be given on the drawing in conformity with BS 308.

Standards publications referred to in this chapter

BS 308 Engineering drawing practice
BS 436 Spur and helical gears
Part 2 Basic rack form, modules and accuracy (1 to 50 metric module)
BS 1134 Assessment of surface texture
BS 3696 Specification for master gears
Part 1 Spur and helical gears

9 Machine tool metrology

Introduction

Tests on machine tools are used to determine the accuracy, position and relative movements of elements which may affect any work produced on them. Machine tool manufacturers use many different and often complex tests, but for the purchaser a number of standardized tests are available. These tests may be used after installation, major maintenance, and reconditioning, or as a periodic check.

British Standards

Tests are covered by two British Standards, namely BS 3800 'Methods for testing the accuracy of machine tools' and BS 4656 'Accuracy of machine tools and methods of tests'. These two standards make use of static tests only, including some practical tests (i.e. the machining of test pieces). Dynamic testing (i.e. the determination of alignment accuracy under dynamic loading conditions) is not included since such tests are still subject to research.

BS 3800, which outlines the basic principles, describes methods of testing the accuracy of machine tools by means of geometrical and practical tests. It provides information on definitions, testing methods, use of checking instruments and tolerances, and describes preliminary checking operations and geometrical checks, as well as some special checks.

BS 4656, which covers the application of the tests to various types of machine tools, makes extensive reference to BS 3800, and specifies both geometrical and practical tests. (Each type of machine tool is covered in a separate part of BS 4656 and these are listed at the end of this chapter.)

The following extract from BS 3800 includes the clauses 'general considerations', 'preliminary checking' and 'practical tests'. The only material from the 'geometrical checks' clause to be included is that needed to explain the five tests selected from various parts of BS 4656. The five give examples of tests for straightness, flatness, parallelism, squareness and rotation. It should be noted that both BS 3800 and BS 4656 give considerably more checks and tests than can be selected for this manual and, if required, reference should be made to the standards. Each part of BS 4656 also gives details of the application of the tests methods which are not included in this publication.

BS 3800

Methods for testing the accuracy of machine tools

. . . .

Methods of test

. . . .

2. General considerations

2.1 Definitions relating to geometrical checks. A distinction should be made between geometrical definitions and those designated in this standard as metrological definitions.

geometrical definitions are abstract and relate only to imaginary lines and surfaces. From this it follows that geometrical definitions sometimes cannot be applied in practice. They take no account of the realities of construction or the possibility of checking.

metrological definitions are concrete, as they take account of lines and surfaces accessible to measurement. They cover in a single result all micro- and macro-geometrical errors. They allow a result to be reached covering all causes of error, without distinguishing them from one another. This distinction should be left to the manufacturers.

Nevertheless, in some cases, geometrical definitions (e.g. definitions of out-of-true running, periodical axial slip, etc.) have been retained in the standard, in order to eliminate any confusion and to clarify the language used, but, when describing testing methods, measuring instruments and tolerances, metrological definitions are taken as a basis.

2.2 Testing methods and use of checking instruments. During the testing of a machine tool, if the methods of measurement only allow verification that the tolerances are not exceeded (e.g. limit gauges) or if the actual deviations could only be determined by high precision measurements for which a great amount of time would be required, it is sufficient, instead of measuring, to ensure that the limits of tolerance are not exceeded.

It should be emphasized that errors of measurement due to the instruments, as well as to the methods used, are to be taken into consideration during the tests. The measuring instrument should not give any error of measurement exceeding a known fraction of tolerance to be verified. The accuracy of the devices used being very variable from one laboratory to another, a calibration sheet should be furnished with each instrument.

Testing operations should be protected from draughts and from disturbing heat radiations (sunlight, electric lamps too close, etc.) and the temperature of the measuring instruments should be stabilized before measuring. The machine itself should be suitably protected from the effects of external heat.

A given test should preferably be repeated, the result of the test being obtained by taking the average of the measurements. However, the various measurements should not show too great deviations from one another. If they do, the cause should be looked for either in the method or the checking instrument or the machine tool itself.

2.3 Tolerances

2.3.1 *Tolerances on measurements when testing machine tools.* Tolerances, which limit deviations to values which are not to be exceeded, relate to the sizes, forms, positions and movements which are essential to the accuracy of working and to the mounting of tools, important components and accessories.

There are also tolerances which apply only to test pieces.

2.3.1.1 *Units of measurement, measuring ranges.* When establishing tolerances, it is necessary to indicate:

(a) the unit of measurement used;
(b) the reference base and the value of the tolerance and its location in relation to the reference base;
(c) the range over which measurement is made.

The tolerance and the measuring range should be expressed in the same unit system. Tolerances, particularly tolerances on sizes, should be indicated only when it is impossible to define them by simple reference to British Standards for machine tool elements. Those relating to angles should be expressed either in units of angle: degree, minute, second or as tangents (microns or millimetres per metre for countries using the metric system or inch per 10 in or inch per foot for countries using the inch-foot system).

When the tolerance is known for a given range, the tolerance for another range comparable to the first one is determined by means of the law of proportionality. For ranges greatly different from the reference range, the law of proportionality cannot be applied: tolerances should be wider for small ranges and smaller for large ranges than those which would result from the application of this law.

2.3.1.2 *Rules concerning tolerances.* Tolerances include errors inherent in the measuring instruments and test methods used. Errors of measurement should consequently be taken into account in the permitted tolerances (see **2.2**).

Example.
Tolerance of run-out: $X\mu$
Inaccuracy of instruments, errors of measurement: $Y\mu$
Maximum permissible difference in the readings during the test: $(X - Y)\mu$

Errors of block gauges, reference discs, etc., inaccuracies arising from comparative laboratory measurements, inaccuracies of form of machine parts used as reference surfaces, including surfaces masked by styli or by support points of measuring instruments should be neglected.

The actual deviation should be the arithmetical mean of several readings taken, neglecting the above causes of error.

Lines or surfaces chosen as *reference bases* should be directly related to the machine tool (e.g. line between centres of a lathe, spindle of a boring machine, slideways of a planing machine, etc.). The direction of the tolerance is defined according to the rules given in **2.3.2.5**.

2.3.2 *Subdivisions of tolerances*

2.3.2.1 *Tolerances applicable to test pieces and to fixed parts of machine tools*

2.3.2.1.1 *Tolerances of dimensions.* The tolerances of dimensions indicated in this standard relate exclusively to the dimensions of test pieces for practical tests and to the fitting dimensions of cutting tools and of checking instruments which may be mounted on the machine tool (spindle taper, turret bores). They are the limits of permissible deviation from the nominal dimensions. They are expressed in length units (e.g. deviations of bearings and bore diameters, for the setting up and the centring of tools.)

For internal and external dimensions of cylindrical and parallelepipedic parts, tolerances should be given in compliance with the rules given in BS 308 'Engineering drawing practice'. The methods of indicating the tolerance by stating the maximum material condition and the deviation or by giving the basic size and the limits in the symbols of BS 1916 'Limits and fits for engineering', are preferred*.

Example. 1.500 / + .0016 or 1.5 H 8

2.3.2.1.2 *Tolerances of form.* Tolerances of form limit the permissible deviations from the theoretical geometric form (e.g. deviations relative to a plane, to a straight line, to a revolving cylinder, to the profile of thread or of tooth). They are expressed in units of length or of angle. Because of the dimensions of the plunger surface or of the support surface, part only of the error of form is detected. Therefore where extreme accuracy is required, the area of the surface covered by the plunger or support should be stated.

In a general way, the plunger surface should be proportional to the precision and to the dimension of the surface to be checked (a surface plate and the table of a heavy planing machine are not checked from the same plunger surface).

NOTE. It should be noted that rules for indicating geometrical tolerances on drawings are given in BS 308 which enable the geometrical accuracy of individual parts to be specified. These rules should be adhered to on manufacturing drawings.

2.3.2.1.3 *Tolerances of position.* Tolerances of position limit the permissible deviations concerning the position of a component relative to a line, to a plane, or to another component of the machine (e.g. deviation of parallelism, of perpendicularity, of alignment, etc.). They are expressed in units of length or angle.

If a tolerance of position is defined by two measurements taken in two different planes, the tolerance should be fixed in each plane when the deviations from those two planes do not affect in the same way the working accuracy of the machine tool.

NOTE 1. When a position is determined in relation to surfaces showing errors of form, these errors of form should be taken into account when fixing the tolerance of position.
NOTE 2. It should be noted that rules for indicating geometrical tolerances on drawings are given in BS 308 which enable the geometrical accuracy of individual parts to be specified. These rules should be adhered to on manufacturing drawings.

2.3.2.1.4 *Rules for the influence of defects of form in determining positional errors.* When positional errors of two surfaces or of two lines (see figure 1, lines *XY* and *ZT*) are being determined, the readings of the measuring instrument automatically include some errors of form. It should be laid down, as a principle, that checking should apply only to the total error, including the errors of form of the two surfaces or of the two lines. Consequently, the tolerance should take into account the tolerance of form of the surfaces involved. (If thought useful,

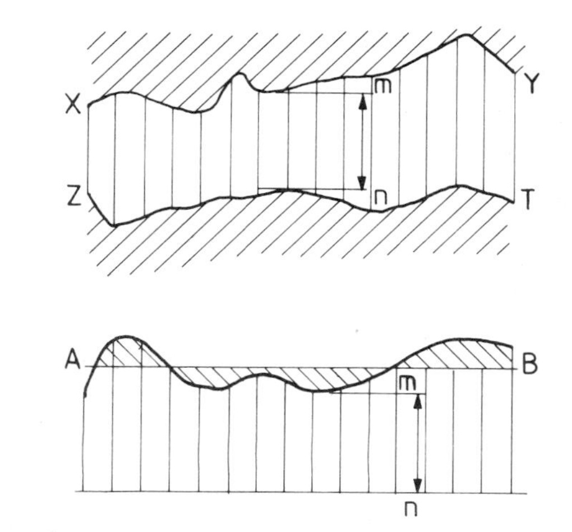

Figure 1.

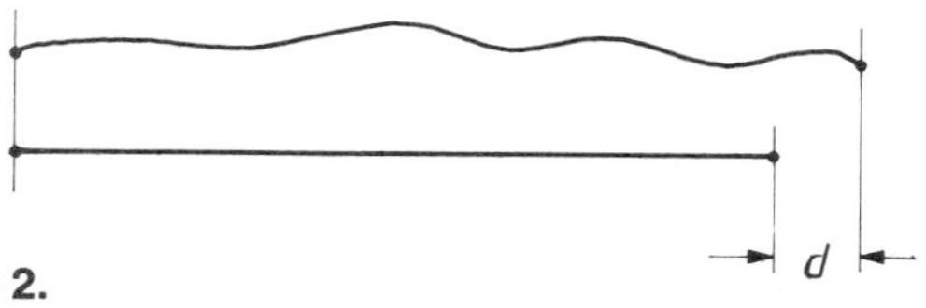

Figure 2.

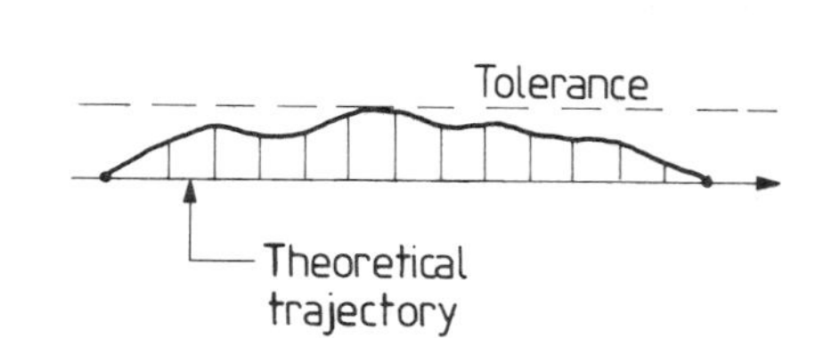

Figure 3.

preliminary checks may ascertain defects of form of lines and of surfaces, of which the relative positions are to be determined.)

When setting out on a graph (see figure 1) the different readings *mn* of the checking instrument, a curve, such as *AB*, is obtained. It is to be accepted, when there is no contradictory stipulation, that the error is to be determined by using, instead of this curve, a line calculated from the minimum squared deviation.

2.3.2.2 *Tolerances applicable to the displacement of a component of a machine tool.*

2.3.2.2.1 *Tolerances of dimensions.* Tolerances of dimensions limit the permissible deviation of the position reached by a point of the moving part from that which it should have reached after moving.

Example 1: Deviation *d*, at the end of the travel, of the position of a lathe cross slide from the position which it should have reached under the action of the lead screw (see figure 2).

Example 2: Angle of rotation of a spindle relative to the angular displacement of a dividing plate coupled to it.

2.3.2.2.2 *Tolerances of form.* These limit the deviation of the actual trajectory of a point relative to the theoretical trajectory (see figure 3). They should be stated in units of length.

* BS 1916 relates to imperial units only. For metric units see BS 4500 'ISO limits and fits', and Chapter 6 of this manual.

2.3.2.2.3 *Tolerances of position.* Tolerances of position limit the permissible deviation between the trajectory of a point on the moving part and the prescribed trajectory (e.g. deviation of parallelism between the trajectory and a straight line or a surface) (see figure 4). They are expressed in units of angle or preferably as successive tangents over a given measurement of length.

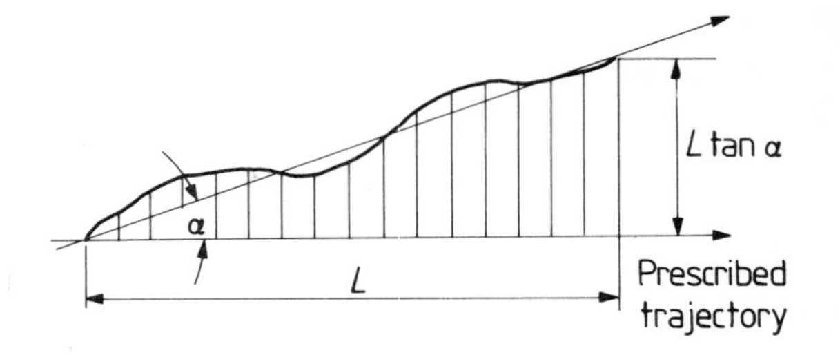

Figure 4.

2.3.2.2.4 *Local tolerances.* Tolerances of form and position are usually relative to the form of position as a whole. e.g. 0.03 per 1000 for a straightness or flatness. It should be observed that checking can show up a deviation (see figure 5) which is not spread over the whole of the form or position, but is concentrated on a short length of the former (e.g. 200 mm). If such defects, seldom met with in practice, are to be avoided, the overall tolerance may be accompanied by a statement of a local tolerance; or it may be simply agreed that the local tolerance, provided that it does not fall below a minimum to be stated (e.g. 0.01 or 0.005 mm) should be proportional to the overall tolerance. In the case under consideration, relating e.g. to straightness, the local error should not in these conditions exceed:

$$\frac{0.03}{1000} \times 200 = 0.006 \text{ mm}$$

If 0.01 mm is accepted as a minimum for any given machine, it is sufficient to check that the local error does not exceed this value.

In practice, local defects are generally imperceptible, as they are covered by the supporting or the feeling surfaces of the measuring instruments. However, when the feeling surfaces are relatively small (plungers of dial gauges or microindicators), the measuring device should be such that the plungers follow a surface of high grade finish (straightedge, test mandrel, etc.). (See measuring device, figure 8.)

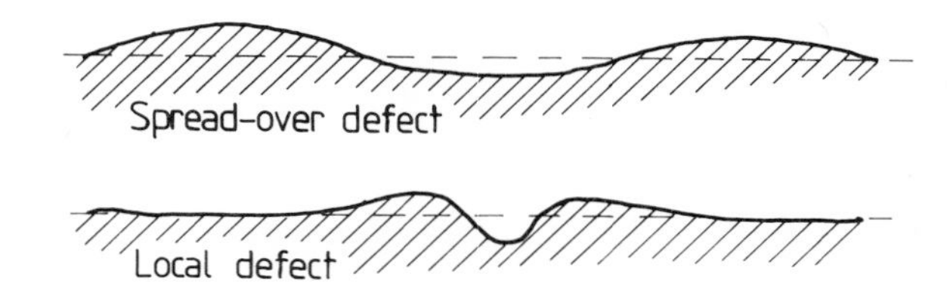

Figure 5.

2.3.2.3 *Cumulative or inclusive tolerances.* The cumulative tolerances are the resultant of several deviations and may be determined by a single measurement, without it being necessary to know each deviation.

Example (see figure 6): The tolerance for the run-out of a shaft is the sum of the tolerance of form (out-of-round of the circumference *ab* on which the stylus is in contact), the tolerance of position (the geometrical axis and the rotating axis of the shaft do not coincide) and the tolerance of out-of-round of the bore of the bearing.

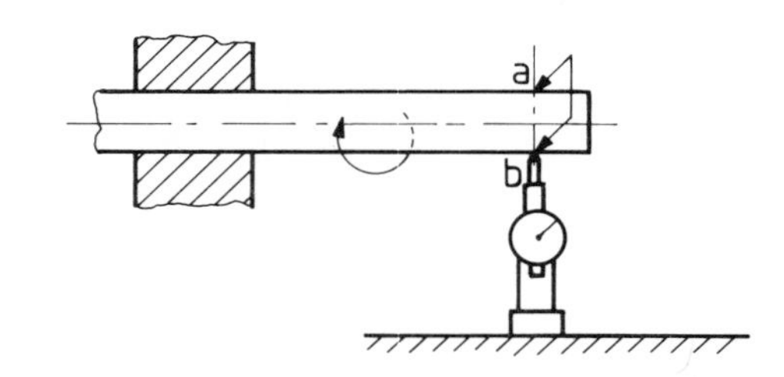

Figure 6.

2.3.2.4 *Symbols and positions of tolerances for relative angular positions of axes, slideways, etc.* When the position of the tolerance in relation to the nominal position is symmetrical, the sign ± may be used. If the position is asymmetrical it should be stated precisely, in words, either in relation to the machine or to one of the components of the machine, or in relation to the operator in his conventional position.

2.3.2.5 *Conventional position of the operator.* For each type of machine a conventional position of the operator is defined. The *front* of a machine is the part which faces the operator. The *right* of a machine is the part which is at his right. The *rear* and the *left* of a machine are the parts opposite to those already defined.

3. Preliminary checking operations

3.1 Installation of the machine before testing. Before proceeding to test a machine tool, it is essential to fix the machine upon suitable foundations and to level it in accordance with the instructions of the manufacturer.

3.1.1 *Levelling.* The preliminary operation of installing the machine involves (see **3.1**) precise levelling and is essentially determined by the particular machine concerned.

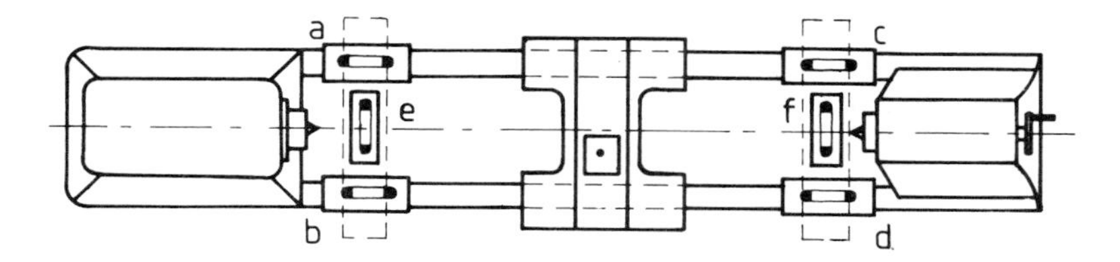

Figure 7.

In the case of a lathe, the plane of the slides (front and rear) is laid horizontally or with a suitable slope. The cross slide is placed in the middle of the bed. When jacks and fixing bolts are used, the extreme ends of the slideways should be placed horizontally and the twisting of the bed should be remedied if necessary. For this purpose, the level is placed in succession (see figure 7) on the longitudinal positions *a*, *b*, *c* and *d*, and the transverse positions *e* and *f*.

After the first installation, checking of the straightness of the slideways (or straightness of the movement of the slide) may be made. It should be noted that this checking is not distinguishable from the setting out of the machine, particularly in the case of large-sized beds. Jacks are often spaced along the bed to effect local corrections as the checking of the slideways progresses.

When installing milling machines, the table of the machine should be set approximately horizontal; the purpose of this is to facilitate the testing operations.

Generally, it is desirable to follow the manufacturer's instructions for the proper setting-out of the machine and for the provision of suitable foundations which, in certain cases, are indispensable.

3.2 Condition of the machine before testing.

3.2.1 *Dismantling of certain elements.* As the tests are carried out, in principle, on a completely finished machine, dismantling of certain elements should only be carried out in exceptional circumstances, in accordance with the instructions of the manufacturer (e.g. dismantling of a grinding machine table in order to check the slideways).

3.2.2 *Temperature conditions of certain elements before testing.* The aim is to check the accuracy of the machine under conditions as near as possible to those of normal functioning as regards lubrication and warming up. During the geometrical and practical tests, elements, e.g. spindles, which are liable to warm up and consequently to change position or shape, should be brought to the correct temperature by running the machine idle in accordance with the conditions of use and the instructions of the manufacturer.

3.2.3 *Functioning and loading.* Geometrical checks are made either when the machine is at a standstill or when it is running idle. When the manufacturer specifies it, e.g. as in the case of heavy-duty machines, the machine should be loaded with one or more test pieces.

4. Practical tests

4.1 Testing. Practical tests should be carried out on pieces, the making of which does not require operations other than those for which the machine has been built. Practical tests to ascertain the precision of a machine tool should be the finishing operation for which the machine has been designed. (It is of primary importance that such tests should be carried out in good faith.)

The number of workpieces or, as the case may be, the number of cuts to be made on a given workpiece, should be such as to make it possible to determine the average precision of working. If necessary, wear on the cutting tool used should be taken into account.

The nature of the workpieces to be made, their dimensions, their material and the degree of accuracy to be obtained and the cutting conditions should be the subject of agreement between the manufacturer and the user.

4.2 Checking of workpieces in practical tests. Checking of workpieces in practical tests should be done by measuring instruments selected for the kind of measurement to be made and the degree of accuracy required.

The tolerances indicated in **2.3.2.1**, particularly in **2.3.2.1.1** and **2.3.2.1.2**, are to be used for these verifications.

4.3 Importance of practical tests. The results of practical tests and geometrical checks can be compared only in so far as these two kinds of tests have the same object. There are cases moreover when, on account of expense or technical difficulties in conducting the tests, the accuracy of a machine is checked only by geometrical checks or only by practical tests.

If the tests by means of geometrical checks and practical tests having the same object do not give the same results, those results obtained by making practical tests should be accepted as the only valid ones.

5. Geometrical checks

5.1 General. For each geometrical check of a given characteristic of shape, position or displacement of lines or surfaces of the machine:

Straightness	see **5.2**
Flatness	see **5.3**
Parallelism, equidistance and coincidence	see **5.4**
Squareness	see **5.5**
Rotation	see **5.6**

a definition*, a method of measurement and the way of determining the tolerance are given.

For each test, at least one method of measurement has been indicated, and only the principles and apparatus used have been shown.

When other methods of measurement are used, their accuracy should be at least equal to the accuracy of those stated in this standard.

NOTE. For the sake of simplicity, the methods of measurement set out in this standard have been chosen systematically from those which employ only the elementary testing instruments most frequently used in engineering workshops, such as straightedges, squares, mandrels, measuring cylinders, spirit levels and dial gauges. It should be observed that other methods, notably those using optical devices, are in fact generally used in machine tool building and in inspection departments. Testing of machine tool parts of large dimensions often requires the use of special devices for convenience and speed.

*See also **2.1**.

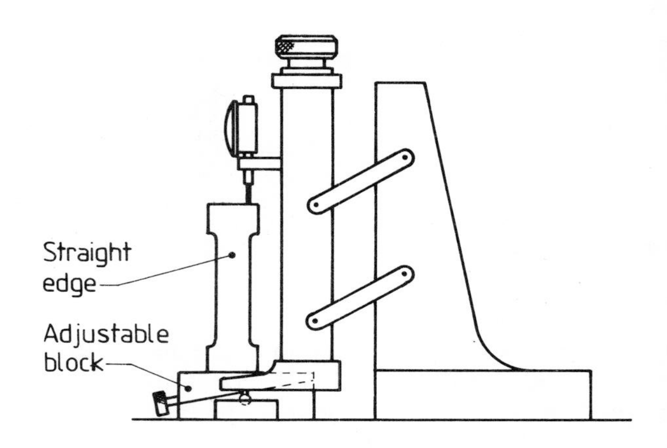

Figure 8.

. . . .

For the purposes of BS 4656, various methods of expressing permissible deviations are employed, each having a particular type of application. The methods are as follows:

---/---	This type of expression applies to deviations of perpendicularity that are ratios.
--- for any length of ---	This type of expression applies to deviations of straightness and parallelism, and is used in fact for local permissible deviations, the measurement length being obligatory.
--- per ---	This type of expression applies to deviations of straightness and parallelism, and is used to recommend a measurement length, but in this case the proportionality rule comes into operation if the measurement length differs from that indicated.

Example of a straightness test

The example on the opposite page is from BS 4656: Part 1 'Lathes, general purpose type'.

Material from BS 3800 cited in the straightness test example.

. . . .

5.2.1.2.2.1 *Spirit-level method.*

(a) *When the line is reasonably horizontal.* The initial straight line of reference is constituted by a straight line *omX*, *o* and *m* being two points on the line to be checked (see figure 11).

The level is placed on *om*, then moved to *mm'*, then to *m'm"*, the distances *om*, *mm'*, being all equal to a value *d*, related to the total length *oA*, which is to be checked (*d* in practice lies between 100 and 500 mm).

Readings of the level on *mm'*, *m'm"* are compared with the reading of the level in its original position *om*. If the level is provided with a regulating device for the bubble, the bubble should be brought to zero in the original position so as to obtain in the operations which follow a direct reading of the

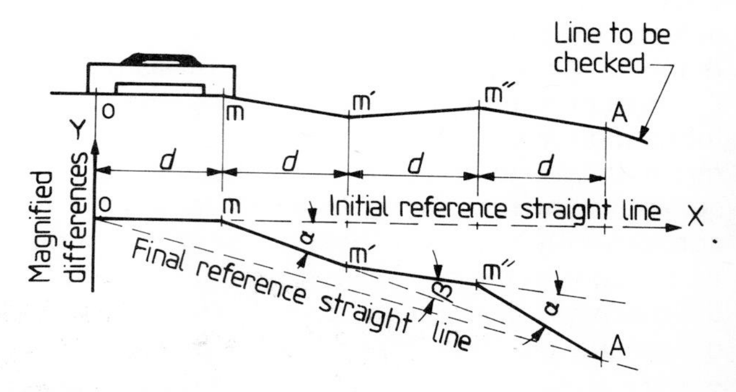

Figure 11.

Part of Table 1. Geometrical tests for centre lathes, general purpose type

All dimensions are in millimetres

Diagram	Object	Permissible deviation			Measuring instruments	Observations and references to the test code BS 3800: 1964
		Precision lathes *$Da \leqslant 500$ †$DC \leqslant 1500$	Other lathes $Da \leqslant 800$	 $800 < Da \leqslant 1600$		
(a)	Checking of straightness of the slideways (a) Longitudinal check: straightness of slideways in the vertical plane	$DC \leqslant 500$ 0.01 (convex) $500 < DC \leqslant 1000$ 0.015 (convex) Local‡ tolerance: 0.005 for any length of 250 $1000 < DC \leqslant 1500$ 0.020 (convex) Local tolerance: 0.005 for any length of 250	$DC \leqslant 500$ 0.01 (convex) $500 < DC \leqslant 1000$ 0.020 (convex) Local‡ tolerance: 0.0075 for any length of 250 $DC > 1000$ 0.02 + 0.01 for each additional 1000 (convex) Local tolerance: 0.015 for any length of 500	$DC \leqslant 500$ 0.015 (convex) $500 < DC \leqslant 1000$ 0.030 (convex) Local‡ tolerance: 0.010 for any length of 250 $DC > 1000$ 0.03 + 0.02 for each additional 1000 (convex) Local tolerance: 0.020 for any length of 500	Precision levels, optical or other methods	**3.11**§, **5.2.1.2.2.1** and **5.2.1.2.2.2** Measurements shall be made at positions equally distributed throughout the length of the bed. The levels shall be placed on the transverse slide. When the slides are not horizontal, use a special straightedge as mentioned in **5.2.1.2.2.1**(b) §See material selected from BS 3800 above. Other references are reproduced below.

* Da means maximum permissible diameter above the bed.
† DC means distance between centres.
‡ See appendix A (not included in this publication).

positions of the lines mm', $m'm''$, in relation to omX. When the distance oA has been traversed, measurements are taken in the opposite direction Ao using the same points, and the average of the results obtained is calculated. All the information needed to trace the profile of the line $omm'm''A$ is then available.

To eliminate local errors in the course of measurement, the level should not be laid on the lines to be tested over the full length of its base. The base should be hollowed out in its central part, or, if this is not possible, the level should be placed on two gauge blocks of equal thickness or on a support with two feet separated by the distance d.

The feet of the level, or its support as the case may be, as well as the surfaces on which the device is to rest in the course of the test, should be very carefully cleaned.

(b) *When the line is not horizontal.* When a line is inclined, the procedure of the last example may be applied, if a support having an angle α equal to the angle of inclination in relation to the horizontal plane is used (see figure 12).

While checking line AB, the level together with its support should keep a constant orientation (e.g. by means of a guiding straightedge R).

NOTE. The level permits checking the straightness only in the vertical plane; for the checking of a line in a second plane, another method should be used (e.g. taut wire and microscope).

5.2.1.2.2.2 *Optical checks.* Numerous methods using optical devices may be used to verify the straightness of a line. The most general are the autocollimation method (measuring the slope) and

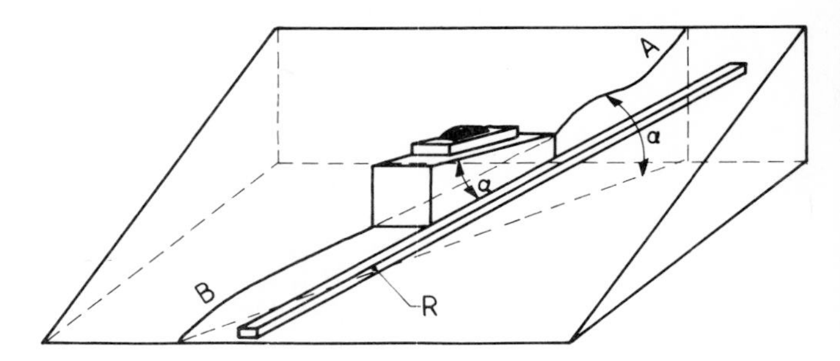

Figure 12.

the method using an alignment telescope (measuring the difference in level).

(a) In the autocollimation method, using a telescope and a microscope mounted coaxially (see figure 13), any rotation of the movable mirror M around a horizontal axis entails a vertical displacement of the image of the reticle in the focal plane; the measurement of this displacement, which may be made with an ocular micrometer, permits the angular deviation α of the mirror holder to be determined.

(b) In the method using an accurate alignment telescope (see figure 14), the measurement of the difference in level a*, corresponding to the distance between the optical axis of the telescope and the mark shown on the target, is read directly on the reticle or by means of the optical micrometer.

. . . .

*i.e. of the deviation a in the measuring plane, whether it is a vertical or a horizontal plane.

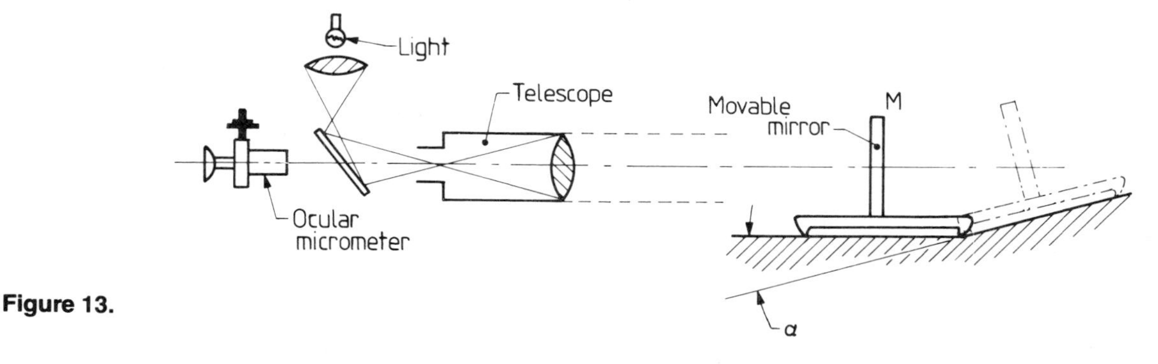

Figure 13.

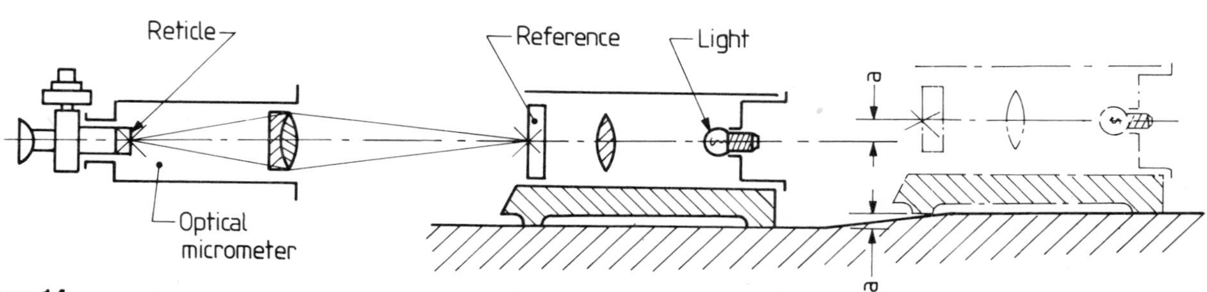

Figure 14.

Example of a flatness test

This example is from BS 4656: Part 10 'Drilling machines, radial type'.

Part of Table 1. Geometrical tests for drilling machines, radial type

All dimensions are in millimetres

Diagram	Object	Permissible deviation	Measuring instruments	Observations and references to test code BS 3800
	Checking of flatness of the base plate	0.1 for a measuring length of 1000 (flat to concave)	Precision level or straightedge and gauge blocks	**5.3.2.2** and **5.3.2.3**

Material from BS 3800 cited in the above example.

. . . .

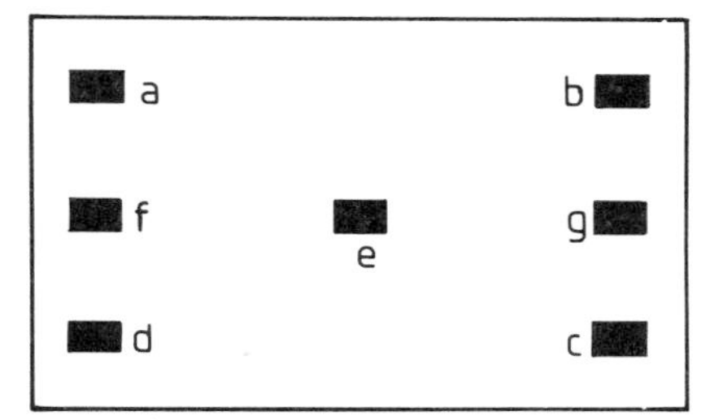

Figure 19.

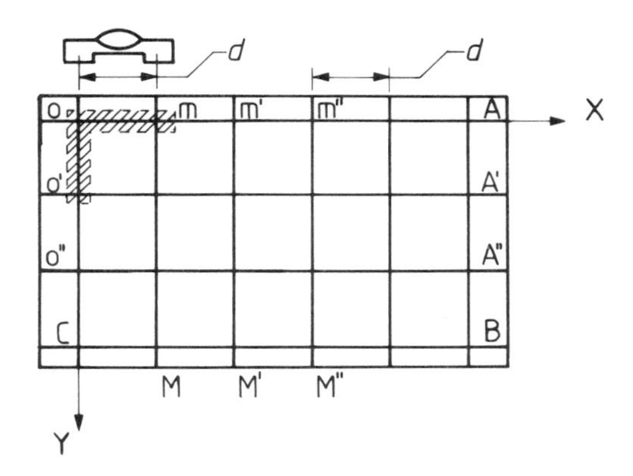

Figure 20.

5.3.2.2 *Checking of flatness by means of a family of straight lines by displacement of a straightedge.* The theoretical plane on which the reference points will be located is first determined. For this purpose, 3 points *a*, *b* and *c* on the surface to be tested are selected as zero marks (see figure 19). Three gauge blocks of equal thickness are then placed on these three points, so that the upper surfaces of the blocks define the reference plane to which the surface is compared. A fourth point *d* lying in the reference plane is then selected in the following manner, using gauge blocks which are adjustable for height; a straightedge is placed on *a* and *c* and an adjustable block is set at a point *e* on the surface and brought into contact with the lower surface of the straightedge. The upper surfaces of the blocks *a*, *b*, *c*, *e*, are therefore all in the same plane. The location of point *d* is then found by placing the straightedge on *b* and *e*; an adjustable block is placed at this point and its upper faces brought into the plane defined by the upper surfaces of the blocks already in position. By placing the straightedge on *a* and *d* and then on *b* and *c*, the locations of all the intermediate points on the surface lying between *a* and *d* and between *b* and *c* may be found. The locations of the points lying between *a* and *b*, *c* and *d*, may be found in the same way (any necessary allowance for sag in the straightedge should be made).

To obtain readings inside the rectangle or square thus defined, it will only be necessary to place at points *f* and *g*, for example, the locations of which will then be known, gauge blocks adjusted to the correct height. The straightedge is placed on these and, with the aid of the gauge blocks, the deviation between the surface to be tested and the straightedge can be measured. It is possible to use an instrument for the testing of straightness as shown, for example, in figure 8.

5.3.2.3 *Checking of flatness by means of a spirit level.* The reference plane is determined by two straight lines *omX* and *oo'Y*, *o*, *m* and *o'* being three points of the surface to be checked (see figure 20).

The lines oX and oY are chosen preferably at right angles and if possible parallel to the sides outlining the surface to be checked. Checking is begun in one of the angles o of the surface and in the direction oX. The profile for each line oA and oC is determined by the method indicated in **5.2.1.2.2.1**. The profile of longitudinal lines $o'A'$, $o''A''$, . . . and CB is determined so as to cover the whole surface.

Supplementary checks may be made following mM, $m'M'$, . . . etc., to check the previous measurements.

When the width of the surface to be checked is important in relation to its length, it is desirable as a cross-check, to take measurements also along diagonals.

. . . .

Example of a parallelism test

This example is from BS 4656: Part 9 'External cylindrical grinding machines with reciprocating table'.

Part of Table 1. Geometrical tests for external cylindrical grinding machines with reciprocating table

All dimensions are in millimetres

Diagram	Object	Permissible deviation	Measuring instruments	Observations and references to the test code BS 3800 : 1964
	Checking of parallelism of the location surfaces for the workhead and tailstock to the longitudinal movement of the table	0.01 up to 1000 For each 1000 increase in length, add: 0.01 Maximum permissible deviation: 0.03 Local tolerance: 0.003 over any length of 300	Dial gauge(s)	**5.4.2.2.2.2** Place a dial gauge(s) on a fixed part of the machine and take measurements successively on the location surfaces for the workhead and the tailstock The table setting carried out during this test shall not be modified for performing tests G 6, G 7 and G 8

Material from BS 3800 cited in the above example.

. . . .

5.4.2.2.2.2 *The plane is not on the moving component itself.* The measuring instrument is attached to the moving component and moved with it by the amount stated; the plunger is at right angles to the surface and slides along it (see figure 33).

If the stylus does not bear directly on the surface (e.g. the edge of a narrow groove), a piece of suitable shape may be used (see figure 34).

5.4.2.2.3 *Parallelism of a trajectory to an axis.* The measuring instrument is fixed to the moving component and is moved with it by the stated amount: the stylus slides over the cylinder or mandrel representing the axis (see figure 35).

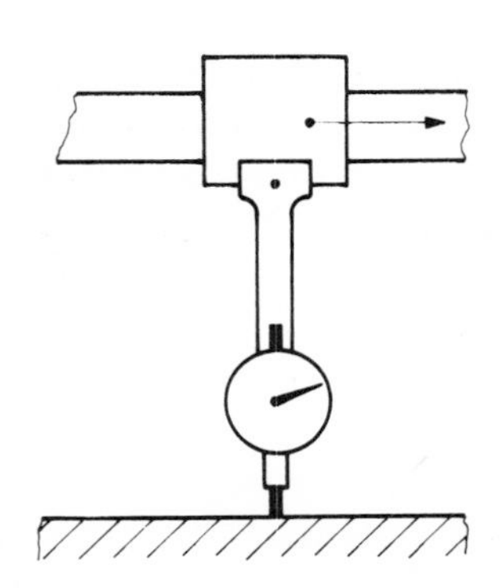

Figure 33.

Unless all planes are of equal importance, the check should be made, if possible, in two perpendicular planes selected as being those most important for the practical use of the machine.

5.4.2.2.4 *Parallelism of a trajectory to the intersection of two planes.* Parallelism between each of the two planes and the trajectory is tested separately, according to **5.4.2.2.2**; the position of the intersecting line is deduced from the position of the planes.

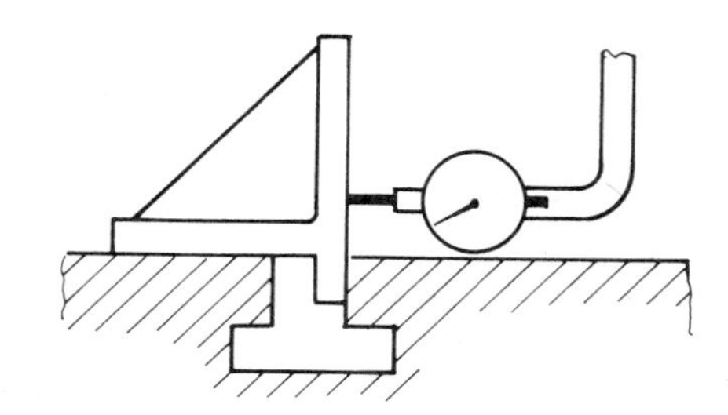

Figure 34.

5.4.2.2.5 *Parallelism between two trajectories.* A dial gauge is attached to one of the moving components of the machine so that its stylus rests on a given point on the other moving part. The two parts are moved together in the same direction by the same amount stated, and the variation in the reading of the measuring instrument is noted (see figure 36).

Unless all planes are of equal importance, this check should be made in two perpendicular planes selected as being those of most importance in the practical use of the machine.

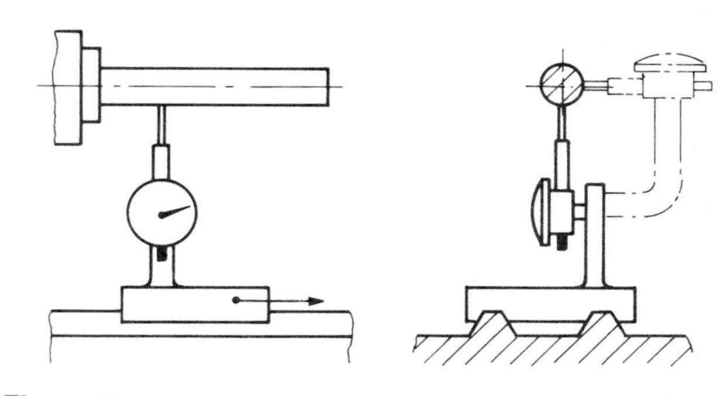

Figure 35.

5.4.2.3 *Tolerance.* Tolerance on parallelism of movement is the permissible variation in the shortest distance between the trajectory of a given point on the moving part and a plane, a straight line or another trajectory within a stated length.

For the method of stating the tolerance, see **5.4.1.3**.

5.4.3 *Equidistance*

5.4.3.1 *Definition.*

Equidistance relates to the distance between the axes and a reference plane. There is equidistance when the plane passing through the axes is parallel to the reference plane. The axes may be different axes or the same axis occupying different positions after pivoting.

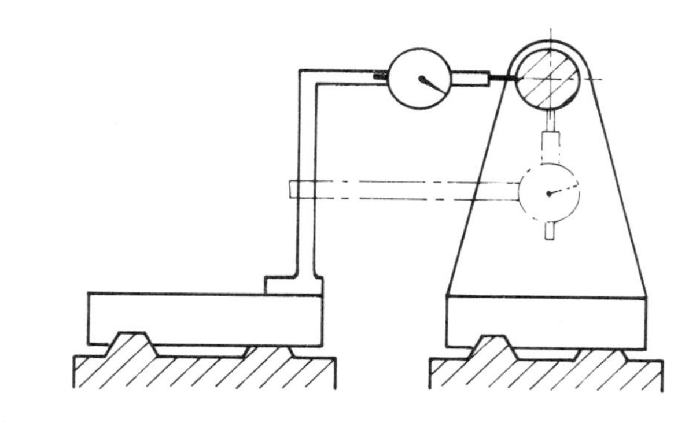

Figure 36.

5.4.3.2 *Methods of measurement.*

5.4.3.2.1 *General.* The problem is identical with that of parallelism between a plane passing through the axes and a reference plane.

Tests for equidistance of two axes, or of a rotating axis, from a plane are, in effect, checks of parallelism (see **5.4.1.2.4**). A test should first be made to check that the two axes are parallel to the plane, then that they are at the same distance from this plane, by using the same dial gauge on the two cylinders representing the axes (see figure 37).

If these cylinders are not identical, it is essential to take into account the difference of radius of the tested sections.

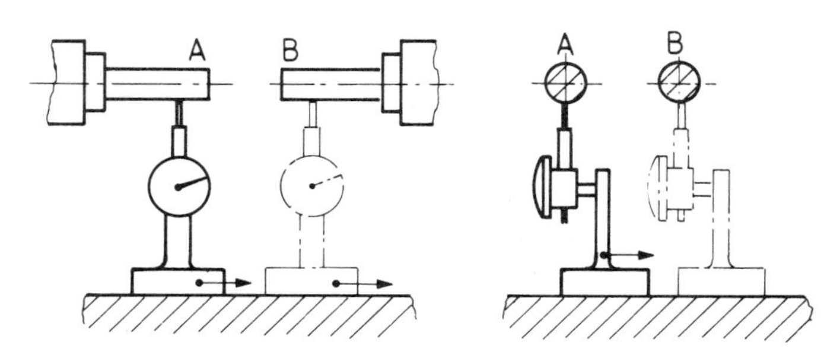

Figure 37.

Example of a squareness test

This example is from BS 4656: Part 3 'Milling machines, knee and column type, horizontal or vertical spindle plain'.

Part of Table 1. Geometrical tests and accuracy requirements for milling machines

All dimensions in millimetres.

Diagram	Test to be applied	Permissible deviation	Measuring instruments	Observations and reference to the test code BS 3800
(a) (b) (a) (b)	Squareness of the table surface to the column ways for knee (in three positions: in the middle and near the extremities of the travel): (a) in the vertical plane of symmetry of the machine (b) in the plane perpendicular to the preceding one	(a) 0.025/300 with $\alpha \leqslant 90°$ (b) 0.025/300	Dial gauge and square	**5.5.2.2.2** Table in central position, saddle and knee locked when taking measurements

Material from BS 3800 cited in the above example.

. . . .

5.5.2.2.2 *Perpendicularity between the trajectory of a point and a plane.* A square is placed on the plane (see figure 51). Parallelism between the motion and the free arm is tested in two perpendicular directions, in accordance with **5.4.2.2.2.2.**

5.5.2.2.3 *Trajectory of a point at 90° to an axis.* A square, with a suitable base, is placed against the cylinder representing the axis (see figure 52). The test for motion parallel to the free arm of the square is made in accordance with **5.4.2.2.2.2.**

. . . .

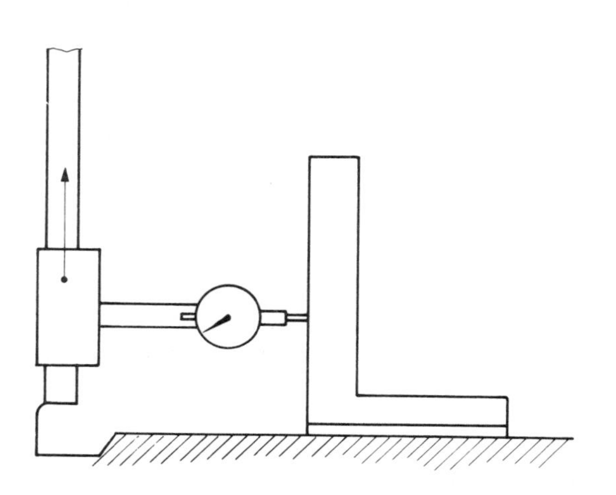

Figure 51.

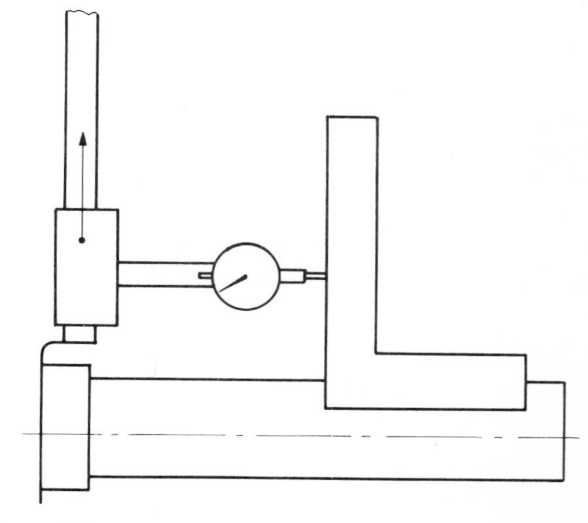

Figure 52.

Example of a rotation (run-out)

This example is from BS 4656: Part 12 'Dividing heads'.

Part of Table 1. Geometrical tests for dividing heads

All dimensions are in millimetres

Diagram	Object	Permissible deviation	Measuring instruments	Observations and references to test code BS 3800
	Measurement of run-out of the internal taper of the spindle: (a) at the mouth of taper; (b) at a distance of 300 mm from the face of the spindle nose	(a) 0.01 FIM (full indicator movement) (b) 0.02 FIM	Dial gauge and test mandrel	**5.6.1.2.3**

Material from BS 3800 cited in the above example.

. . . .

5.6.1.2.3 *Internal surface.* If the dial gauge cannot be used directly on a cylindrical or tapered bore, a test mandrel is mounted in the bore. The projecting cylindrical part of this mandrel is used for the test, in accordance with the previous clause. However, if the test is made at one section only of the mandrel, the position of only one circle of measurement in relation to the axis would be determined. As the axis of the mandrel may cross the axis of rotation in the plane of measurement, checking should be done on two sections *A* and *B* which are a specified distance apart (see figure 58).

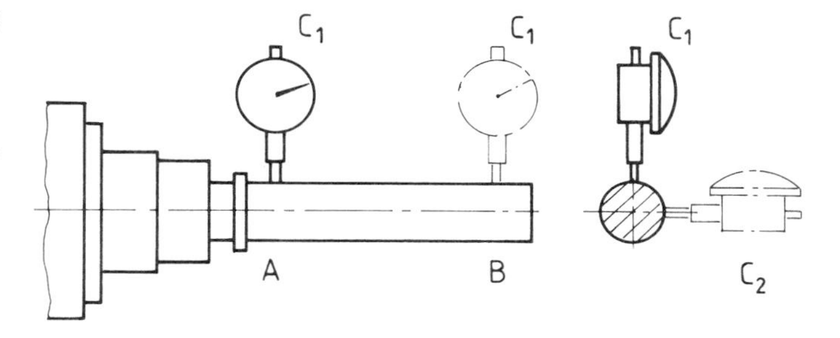

Figure 58.

For instance, one test should be made close against the housing of the mandrel and another at a specified distance from it. To provide for any lack of accuracy in inserting the mandrel into the bore, particularly with tapered bores, *these operations should be repeated at least four times, the mandrel being turned through 90 ° in relation to the spindle. The average of the readings should be taken.*

In each case, the run-out should be measured in a vertical axial plane and then in a horizontal axial plane (positions C_1 and C_2 in figure 58).

The above methods require the following comments:

As the movement of the stylus may vary directionally during the checking of run-out, every guarantee should be afforded for the accuracy of the measuring instrument (minimum drag).

When checking with a test mandrel, the exact shape of the bore is not checked.

A check of the run-out of the spindle by machining and testing a cylindrical workpiece will take into account only errors in the bearings of the spindle. This turning test therefore gives no information of the exact shape of the cylindrical or conical bore, or of the actual position of the bore in relation to the axis of rotation.

The above methods apply only to spindles in plain bearings or ball and roller bearings. Spindles which are automatically centred during rotation (e.g. by hydraulic pressure) can only be tested when running at normal speed. In such cases instruments involving no contact should be used, e.g. a capacitative pick-up, an electromagnetic pick-up or any other suitable instrument.

Standards publications referred to in this chapter

BS 3800 Methods for testing the accuracy of machine tools

BS 4656 Accuracy of machine tools and methods of test

- Part 1 Lathes, general purpose type
- Part 2 Copying lathes and copying attachments
- Part 3 Milling machines, knee and column type, horizontal or vertical spindle plain
- Part 4 Milling machines, bed type, horizontal or verticle spindle
- Part 5 Milling machines, knee and column type, horizontal spindle, universal
- Part 6 Surface grinding machines with vertical grinding wheel spindle and reciprocating table
- Part 7 Surface grinding machines with horizontal grinding wheel spindle and reciprocating table
- Part 8 Internal cylindrical grinding machines with horizontal spindle
- Part 9 External cylindrical grinding machines with reciprocating table
- Part 10 Drilling machines, radial type
- Part 11 Drilling machines, vertical floor mounted column and pillar types
- Part 12 Dividing heads
- Part 13 Broaching machines, vertical surface type
- Part 14 Broaching machines, vertical internal type
- Part 15 Drilling machines, turret and single spindle co-ordinate types
- Part 17 Broaching machines, vertical universal type
- Part 18 Broaching machines horizontal internal type
- Part 19 Gear hobbing machines
- Part 20 Machining centres, vertical spindle type
- Part 21 Boring and milling machines. Horizontal spindle table type and rotary table type
- Part 22 Vertical boring and turning lathes, single and double column types
- Part 23 Surface grinding machines two columns – slideway grinding machines
- Part 24 Cylindrical external centreless grinding machines
- Part 25 Gear planing machines
- Part 26 Gear shaping machines
- Part 27 Machining centres, horizontal spindle type
- Part 28 Numerically controlled turning machines up to and including 1500 mm turning diameter
- Part 29 Automatic lathes, multi-spindle (indexing drum) type
- Part 31 Capstan, turret and automatic lathes, single spindle type greater than 25 mm diameter capacity

10 Surface texture

Introduction

In practice, all manufactured surfaces depart to some extent from absolute perfection. It has long been recognized that careful finishing of components can give rise to longer life and improved fatigue resistance, bearing properties, functional interchangeability and wear resistance. This is so in all fields of industry.

At one time the terms 'rough machine', 'medium machine' and 'fine machine', or equivalent symbols, were used on engineering drawings, leaving the surface to be controlled by limitations of the machining process involved and arbitrary opinions of operator and inspector which, all too often, did not coincide. The problems, which became increasingly acute as the demand for a more comprehensive specification increased to keep pace with technological developments, called for a method of assessing the texture of the surface under consideration.

Geometry of surfaces

The imperfections of any surface take the form of a series of peaks and valleys which may vary both in height and spacing and result in a texture which, in feel or appearance, and in properties generally, is often characteristic of the process employed in its production. The complete texture of any surface is a combination of irregularities of various kinds and magnitudes arising from different causes. An ideally complete assessment would involve the measurement of each and every departure from the ideal surface plus an assessment of the effect of the combined texture on the functioning of the component surface.

The irregularities on any machined surface commonly represent the joint effects of roughness arising from the inherent action of the cutting process and waviness attributable to vibration and machine deflection. These may be superimposed on departures from true geometrical form. It is normally only the roughness and waviness which are considered to make up the surface texture, although form errors are often measured in a similar manner.

In addition to the irregularities mentioned above, the direction of the predominant surface pattern, ordinarily determined by the production method, is evident on most surfaces. This is known as a *lay* (see Figure 10.1).

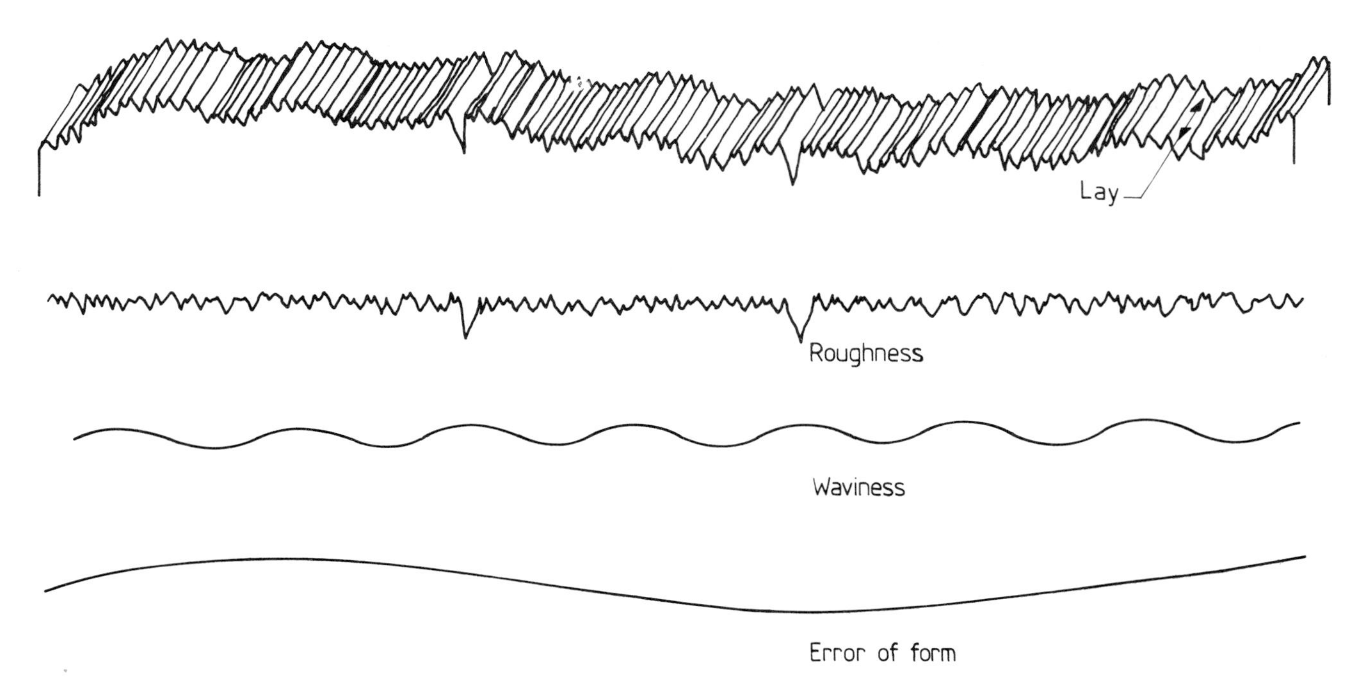

Figure 10.1 *Lay, roughness, waviness and error of form*

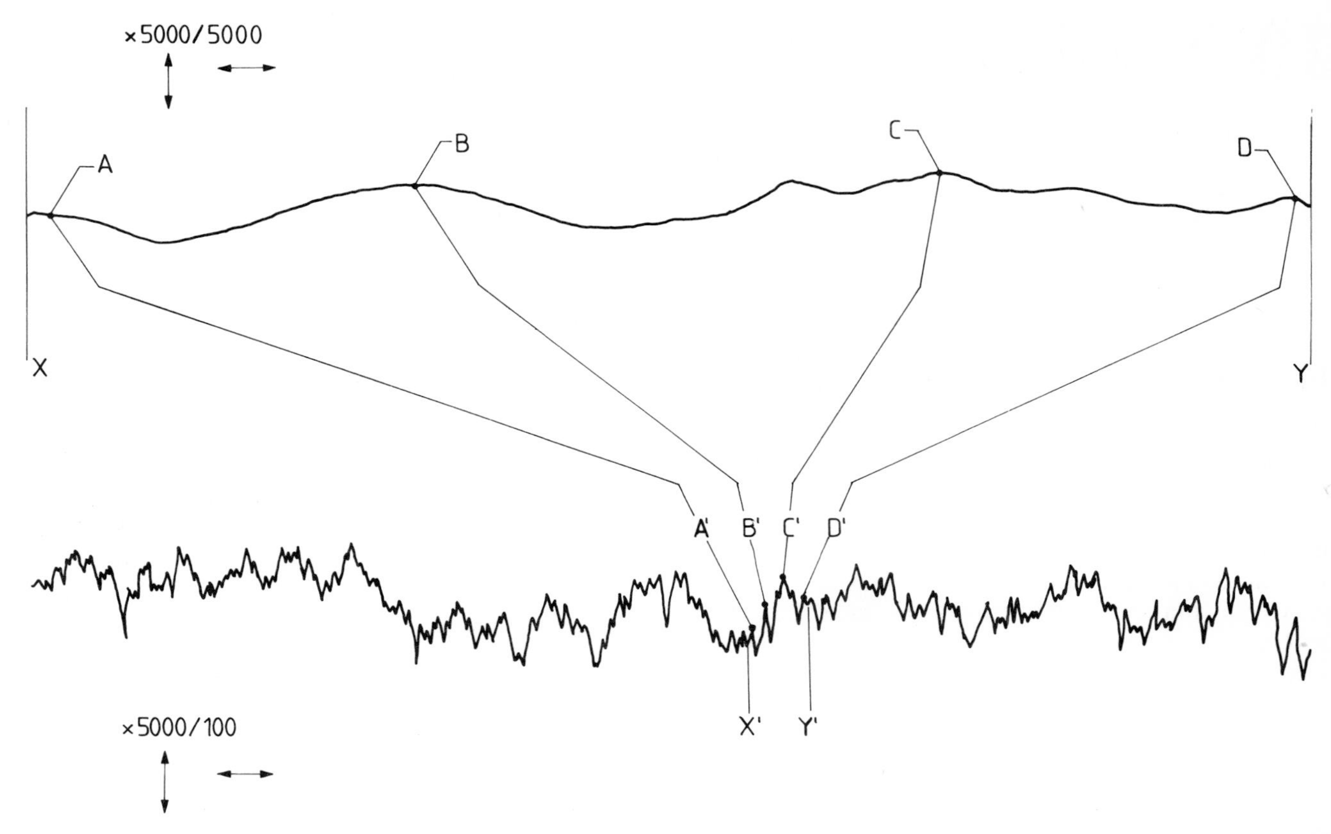

Figure 10.2 *Effect of different horizontal magnification*

Measurement of surface texture

The obvious problems encountered in endeavouring to make complete assessments of three-dimensional surfaces are overcome by confining measurements to profiles of plane sections taken through the surfaces (i.e. single line traces).

Tomlinson surface recorder

An instrument designed in 1939 by Dr G. A. Tomlinson of the National Physical Laboratory recorded the movement of a stylus passing over the surface under inspection on to a smoked glass disc and gave a vertical magnification of × 100, there being no horizontal magnification. The trace was then further magnified by an optical projector before a hand or photographic record was taken for analysis.

Modern stylus instruments

Stylus instruments are still the most widely used method of surface texture measurement and the need for differing magnifications of vertical and horizontal movements of the stylus is still necessary to avoid inconveniently long traces (see Figure 10.2).

Stylus instruments are made-up of the following units:

(a) a skid which, when drawn over the surface, provides a datum;
(b) a stylus;
(c) an amplifying device for magnification of the stylus movement;
(d) a recording device to provide a trace and/or a meter to provide a direct read-out.

Figure 10.3 shows the main parts of a stylus surface texture measuring instrument.

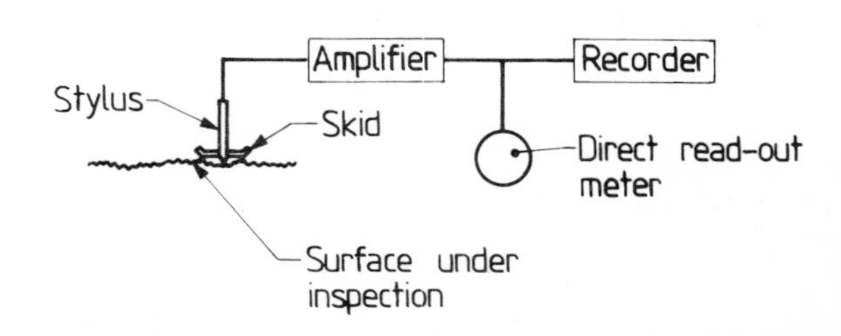

Figure 10.3 *Main parts of a stylus surface texture measuring instrument*

Quantifying the profile graph

Having obtained profile graphs (traces) of a number of different surfaces they can be compared, but still only in a subjective manner. A system which interprets a complex surface texture and presents a code or number to indicate *all* the characteristics which will affect its performance is not possible. Consider, for example, Figure 10.4 which shows, diagramatically, a surface containing roughness, waviness and error of form. If the measurement is confined to a short sampling length L_1 of the surface, the value obtained for the total height will be H_1 (total height being considered here for reasons of demonstration). This is a measure of the roughness neglecting the occasional deep scratches; it is nominally the same for all parts of the surface and it suppresses the irregularities of greater spacing. As the sampling length is increased, however, the height will eventually increase until, for the length L_2, it reaches a new value H_2 which takes into account the waviness but, which still ignores errors of form. Finally, for the whole surface, sampling length L_3 would give a value H_3, including all irregularities and errors.

Thus, various values can be obtained according to the length of surface selected for assessment. It might seem that the length of surface which most fairly expresses the quality of the surface texture is the greatest, i.e. L_3. That this may not be so is demonstrated by Figure 10.5, in which profile A represents a finely finished surface having a large error of form, while profile B represents a more coarsely finished surface having no error of form. Although both surfaces have the same total height they are obviously different in quality.

An obvious method of measuring the roughness regardless of the other irregularities is to limit the measurement to a sufficiently short length of the surface. In the case of some surfaces, this may become very small. The difficulty is then encountered that the measured value over a sampling length may vary considerably from point to point. Such statistical variations must not be confused with true variations of texture in different parts of the surface, as they are merely the incidental result of the method of analysis and have to be smoothed out. As shown in Figure 10.6, this can be done by taking, as the true value, the mean of a number of observations.

The observations may conveniently be taken in a row along a short length of the surface. Thus, if the roughness does not exceed some given spacing, say, 0.8 mm, the graph may be divided into successive sections L_1, L_2, L_3, etc., each 0.8 mm long and the average height of each section found separately. Taking the mean

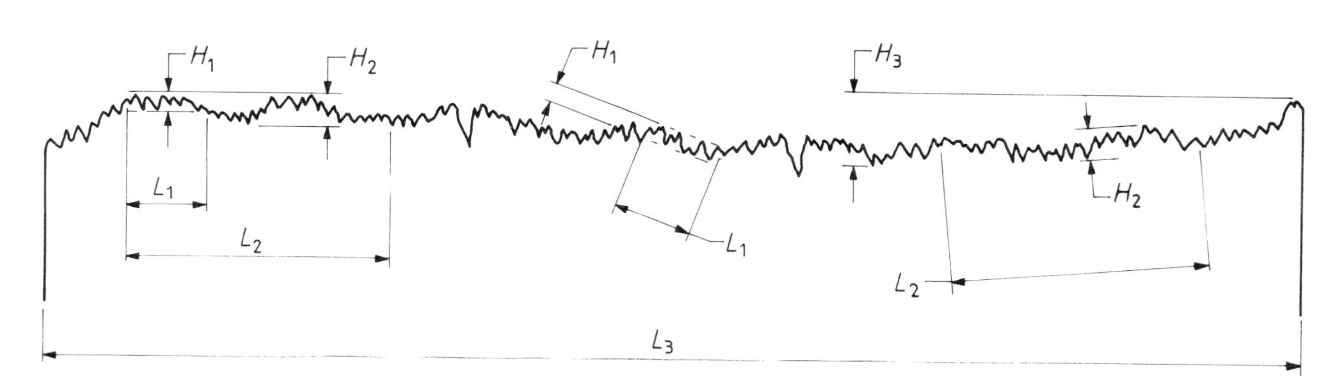

Figure 10.4 *Examples of heights and sampling lengths on a surface containing roughness, waviness and error of form*

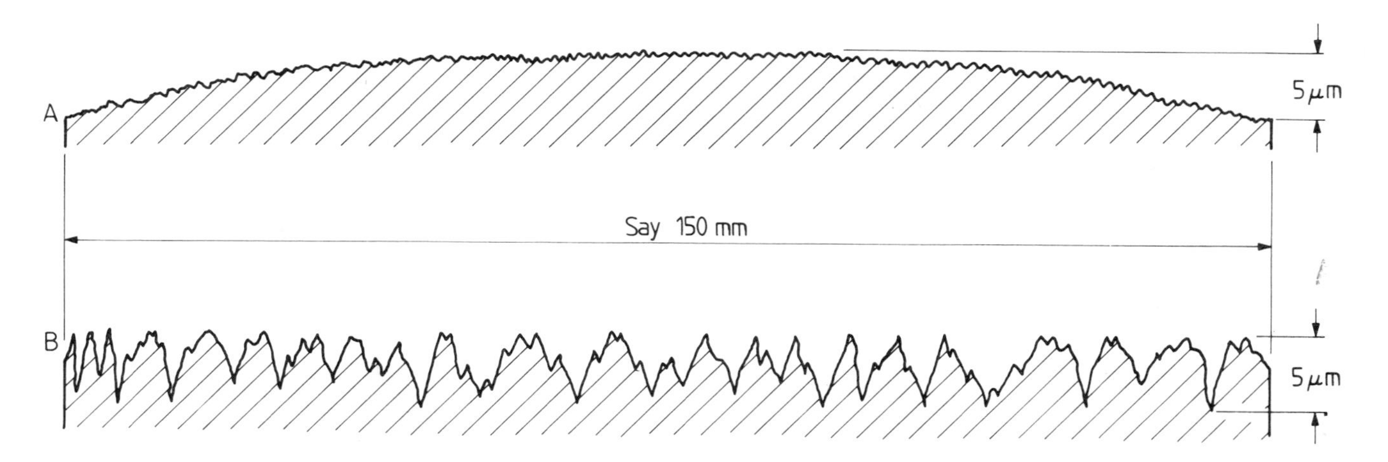

Figure 10.5 *Surface textures having the same total height but differing departures from geometrical profile*

value from a few (say five) consecutive sections will usually suffice to eliminate the effect of variations between individual sections.

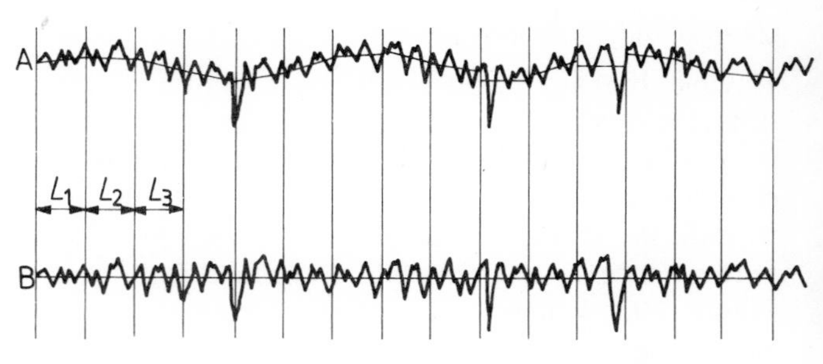

Figure 10.6 *Analysis of profile graph showing the elimination of longer spacing errors*

Parameters

There are a number of ways (parameters) by which roughness can be expressed numerically, but the most widely used is the arithmetical mean deviation (designated R_a). Once known as the centre-line-average (CLA), this parameter is defined as: the arithmetical average value of the departure of the profile above and below the reference line (centre or electrical mean line) throughout the prescribed sampling length. See material from BS 1134 'Method for the assessment of surface texture', Part 1 'Method and instrumentation' below. BS 1134 also describes the ten-point height (R_z) of irregularities but it should be noted that many other parameters exist (see PD 7306 'Introduction to surface texture' published by BSI, and *Exploring surface texture* by H. Dagnall published by Rank Taylor Hobson, 1980).

BS 1134

Assessment of surface texture

Part 1. Method and instrumentation

. . . .

1.2 Terminology. For the purposes of this British Standard the following terms and definitions have been adopted (see also figure 1):

(1) real surface. The surface limiting the body, separating it from surrounding space.

(2) real profile. The contour that results from the intersection of the real surface by a plane conventionally defined with respect to the geometrical surface.

(3) geometrical surface (nominal surface). The surface determined by the design, neglecting errors of form and surface roughness.

(4) geometrical profile (nominal profile). The profile that results from the intersection of the geometrical surface by a plane conventionally defined with respect to this surface.

(5) effective surface (measured surface). The close representation of a real surface obtained by instrumental means.

(6) effective profile (measured profile). The contour that results from the intersection of the effective surface by a plane conventionally defined with respect to the geometrical surface.

(7) irregularities. The peaks and valleys of a real surface.

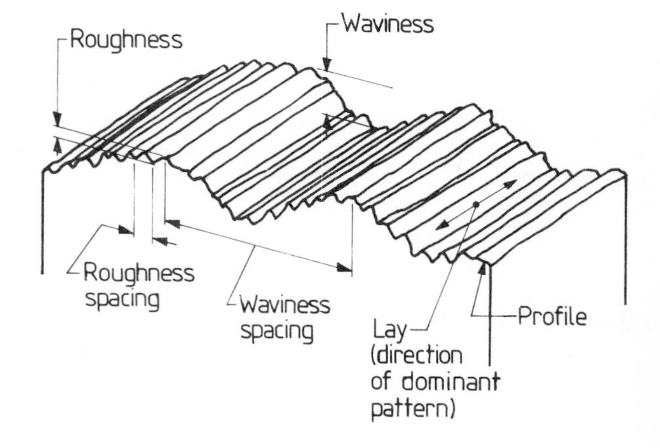

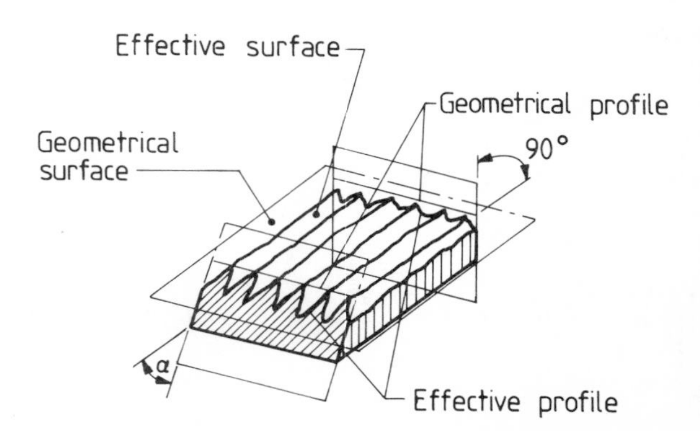

Figure 1. Surface characteristics and terminology

(8) spacing. The average distance between the dominant peaks on the effective profile.
(9) surface texture. Those irregularities with regular or irregular spacing which tend to form a pattern or texture on the surface. This texture may contain the following components:
(a) roughness. The irregularities in the surface texture which are inherent in the production process but excluding waviness and errors of form.
(b) waviness. That component of surface texture upon which roughness is superimposed. Waviness may result from such factors as machine or work deflections, vibrations, chatter, heat treatment or warping strains.
(10) lay. The direction of the predominant surface pattern, ordinarily determined by the production method used.
(11) sampling length. The length of profile selected for the purpose of making an individual measurement of surface texture.
(12) reference line. The line chosen by convention to serve for the quantitative evaluation of the roughness of the effective profile.
(13) centre line. A line representing the form of the geometrical profile and parallel to the general direction of the profile throughout the sampling length, such that the sums of the areas contained between it and those parts of the profile which lie on each side of it are equal.
(14) least-squares mean line. A reference line representing the form of the geometrical profile within the limits of the sampling length, and so placed that within the sampling length the sum of the squares of the deviations of the profile from the mean line is a minimum.
NOTE. The least-squares mean line is unique in position and direction, but its graphical determination can be somewhat laborious.
(15) meter cut-off (B_{max}). In a profile meter instrument, the conventionally defined wavelength separating the transmitted from the attenuated components of the effective profile.
NOTE. The meter cut-offs of profile meter instruments are made equal to the desired sampling lengths. Electrical integrating instruments indicate automatically the average results from several consecutive sampling lengths.
(16) electrical mean line. In an electric meter instrument, a reference line established by the circuits determining the meter cut-off, which line divides equally those parts of the modified profile lying above and below it.
(17) modified profile. The effective profile modified by such defined filter means as are used for suppressing those undulations of the real profile that are not or are not fully to be included in the measured roughness parameters of the surface.
(18) arithmetical mean deviation (R_a). The arithmetical average value of the departure of the profile above and below the reference line (centre or electrical mean line) throughout the prescribed sampling length (see figure 2).
(19) ten point height (R_z) of irregularities. The average distance between the five highest peaks and the five deepest valleys within the sampling length, measured from a line parallel to the reference line and not crossing the profile (see figure 3).
(20) profile recording instrument. An instrument recording the co-ordinates of the profile of the surface.
(21) recording traversing length. The maximum recording movement of the stylus along the surface.
(22) profile meter instrument. An instrument used for the measurement of surface texture parameters.
(23) measuring traversing length. The length of the modified profile used for measurement of surface roughness parameters. (It is usual for the measuring traversing length to contain several sampling lengths.)
(24) profile meter instrument with predetermined measuring traversing length. An instrument in which the length used for measurement has a defined beginning and end determined by switches or other instrumental means.
(25) profile meter instrument with 'running' measuring traversing length (giving a running average). An instrument in which the length used for measurement results from the characteristics of the profile meter and moves along the surface with the pick-up.
NOTE. In instruments of this type the reading may fluctuate according to local variations of the profile.
(26) static measuring force. The force which the stylus exerts along its axis on the examined surface without taking into account the dynamic components arising in the process of traversing the surface by the stylus.
(27) rate of change of measuring force. The change of static measuring force per unit displacement of the stylus along its axis.
(28) vertical magnification of a profile recording instrument (V_v). The ratio of the movement of the indicating device of the recorder to the displacement of the stylus in a direction normal to the surface.
(29) horizontal magnification of a profile recording instrument (V_h). The ratio of the movement of the recorder chart to that of the stylus along the surface.
(30) error of vertical magnification of a profile recording instrument. The difference between the nominal and the actual values of the vertical magnification referred to the nominal value and expressed as a percentage.
(31) error of horizontal magnification of a profile

recording instrument. The difference between the nominal and the actual values of the horizontal magnification referred to the nominal value and expressed as a percentage.

(32) basic error of a profile meter reading. The percentage difference between the instrument reading and the value of the surface texture parameter as defined by the stylus and meter cut-off (without skid) of the instrument.

(33) method divergence of the instrument reading. For a given measured profile the percentage difference between the value of a surface texture parameter determined with respect to the mean line of the defined wave filter and a succession of straight centre lines each equal in length to the meter cut-off, both determinations being referred to the same part and overall length of the same cross-section (see appendix A (not included in this manual)).

1.3 Sampling lengths

. . . .

Table 1. Standard sampling lengths

Millimetres	Inch
0.08	0.003
0.25	0.01
0.8	0.03
2.5	0.1
8.0	0.3
25.0	1.0

1.4 R_a values

. . . .

Table 2. Preferred R_a values

Nominal R_a values		Roughness grade number
μm	μ in	
50	2000	N 12
25	1000	N 11
12.5	500	N 10
6.3	250	N 9
3.2	125	N 8
1.6	63	N 7
0.8	32	N 6
0.4	16	N 5
0.2	8	N 4
0.1	4	N 3
0.05	2	N 2
0.025	1	N 1
0.0125	0.5	—

NOTE. The values given in table 2 are expressed as 'preferred' in order to discourage unnecessary variation of the values expressed on drawings. It should be realized that in some circumstances, warranting departure from the preferred values, other values may be specified.

1.5 Graphical determination of R_a values. The following procedure shall be observed when determining R_a values from graphical recordings. It will be assumed for the moment that the surface is nominally flat, and that the record is produced in rectilinear co-ordinates in which a truly flat surface is represented as a straight line.

It is necessary first to determine the centre line of each successive sampling length contained within the traversing length of the record. Consider figure 2*a*: draw a straight line X . . . X, parallel to the general course of the record over the length (L) that represents a sampling length, and for convenience through the lowest valley. The direction can usually be determined with sufficient accuracy by eye. In practice, this is done almost instinctively. Where the texture has a distinguishable periodicity it is essential that the sampling length should be chosen to include a whole number of wavelengths.

The area P between the profile and X . . . X is then determined either by measuring equally spaced ordinates or by the use of a planimeter, through the chosen sampling length. The total area P, when

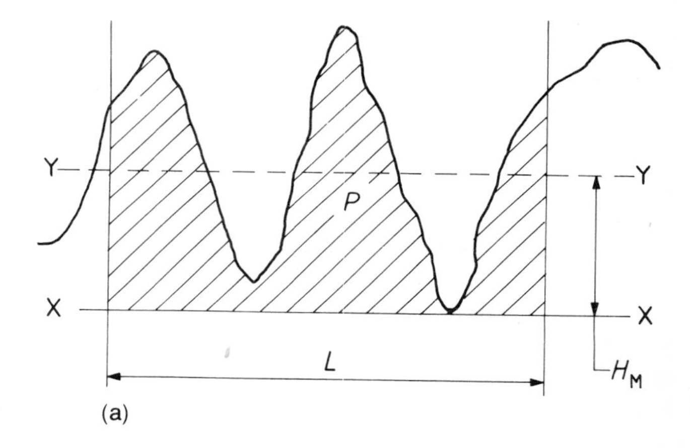

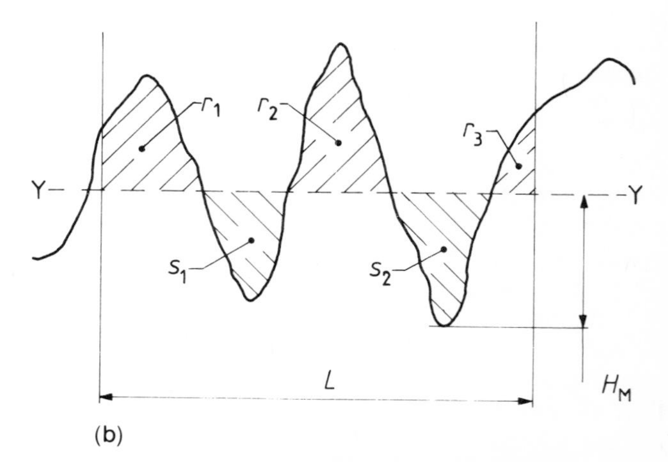

Figure 2. Graphical determination of R_a values

divided by L, gives the height H_M of the centre line Y . . . Y above X . . . X.

$$\text{Thus } H_M = \frac{\text{area } P}{L}$$

The centre line Y . . . Y can now be drawn in parallel to X . . . X and at a height H_M above it.

The areas r_1, r_2, r_3 . . . and s_1, s_2 . . . above and below the centre line shown in figure 2*b* can then be determined. Taking all areas positively:

$$R_a \text{ in } \mu\text{m} = \frac{\text{sum of areas } r + \text{sum of areas } s}{L} \times \frac{1000}{V_v}$$

where L is expressed in millimetres and areas are expressed in square millimetres on the record, and V_v = vertical magnification. Finally, the required value of R_a over the measuring traversing length is taken as the mean of the successive values of the sampling length.

If the surface is intentionally curved, the curvature will generally be neutralized, prior to recording, by some form of guiding or filter device.

1.6 Graphical determination of R_z values. For some purposes it is convenient to have an assessment of average peak-to-valley height of surface irregularities. The R_z or 'ten point height' method (see figure 3) which, in essence, is an arbitrary way of avoiding the effect of exceptional peaks and valleys in the final computation, shall be used in determining average peak-to-valley values. R_z values are generally from 4 to 7 times the corresponding R_a values, the ratio depending upon the shape of the profile.

The five highest peaks and five deepest valleys are conveniently measured from an arbitrary base line A' . . . B' drawn parallel to the centre line A . . . B of the chosen sampling length L. R_z is then determined from:

$$R_z = \frac{(R_1 + R_3 + \ldots R_9) - (R_2 + R_4 + \ldots R_{10})}{5}$$

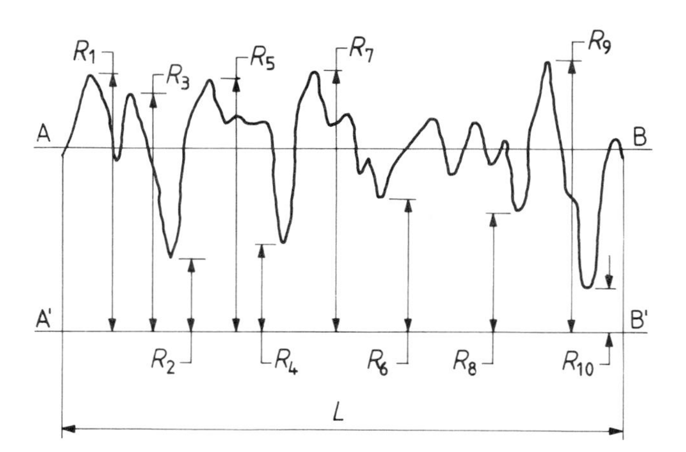

Figure 3. Graphical determination of R_z values

. . . .

2.4 Traverse. Electrical integrating instruments giving direct scale and pointer indications of R_a values take, in effect, an automatic average over a succession of sampling lengths determined by the meter cut-off.

In profile meters with predetermined or running traversing lengths, the measuring traversing length shall depend on the cut-off value (B_{max}) within the limits given in table 4.

Table 4. Measuring traversing lengths

Type of profile meter	Cut-off (B_{max})	Measuring traversing length min.	max.
	mm	mm	mm
Predetermined traversing length	0.08	0.4	2
	0.25	1.25	5
	0.8	2.4	8
	2.5	5.0	15
	8.0	16.0	40
Running traversing length	0.25	2.5	16
	0.8	5.0	16

2.5 Values of vertical and horizontal magnification. The values of vertical and horizontal magnification for profile recording instruments shall be selected from the following preferred series:

Vertical (V_v): 100, 200, 500, 1000, 2000, 5000, 10 000, 20 000, 50 000, 100 000.
Horizontal (V_h): 10, 20, 50, 100, 200, 500, 1000, 2000, 5000.

2.6 Transmission characteristics in the long wavelength

2.6.1 Meter cut-off values. The standard meter cut-off values to be used in instrument construction shall be selected from those in table 5.

Table 5. Standard meter cut-off values

Millimetres
0.08
0.25
0.8
2.5
8.0

NOTE. A meter cut-off of 0.8 mm is found adequate for most of the finer surfaces.

Surface roughness values produced by common production processes and materials

Table 10.1 shows results in terms of R_a values that can be expected from various common production processes and materials, but they should be regarded as a guide only.

Table 10.1 *Surface roughness values produced by common processes and materials*

Process	Roughness value, R_a (in µm): 50–25	25–12.5	12.5–6.3	6.3–3.2	3.2–1.6	1.6–0.8	0.8–0.4	0.4–0.2	0.2–0.1	0.1–0.05	0.05–0.025	0.025–0.0125
Flame cutting	▨	■	▨									
Snagging	▨	■	■	▨								
Sawing	▨	■	■	■	▨							
Planing, shaping	▨	■	■	■	■	■	▨					
Drilling			▨	■	■	▨						
Chemical milling			▨	■	■	▨						
Electro-discharge machining			▨	▨/■	■	▨						
Milling		▨	▨	■	■	■	▨	▨				
Broaching				▨	■	■	▨					
Reaming				▨	■	■	▨					
Boring, turning		▨	▨	■	■	■	■	▨	▨	▨		
Barrel finishing					▨	▨	■	■	▨	▨		
Electrolytic grinding							▨/■	■	▨			
Roller burnishing							▨	■	▨			
Grinding				▨	▨	■	■	■	■	▨	▨	
Honing						▨	■	■	■	▨	▨	
Polishing							▨	■	■	▨	▨	▨
Lapping							▨	■	■	■	▨	▨
Superfinishing							▨	▨	■	■	▨	▨
Sand casting	▨	■	▨									
Hot rolling	▨	■	▨									
Forging		▨	■	■	▨							
Permanent mould casting				▨	■	▨						
Investment casting				▨	■	▨	▨					
Extruding			▨	▨	■	■	▨					
Cold rolling, drawing				▨	■	■	▨	▨				
Die casting					▨	■	▨					

Key

■ Average application

▨ Less frequent application

Process designations and guidance on suitable meter cut-offs

In general each finishing process has its characteristic surface texture. The process may be stated in conjunction with the R_a value and the sampling length in order to define the required surface more fully. Table 10.2 gives typical process designations and indicates the meter cut-offs (equal to the sampling lengths) found by experience to be suitable for each process listed.

Table 10.2 *Process designations and suitable meter cut-off values for various finishing processes*

Typical finishing process	*Designation*	*Meter cut-off (mm)*				
		0.25	*0.8*	*2.5*	*8.0*	*25.0*
Milling	Mill		×	×	×	
Boring	Bore		×	×	×	
Turning	Turn		×	×		
Grinding	Grind	×	×	×		
Planing	Plane			×	×	×
Reaming	Ream		×	×		
Broaching	Broach		×	×		
Diamond boring	D. Bore	×	×			
Diamond turning	D. Turn	×	×			
Honing	Hone	×	×			
Lapping	Lap	×	×			
Superfinishing	S. Fin.	×	×			
Buffing	Buff	×	×			
Polishing	Pol.	×	×			
Shaping	Shape		×	×	×	
Electro-discharge machining	EDM	×	×			
Burnishing	Burnish		×	×		
Drawing	Drawn		×	×		
Extruding	Extrude		×	×		
Moulding	Mould		×	×		
Electro-polishing	El-Pol.		×	×		

Note: It will be appreciated that while the range of meter cut-off values associated with the processes listed in this table will generally be found suitable, the proper cut-off value is in fact determined not by the process but by the dominant peak spacing the process has produced or will produce. The appearance of 0.8 mm cut-off in all but one of the range does not mean that this is the comprehensive value that will serve for all purposes. It will lead to a realistic assessment of the hills and valleys in question only when it embraces their dominant peak spacing.

The indication of surface texture on engineering drawing

Information arranged as shown in Figure 10.7, and illustrated in Figure 10.8, may be indicated on the drawing in conjunction with the machining symbols.

When it is necessary to indicate lay the symbols in Figure 10.9, from BS 308 'Engineering drawing practice', Part 2 'Dimensioning and tolerancing of size', should be used.

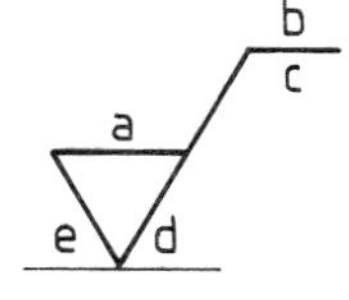

a denotes roughness value(s)
b denotes production method, treatment or coating
c denotes sampling length
d denotes direction of lay
e denotes machining allowance

Figure 10.7 *Additional information applied to machining symbol*

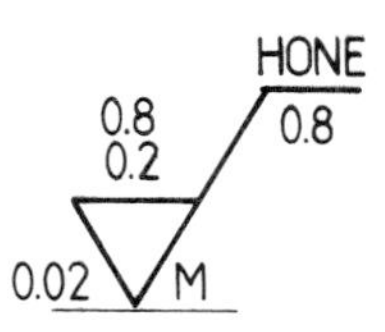

Figure 10.8 *Example of additional information applied to machining symbol*

Summary

Roughness values should not be placed on drawings as the arbitrary opinion of the designer but preferably should result from investigation taking into consideration all factors, including load, lubrication, materials, direction of movement, speeds, temperature, cost, etc.

Two schools of thought exist on whether or not all surfaces should be specified and controlled. It should be recognized that to specify a surface texture value is an instruction for the surface to be inspected and that to specify an unnecessarily smooth surface can be unsound economically (see Figure 10.10).

In the absence of full information, experience with similar designs and processes is often a useful starting-off point. Examination of roughness comparison specimens may also be helpful.

Roughness comparison specimens

Roughness comparison specimens are specimen surfaces of known average roughness height, R_a, representative of particular machining, or other, process. The specimen is used to give guidance on the feel and appearance of the particular production process and roughness grade (see Figure 10.11).

Details of roughness comparison specimens are given in BS 2634 'Roughness comparison specimens'. No part of BS 2634 is reproduced in this manual.

Lay symbol	*Indication on the drawing*	*Interpretation*
=	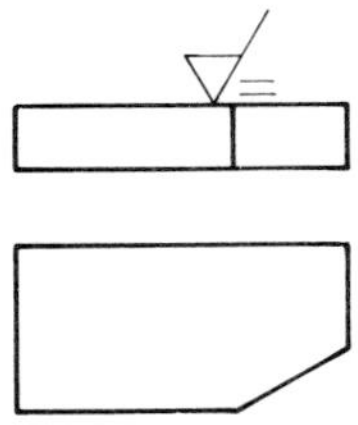	Parallel to the plane of projection of the view in which the symbol is used Direction of lay 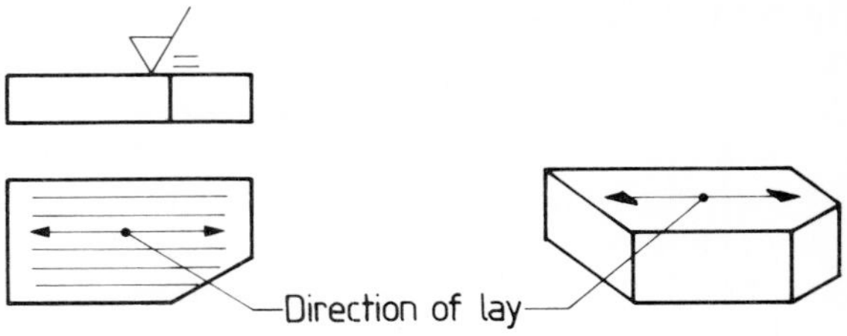
⊥	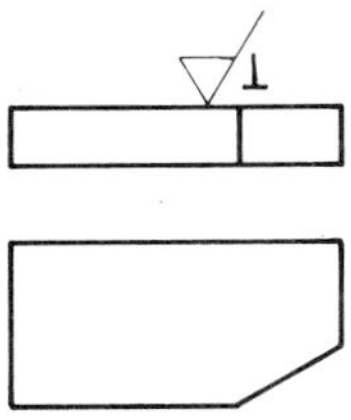	Perpendicular to the plane of projection of the view in which the symbol is used Direction of lay 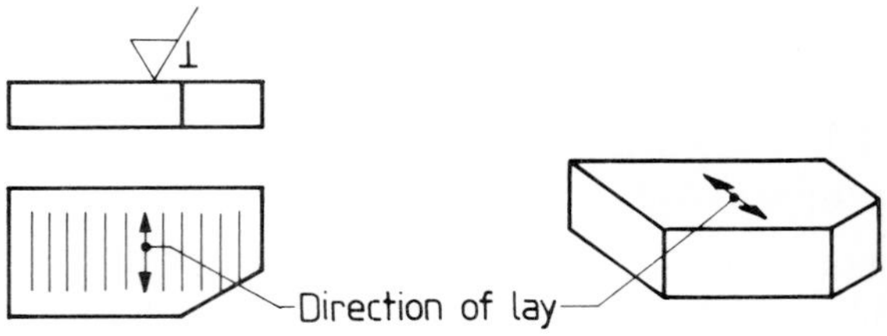
X	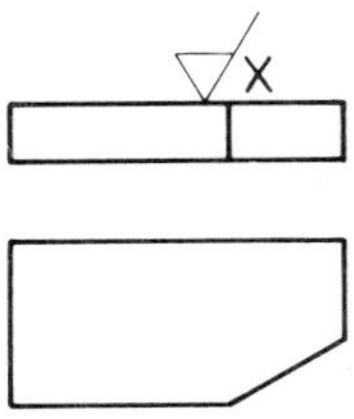	Crossed in two slant directions with regard to the plane of projection of the view in which the symbol is used Direction of lay 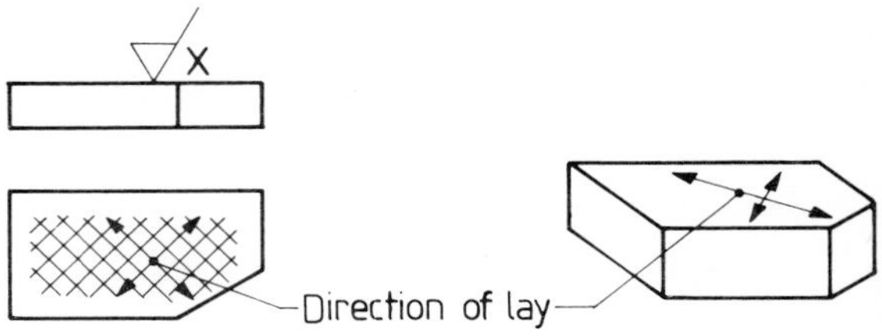
M	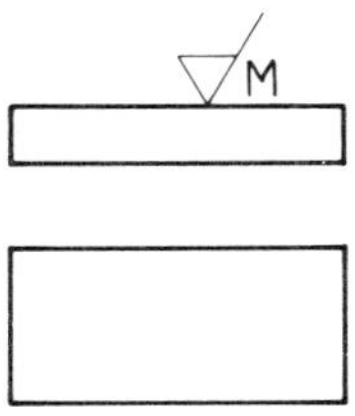	Multi-directional 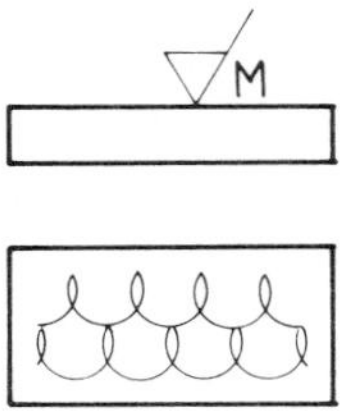
C	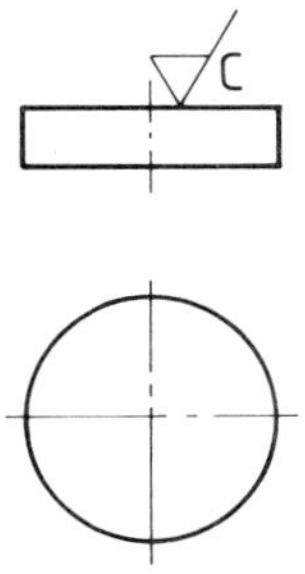	Approximately circular relative to the centre of the surface to which the symbol is applied 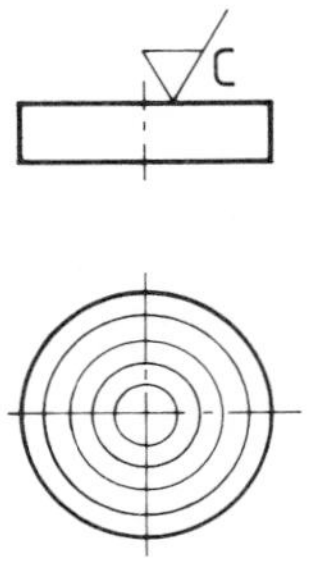
R	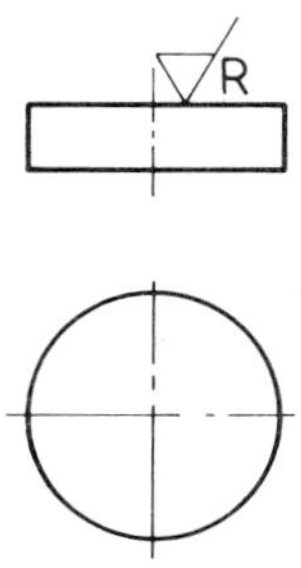	Approximately radial relative to the centre of the surface to which the symbol is applied 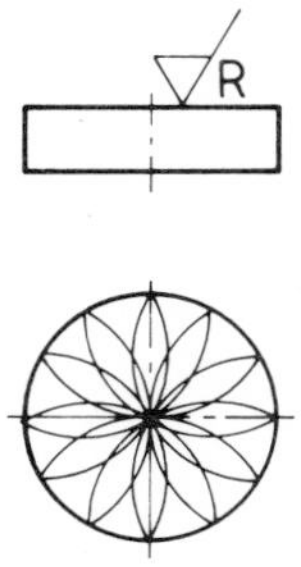

Figure 10.9 *Symbols for the direction of lay*

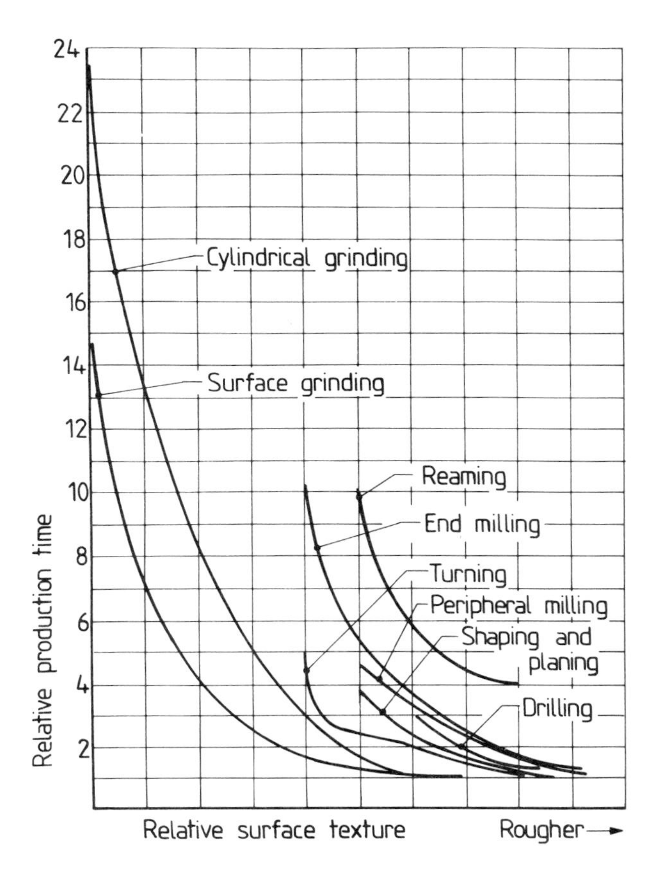

Figure 10.10 *Typical relationships of surface texture to production time*

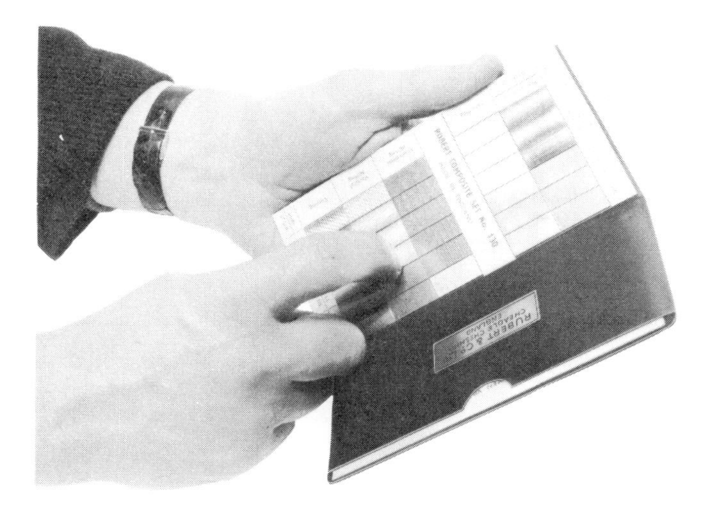

Figure 10.11 *Roughness comparison specimens in use*
Reproduced by permission of Rubert & Co. Ltd

Standards publications referred to in this chapter

BS 308 Engineering drawing practice
Part 2 Dimensioning and tolerancing of size

BS 1134 Method for the assessment of surface texture
Part 1 Method and instrumentation
Part 2 General information and guidance

BS 2634 Roughness comparison specimens (in 3 parts)

PD 7306 Introduction to surface texture

Other publication referred to in this chapter

Exploring surface texture by H. Dagnall published by Rank Taylor Hobson (1980)

11 Roundness

Introduction

Technical advancement which called for increased dimensional control also required a greater control of geometrical form and, in particular, roundness.

The importance of roundness can be illustrated by considering a shaft in a plain bearing. If both shaft and bearing are truely round and of an appropriate fit, the shaft will run smoothly, particularly if correctly lubricated. If, however, the shaft is oval, as shown in Figure 11.1, or lobed as in Figure 11.2, then smooth running will not result. The problem will be increased if *both* components of the fit suffer from out-of-roundness, and even the use of lubricants will not overcome these shortcomings. Measurement of the diameter will not necessarily identify the problems associated with out-of-roundness. A lobed part will give the same reading for diameter when measured between a pair of parallel faces as is employed by a micrometer (see Figure 11.3), but a lobed part is not round.

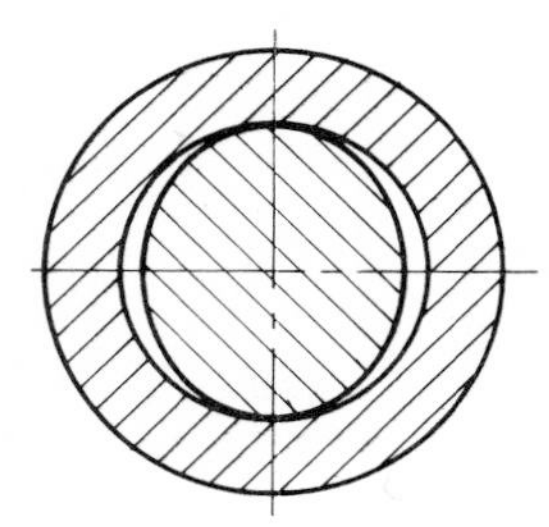

Figure 11.1 *Oval shaft*

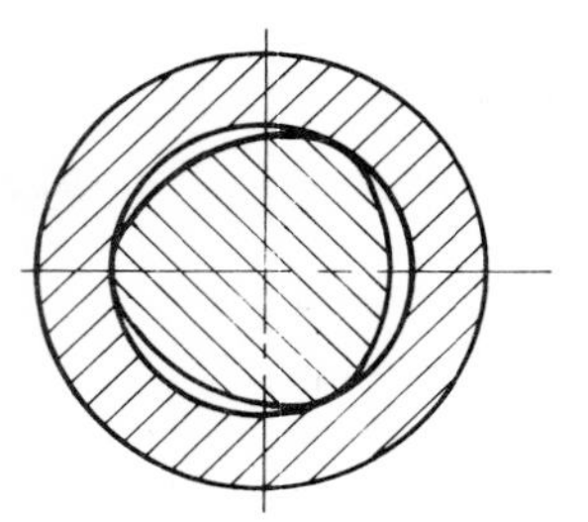

Figure 11.2 *Lobed shaft*

BS 3730

BS 3730 'Assessment of departures from roundness' is in three parts as follows:

Part 1 Glossary of terms relating to roundness measurement

Part 2 Specification for characteristics of stylus instruments for measuring variations in radius (including guidance on use and calibration)

Part 3 Methods for determining departures from roundness using two- and three-point measurement

BS 3730: Part 1 defines metrological terms used in the determination of deviations from roundness and includes appendices illustrating the sequential steps involved in roundness measurement.

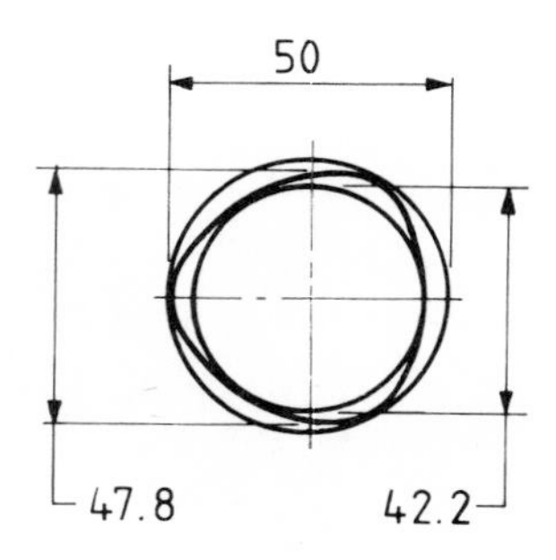

Figure 11.3 *Example of maximum and minimum circles related to a lobed part*

BS 3730

Assessment of departures from roundness

Part 1. Glossary of terms relating to roundness measurement

1. Scope

This Part of this British Standard defines metrological terms used in the determination of deviations from roundness.

Section one. General terms (surfaces, planes, axes)

real surface. A surface limiting the body separating it from the surrounding medium (see figure 3).

nominal axis of rotation. The theoretically exact axis about which the spindle of a perfect instrument rotates.

instantaneous axis of rotation. The axis about which the spindle of an instrument actually rotates at any instant.

NOTE. The instantaneous axis of rotation may be continuously varying within the confines of the bearings.

reference axis of rotation. The mean of the *instantaneous axes of rotation* about which the spindle of an instrument rotates.

instantaneous error of. The difference between the position of the *instantaneous axis of rotation* and the *reference axis of rotation*.

NOTE. Errors of rotation of the instrument may comprise radial, axial and tilt components.

nominal plane of measurement. A plane perpendicular to the *nominal axis of rotation* of the instrument.

plane of measurement. A plane perpendicular to the *reference axis of rotation* and passing through the point of contact of the detecting element of the instrument with the workpiece.

direction of measurement. The direction along which radial variations are determined. It substantially intersects the axis of rotation of the instrument and it generally lies in the plane of measurement.

axis of workpiece. A defined straight line about which the relevant part of the workpiece is considered to be round (see figure 2).

NOTE. There may be several ways in which the axis can be defined, including the following.

(a) Straight line such that the root-mean-square value of the distances from it of the defined centres of a representative number of cross sections has a minimum value.
(b) Straight line passing through the defined centres of two separated and defined cross sections.
(c) Straight line passing through the centre of one defined cross section and perpendicular to a defined shoulder.
(d) Straight line passing through two support centres. This axis is independent of the surface of the workpiece.
(e) Axis of two co-axial surfaces of revolution just enclosing the surface irregularities of the workpiece.

setting-up eccentricity. The distance in the plane of measurement between the point of intersection therewith of the *reference axis of rotation* of the instrument and the defined centre of the workpiece profile.

magnification of the instrument. The ratio of the output value of the instrument to the displacement of the stylus in the direction of measurement.

Section two. Profiles

real roundness profile. The profile resulting from the intersection of the real surface of a round workpiece by a plane perpendicular to its defined axis.

profile transformation. The action of transforming a profile at any stage, as for example by a stylus, filter, recorder, etc.

traced profile. The profile traced by the stylus (see figure 3).

NOTE. It may tend to include or exclude surface roughness according to the dimensions of the stylus.

modified profile. The *traced profile* intentionally modified by an analogue or digital wave filter that has defined characteristics.

displayed profile. The representation of the *traced or modified profile* displayed by the instrument, for example as a trace, oscilloscope presentation or a data log.

Section three. Circles

workpiece reference circle. A circle fitting the *traced profile* of the workpiece in a defined way, to which the departures from roundness and the geometric roundness parameters are referred.

display reference circle. A circle representing the *workpiece* reference circle, fitted to the displayed representation of the *traced profile* in the defined way.

NOTE 1. Residual eccentricity left after setting up will result in slight distortion of the displayed representation of the workpiece profile and in principle the reference circle of the display should be distorted correspondingly, approximately to the shape known as limaçon. Electrical

methods of plotting the reference circle on a polar graph and digital representations often do this automatically. The distortion, having a maximum value given by $E^2/2R$, can generally be neglected if the residual eccentricity, E, measured at the display is kept within about 15 % of the mean radius, R, of the profile for general testing and within 7 % for more critical applications.
NOTE 2. Other reference profiles, e.g. elliptical, tri-lobe or electric wave filter mean line, may find auxiliary use for analytical purposes.

least squares mean circle.* A circle such that the sum of the squares of the departures from it of the *traced or modified profile* of the workpiece is a minimum.

NOTE. The circle is generally determined with sufficient accuracy from a finite number of radial ordinates suitably spaced.

minimum circumscribed circle.* Smallest possible circle that can be fitted around the *traced or modified profile* of a shaft.

maximum inscribed circle.* Largest possible circle that can be fitted within the *traced or modified profile* of a hole.

minimum zone circles.* Two concentric circles enclosing the *traced or modified profile* and having the least radial separation.

Section four. Circumference

undulations per revolution (upr). The number of complete periodic undulations contained in the periphery of the workpiece.

NOTE. There cannot be less than 1 complete undulation per 360° or 2π radians.

sinusoidal undulation number (n_s). The dominant or superimposed periodic sinusoidal waves measured during one revolution of the workpiece.

sinusoidal undulation frequency. The *sinusoidal undulation number* multiplied by revolutions per second of the instrument (expressed in hertz).

angular wavelength. The angle, θ, subtended at the centre of the workpiece by one complete periodic undulation (see figure 1) expressed in degrees or radians as follows:

$$\theta \text{ (in degrees)} = \frac{360}{\textit{undulations per revolution}}$$

$$\theta \text{ (in radians)} = \frac{2\pi}{\textit{undulations per revolution}}$$

circumferential wavelength. The circumference of the workpiece divided by the *sinusoidal undulation number*.

*These terms define reference circles to be fitted to the traced profile of the workpiece and to the displayed representation of the traced profile when adequately centred on the axis of rotation.

Section five. Filter function of the apparatus

wave filter. A system transmitting a range of sinusoidal frequencies for which the ratio of output to input amplitude is nominally constant, while attenuating (i.e. reducing) the ratio for frequencies lying outside the range at either or both ends.

NOTE. A characteristic of the electric wave filter is that the ratio is dependent only on frequency and is independent of amplitude, in contrast to mechanical filtering methods (e.g. by the stylus) which are influenced by frequency and also by amplitude.

amplitude transmission characteristic. The ratio of output amplitude to input amplitude plotted for each of a range of sinusoidal frequencies covering the operative range of apparatus.

NOTE. The ratio may be expressed as a percentage or in decibels.

rate of attenuation of filter. The maximum slope of the transmission characteristic.

NOTE 1. The attenuation rate is determined by the design of the filter and is expressed in decibels per octave.
NOTE 2. The rate of attenuation at the 75 % transmission cut-off may also be significant and be expressed.

phase shift. Displacement, expressed in time or space, between sinusoidal output and input signals of given frequency.

NOTE. The phase shift produced by a filter, such as the 2–CR filter, is generally dependent on the rate of attenuation at each frequency considered. Phase shift through a phase corrected (digital) filter can be zero or the same for all frequencies.

undulation cut-off. The number of sinusoidal undulations per 360° at the upper or lower end of the pass band, where the transmission has been attenuated to 75 % of its maximum (except for 1 undulation per revolution).

undulation range of the filter. The range of undulations lying between the upper and lower undulation cut-off.

NOTE. Undulation range may be expressed in spatial terms (undulations per 360°) or in temporal means (frequency in hertz).

Section six. Additional terms

method divergence. The numerical difference between the results of two methods of

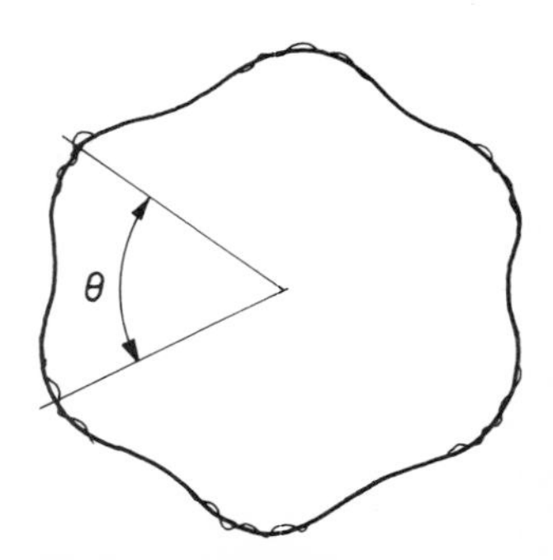

Figure 1. Angular wavelength

measurement that are both standardized but give values that are nominally but not precisely equal.

measurement between support centres. A method in which the axis used for measurement is the common axis between the support centres formed in the workpiece itself.

NOTE. In this method the real axis of rotation of the workpiece can lead to differences in the measurement that can be caused by:

(a) defects of form, orientation, alignment of the centres and the bench marks;
(b) possible eccentricity of the section being checked.

Appendix A

Coordinate system for roundness measurement

The coordinate system for roundness measurement is shown in figure 2.

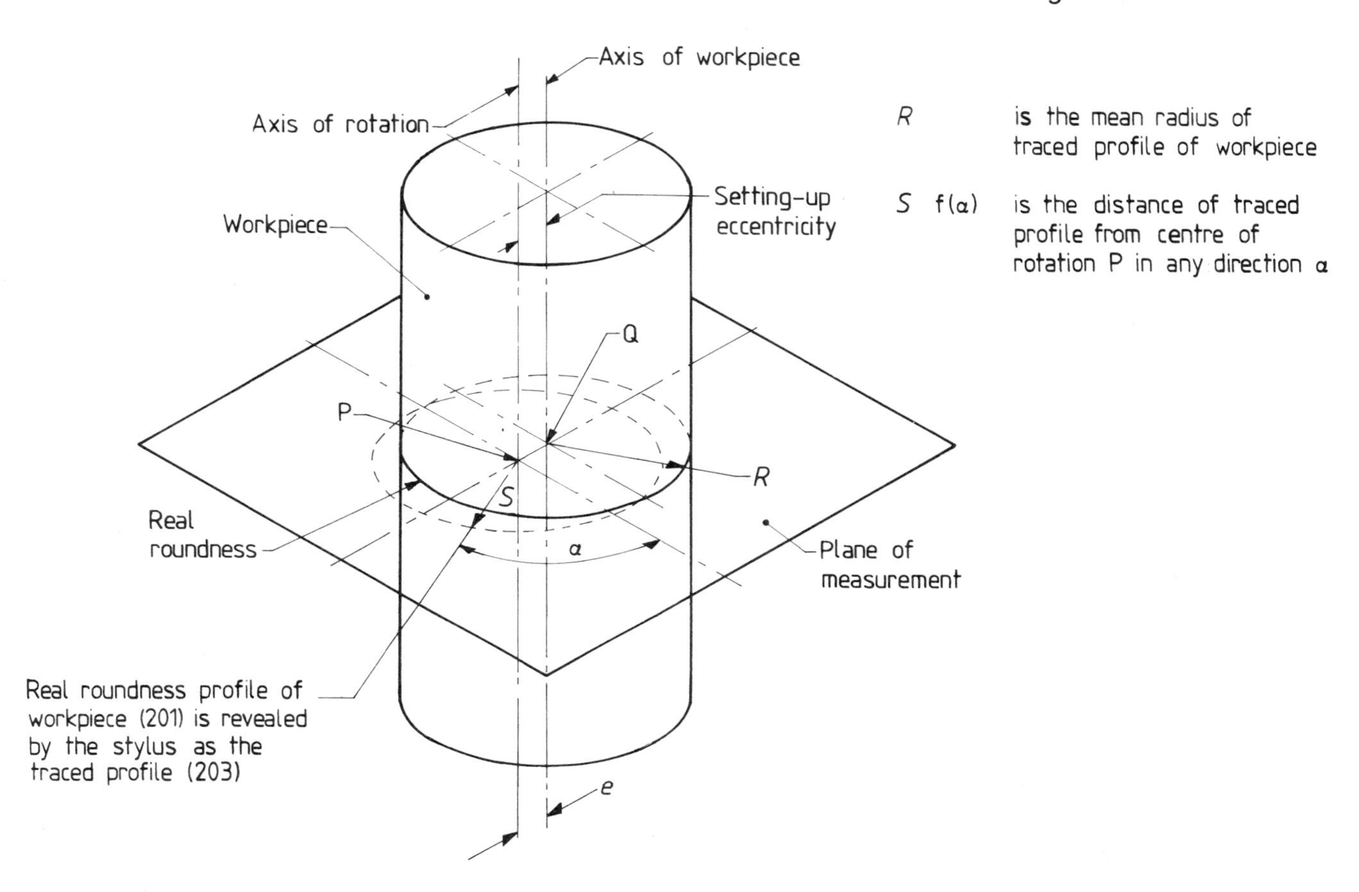

Figure 2. Coordination system for roundness measurement

Appendix B

Evaluation of workpiece parameters

The evaluation of workpiece parameters is shown in flowchart form in figure 3.

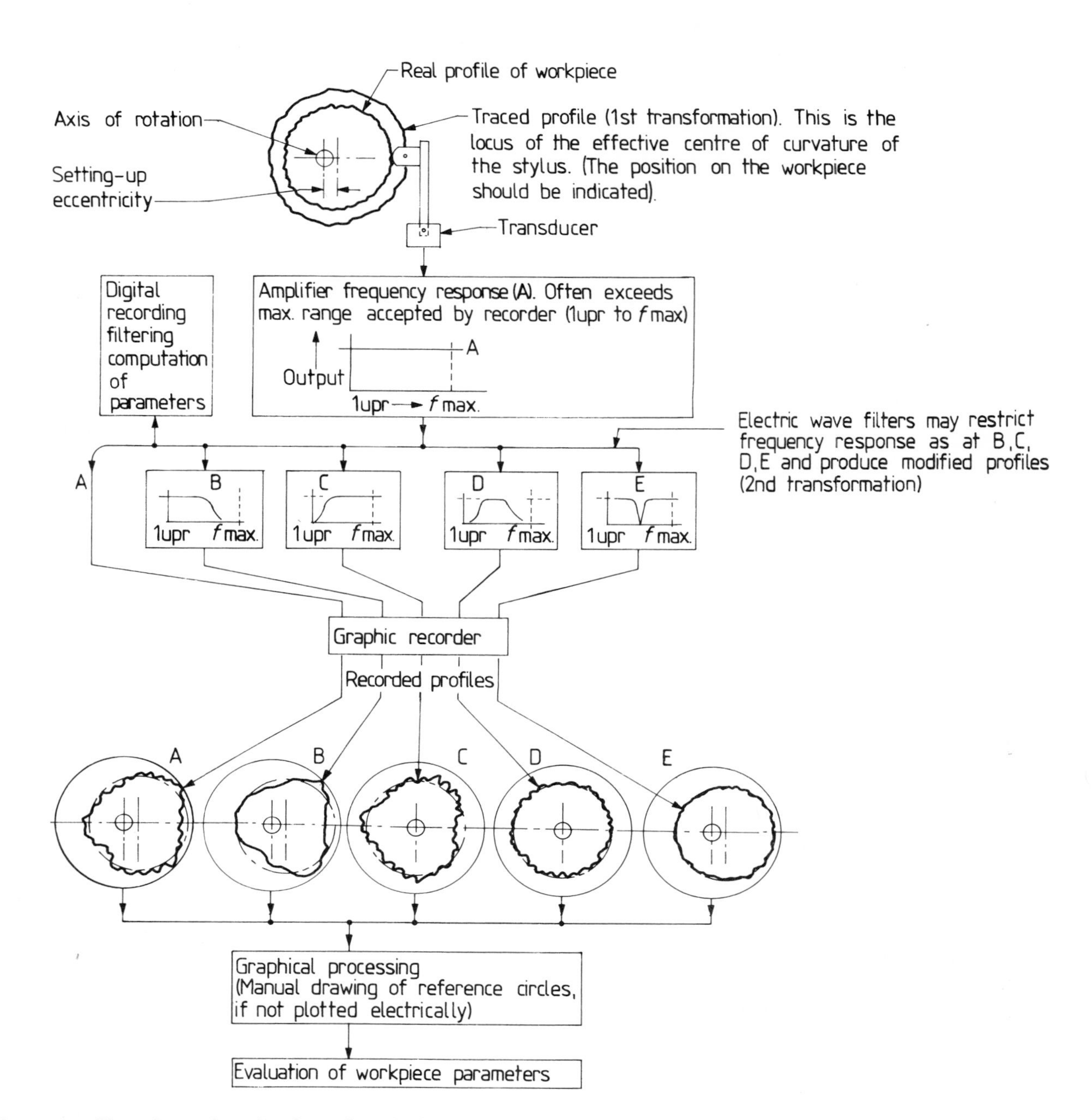

Figure 3. Flowchart of evaluation of workpiece parameters

NOTE. Channels A, B and E transmit 1 undulation per revolution and show eccentricity. C and D reject 1 undulation per revolution and show the profile with its least squares circle centred on the chart.

BS 3730: Part 2 specifies the characteristics of contact (stylus) instruments based on the method of determining departures from roundness by measuring variations in radius. It relates to the assessment of the departures from ideal roundness of a workpiece through the medium of a profile transformation, obtained under reference conditions, expressed as any one of the following centres:

(a) centre of the least squares circle;
(b) centre of the minimum zone circle;
(c) centre of the minimum circumscribed circle;
(d) centre of the maximum inscribed circle.

Each of the above centres may have its particular field of application.

Appendices deal with the departure of roundness from the measured profile, use of instruments, calibration and determination of systematic errors of rotation, and rules for plotting the position of the least squares centres.

BS 3730

Assessment of departures from roundness

Part 2. Specification for characteristics of stylus instruments for measuring variations in radius (including guidance on use and calibration)

. . . .

2. Definitions

For the purposes of this Part of BS 3730 the definitions given in BS 3730: Part 1 apply, together with the following.

overall instrument error. The difference between the value of the parameter indicated, displayed or recorded by the instrument and the true value of the parameter. The value of this error is determined by measuring a test piece.

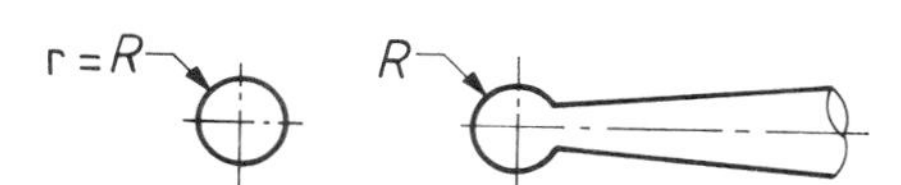

Figure 1. Spherical stylus

3. Instruments

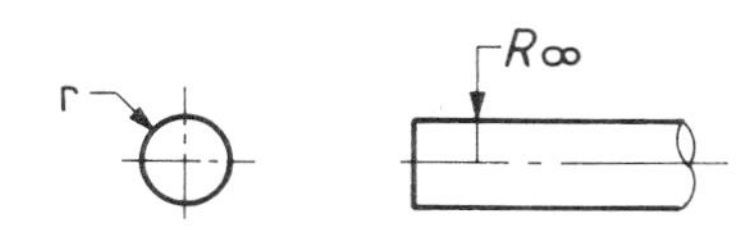

Figure 2. Cylindrical stylus

3.1 Instrument types and general requirements

3.1.1 *General.* Instruments of the stylus type employed for the determination of departures from ideal roundness shall be one of the following types:

(a) a stylus and transducer rotating round a stationary workpiece; *or*

(b) a rotating workpiece engaged by a stationary stylus and transducer.

NOTE. By the character of the output information, instruments for the measurement of roundness fall into two groups:

(a) profile recording;
(b) direct display of the values of the parameters.

Both groups can be combined in one instrument.

Stylus instruments shall comply with **3.1.2** to **3.1.4**.

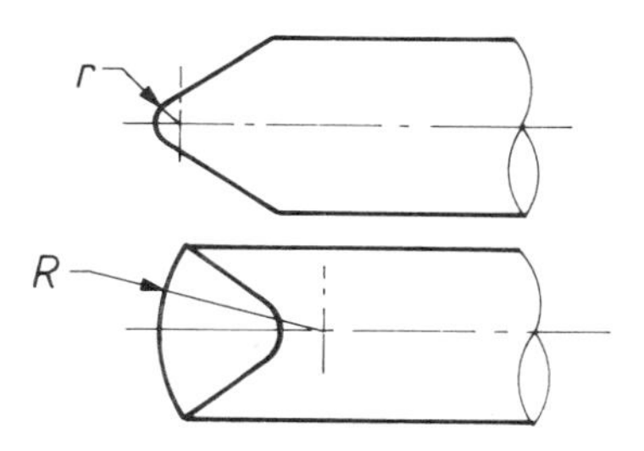

Figure 3. Toroidal (hatchet) stylus

3.1.2 *Stylus types and dimensions.* The surface characteristics of the part under examination are of primary importance in the choice of stylus and variations to comply with different requirements, depending upon the nature and magnitude of the irregularities that are to be taken into account, shall be as shown in figures 1 to 4 (see appendix A).

NOTE. No order of preference is implied by the order of figures 1 to 4.

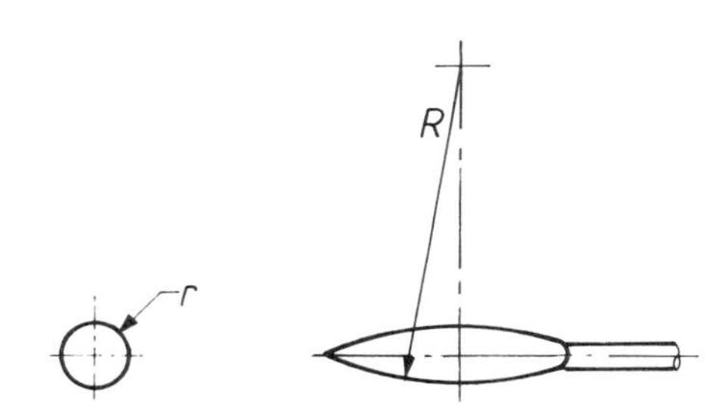

Figure 4. Ovoidal stylus

The dimensions *r* and *R* of the various styli shall be selected from the following values:

0.25 mm, 0.8 mm, 2.5 mm and 8 mm.

. . . .

3.2 Instrument error

3.2.1 *Overall instrument error.* The overall instrument error, comprising systematic and random components from the spindle error, electric noise, vibration, magnification, etc., shall be expressed as a percentage of the upper limiting value of the measuring range used.

3.2.2 *Errors of rotation of the instrument*

3.2.2.1 *General.* The following errors of rotation shall be determined under reference conditions at assigned positions of measurement:

(a) *Radial instrument error:* the value of roundness parameter that would be indicated by the instrument when measuring a perfectly round and perfectly centred section of a test piece, in a direction perpendicular to the reference axis of rotation;

(b) *Axial instrument error:* the value derived from the zonal parameter displayed by the instrument when measuring on a perfectly flat test piece set perfectly perpendicular to the reference axis of rotation.

NOTE. The components of errors of rotation are vector quantities and should not therefore be algebraically added to the measured value of a roundness parameter in an attempt to allow for errors of rotation.

3.2.2.2 *Statements of errors of rotation.* The displacements that the rotating member can exhibit, within the confines of its bearings, shall be divided into combinations of:

(a) radial displacements parallel to itself;
(b) axial displacements parallel to itself;
(c) tilt.

As the magnitude of the radial instrument error measured at the stylus depends on the position of the measurement plane along the axis of rotation and the magnitude of the axial instrument error depends on the radius at which the flat test piece is measured, the axial and radial positions selected for test shall be stated.

The radial instrument error shall be expressed at two stated and well separated positions along the axis, or at one position together with the rate of change of the radial instrument error along the axis.

The axial instrument error shall be expressed on the axis and at one stated radius.

Appendix A

Departure from roundness of the measured profile of the workpiece

In this Part of this standard, the departure from ideal roundness is assessed as the difference between the largest and the smallest radii of the measured profile of the workpiece, measured from one or other of the following centres.

(a) *Least squares centre (LSC):* the centre of the least squares mean circle (see figure 6).
(b) *Minimum zone centre (MZC):* the centre of the minimum zone circle (see figure 7).
(c) *Minimum circumscribed circle centre (MCC):* the centre of the minimum circumscribed circle for external surfaces (see figure 8).

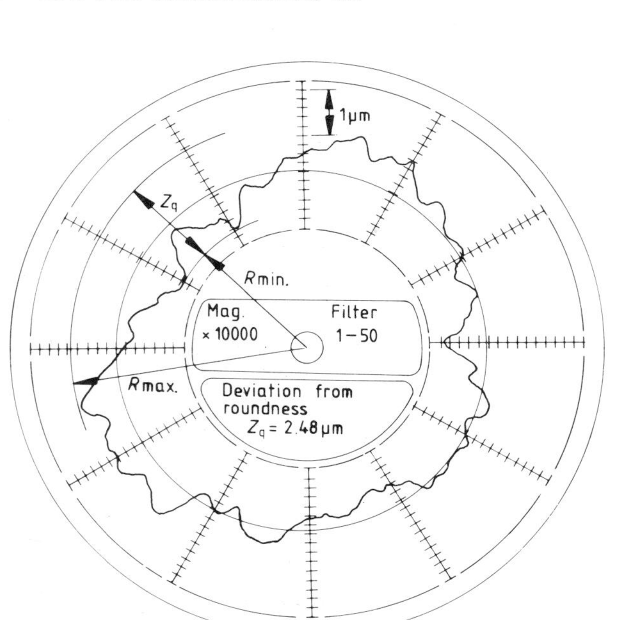

Figure 6. Assessment of roundness from least squares centre (Z_q)

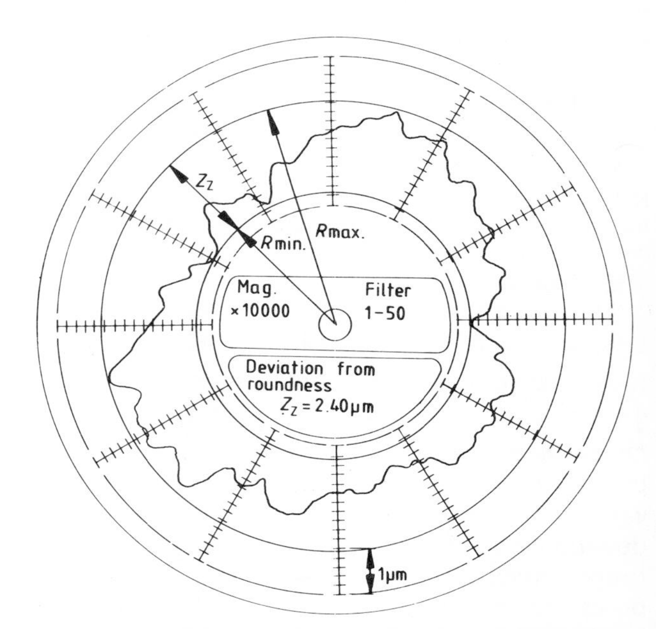

Figure 7. Assessment of roundness from minimum zone centre (Z_z)

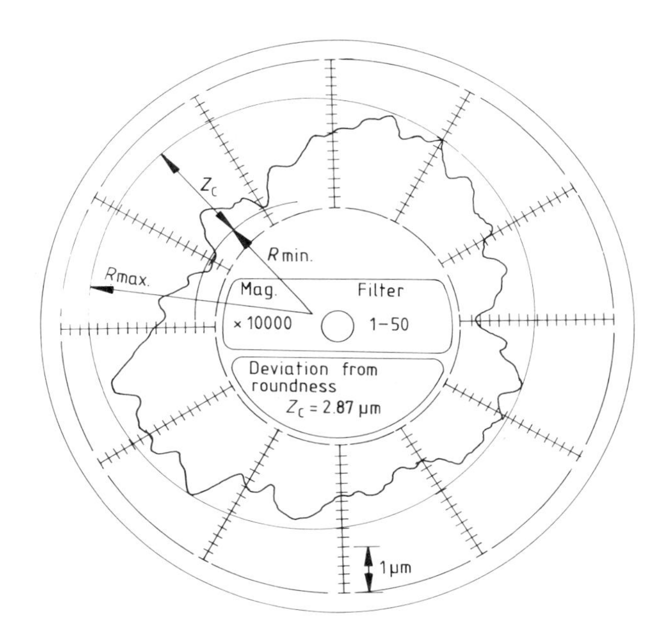

Figure 8. Assessment of roundness from centre of minimum circumscribed circle (Z_c)

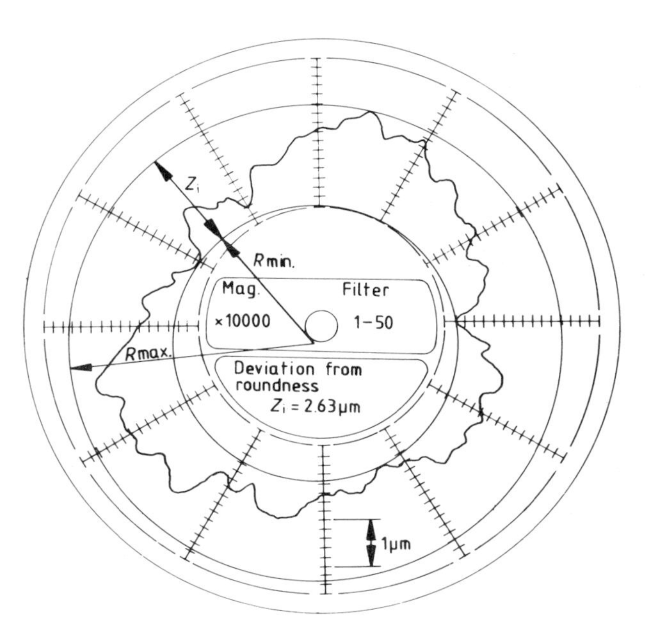

Figure 9. Assessment of roundness from centre of maximum inscribed circle (Z_i)

(d) *Maximum inscribed circle centre (MIC):* the centre of the maximum inscribed circle for internal surfaces (see figure 9).

The largest and smallest radius, in each case, is commonly used to define a concentric zone. The kind of zone may be designated by the letter z, together with a suffix denoting its centre. For the purposes of this standard the following suffices are used:

(e) least squares: suffix q, thus Z_q;
(f) minimum width: suffix z, thus Z_z;
(g) minimum circumscribed: suffix c, thus Z_c;
(h) maximum inscribed: suffix i, thus Z_i;

NOTE. The use of circles drawn on the chart to represent circles fitting the profile of the workpiece assumes that the workpiece is sufficiently well centred on the axis of the instrument (see **B.1.1**, figure 10 and appendix F).

Appendix B

Use of the instrument

B.1 General

B.1.1 This appendix gives general guidance on setting up and measurement. The workpiece is set up so that the section to be measured is sufficiently well centred on the axis of rotation to avoid excessive distortion due to eccentricity and with its axis sufficiently parallel to the axis of rotation to avoid excessive inclination errors.

Several kinds of distortion result from polar plotting because on the chart only the variations in radius of the workpiece, together with the eccentricity, and not the radius itself, are highly magnified.

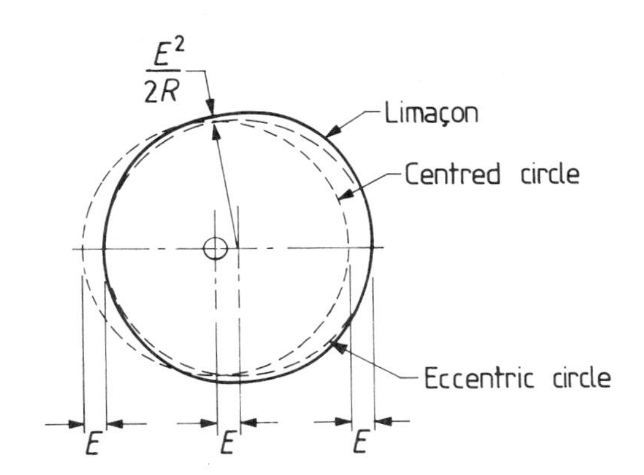

Figure 10. Effect of eccentricity

B.1.2 In the direction of eccentricity, the radius of the eccentric plot is independent of the eccentricity, but in the perpendicular direction the radius is slightly increased in proportion to the square of the eccentricity (see figure 10). Strictly, the eccentric plot of a perfect circle has the form of a limaçon, which however is hardly perceptible as such when the eccentricity is very small. Graphical compensation is sometimes possible, compensation by electrical methods is widely practised and elimination of the distortion by digital correction is being realized.

B.1.3 The circumferential separation of the peaks round a periodic profile is greater than that of the valleys even though the difference on the workpiece is negligibly small and, to avoid giving a misleading impression, the ratio of peak to valley radius measured from the centre of the chart should not be too great.

B.1.4 Inclination of the axis of the workpiece to the axis of rotation will cause a perfectly round cylinder to appear elliptical. The diametral difference on the chart will be given by:

$MD\ (1-\sec\theta)$

where

D is the diameter of the workpiece (see figure 11);
θ is the angle of inclination (see figure 11);
M is the magnification.

Conversely, appropriate inclination can cause an elliptical cylinder to appear to be round.

B.1.5 Some guiding rules for plotting and reading polar graphs are given in appendix F.

B.2 Direction of measurement

B.2.1 When the workpiece is a cylinder, its roundness will be assessed in a cross section perpendicular to the axis of rotation of the instrument, the direction of measurement will be perpendicular to this axis and the traced profile that is plotted and measured will be that of the cross section. This forms the normal basis of roundness measurement and assessment.

B.2.2 When the workpiece is conical or toroidal, the question of which is the more significant functional direction has to be determined by details of the application and the direction in which the surface is likely to be operative. Also, the question can arise whether the direction of measurement should be perpendicular to the axis, or normal to the surface (see figure 12). If the direction of measurement is normal to the surface, the profile will be that formed by the intersection of the workpiece with a perfect, nominally coaxial cone of the complementary semi-angle, along the generators of which the variations in the profile will be measured. On the profile graph, however, these variations will be displayed as though they were normal radial variations and their zone width will have to be multiplied by the secant of the semi-angle of the workpiece cone if the radial value expressed in the normal cross section is required. A ball bearing ring raceway (see figure 13), which is a portion of a toroid, can be treated as a conical surface formed by the tangents to the zone contacted by the stylus.

B.3 Considerations regarding roughness texture and interdependance of roundness, roughness and stylus radius

B.3.1 The question will arise of whether closely spaced irregularities of the cross section, which can generally be traced to circumferential components of the roughness texture, should be included in or excluded from the zonal assessment of roundness (the roundness parameters as so far defined).

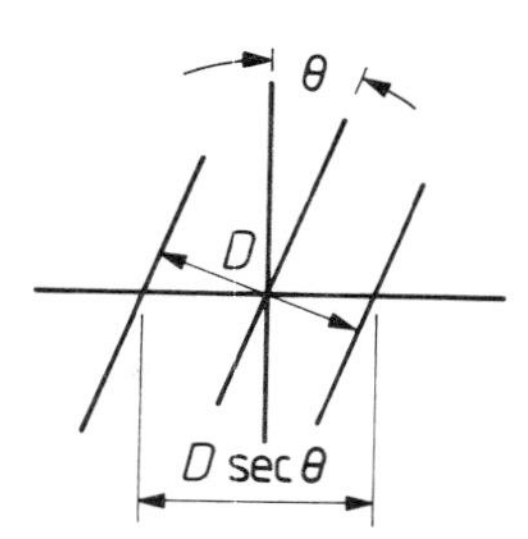

Figure 11. Effect of inclination of workpiece axis

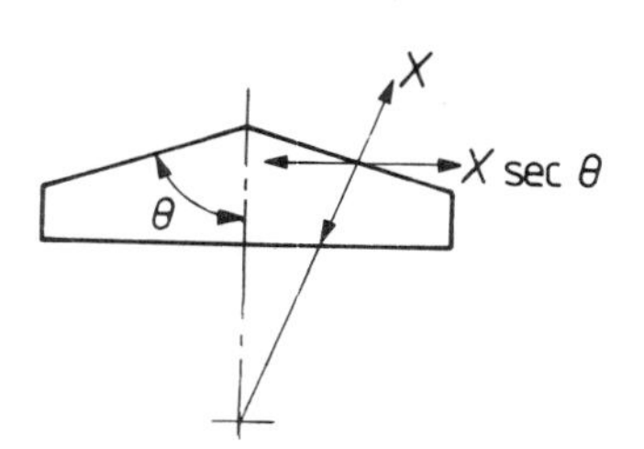

Figure 12. Conical surface

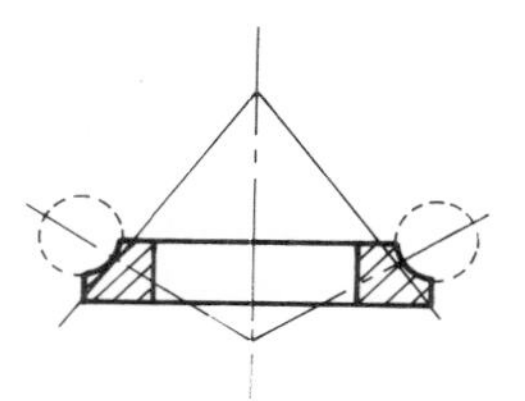

Figure 13. Toroidal surface

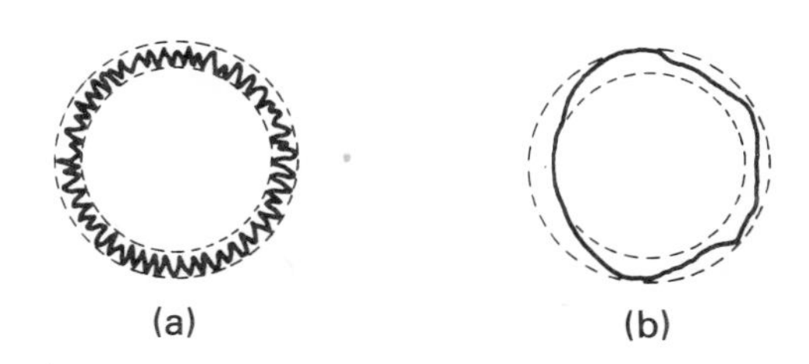

Figure 14. Effect of roughness texture

The decision has to reflect the intended use of the information obtained and the intended use of a workpiece. For example, sliding contact with another surface of similar form can be distinguished from rolling contact with balls and rollers. The inclusion or exclusion of the effects of roughness texture by instrumental means can greatly affect the value of the roundness parameter.

Consider the profiles in figures 14(a) and 14(b). They have the same value of zonal parameter but their very different characteristics are traceable to different causes and they are unlikely to be equal functionally.

If the two profiles are those of ball bearing raceways, figure 14(a) would give rise to high frequency vibration and noise and figure 14(b) might be preferred; but if they are the profiles of shafts, mandrels, piston, etc., it is likely that figure 14(a) will be preferred.

If the point of interest is the geometry of the workpiece or of the machine that made it, the geometry generally being characterized by a relatively small number of peripheral undulations, it is likely that the most meaningful assessment will be made by excluding the roughness, which could sometimes be large enough to mask the departure from roundness. The roughness may then have to be considered separately.

B.3.2 The extent to which the circumferential components of the roughness texture are taken into account depends on the characteristics of the texture (lay, height, spacing) and on the dimensions of the stylus in combination with the frequency response of the instrument.

B.3.3 Experimental evidence has indicated that a stylus of about 10 mm radius engaging a workpiece with straight generators will suppress most of the axial component of the grinding and turning marks normally encountered, but is less effective in suppressing residual circumferential components or roughness with an axial lay (extrusion, broaching) because of difficulty in securing a small enough difference of circumferential curvature.

Figure 15 shows diagrammatically how styli of short and long radii react to the tool marks on a turned cylinder. The short-radius stylus will move from the crest on the one side to a valley on the other side and back to the crest again and in doing so will follow a truly circular path only if the shape of a tool mark happens to be truly sinusoidal, which it rarely is. On the other hand, if a hatchet-like stylus of long radius is used, the record will be representative of the roundness of the envelope of the part. It will be substantially circular despite the presence of the tool marks.

The principle is illustrated in figure 16 which shows the envelope A traced with a hatchet stylus and the cross section B traced with a sharply pointed stylus of a part turned in an ordinary toolroom lathe, the tool producing a tool mark as shown separately. The styli were adjusted so that they would contact a smooth cylinder in the same transverse plane. Thus the trace of the sharp stylus should lie everywhere inside that of the hatchet, except at the highest crest where the two traces could touch.

The envelope traced by the hatchet is as round as can be assessed at the low magnification but the lack of roundness in the cross section traced by the sharp stylus is obvious.

A spherical radius of less than 0.8 mm, for example

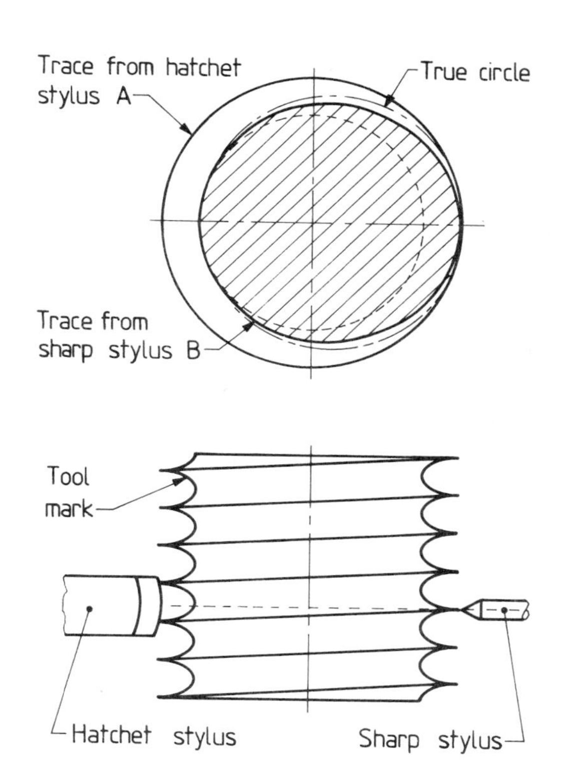

Figure 15. Effect of stylus radius when in contact with surface

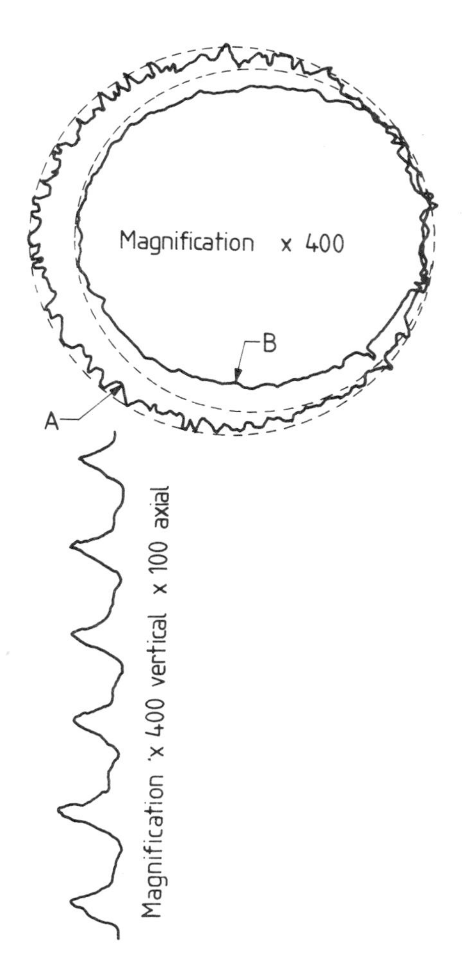

Figure 16. Traces by hatchet and sharp styli

0.25 mm, would fully enter a turned texture produced by a tool having the widely used radius of 0.8 mm and would largely enter many scratch marks produced by grinding but could still suppress the finest texture as produced by lapping, honing and the finest grinding.

There are advantages in using a small radius for the circumferential direction combined with a large axial radius: hence the often used toroidal (hatchet) form, which facilitates measurement in holes.

B.3.4 High frequency circumferential components, whether found by a sharp or by a blunt stylus, are best suppressed by means of an electric wave filter having a suitable cut-off.

B.3.5 The choice of stylus radius for the measurement of grooves (e.g. ball bearing raceways) involves not only the question of roughness but also that of positioning the stylus in the groove.

It will be seen from figure 17 that if the centre of the stylus is offset from the direction of measurement XX, errors in the measurement will result if the offset *y* varies as the stylus rotates and that the probability of error will increase as the difference in radii of stylus and workpiece is reduced.

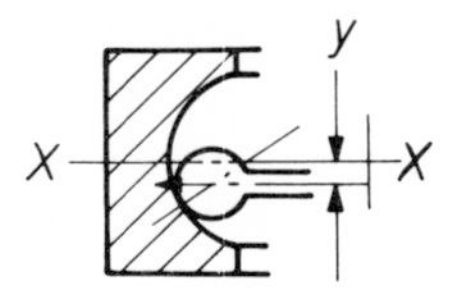

Figure 17. Positioning of a stylus in a groove

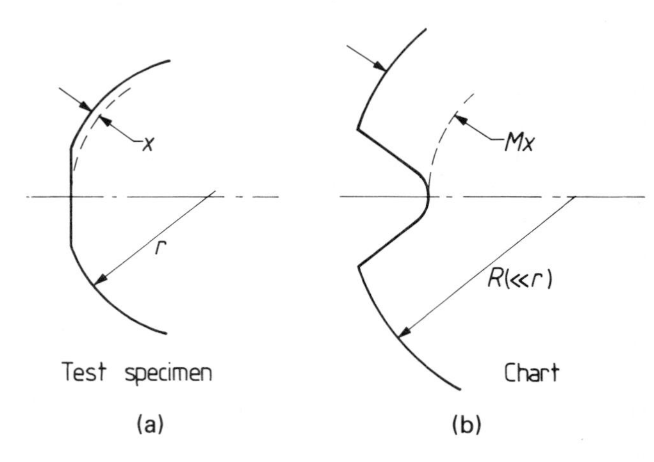

Figure 18. Error due to stylus displacement

Appendix C

Calibration

C.1 Calibration of radial magnification. Static calibration can be effected by displacing the stylus in any convenient and precise way, for example by means of a screw-driven reducing lever or by gauge blocks.

Dynamic calibration can be effected by means of a cylinder having one or more small flats round the periphery (see figure 18). Because the angular subtense of a flat is generally small and may come near the high frequency response of the instrument, such specimens need to be proportioned and calibrated for the particular characteristics of the instrument with which they can be used.

C.2 Calibration of axial magnification. Static calibration can generally be effected with the same devices as are used for the calibration of radial magnification.

Appendix D

Determination of systemmatic errors of rotation

D.1 General. Provided that the errors of a given spindle are sufficiently repeatable, their polar values can be determined and applied as corrections to the profile. In some instruments this is done automatically, so that the recorded profiles and output data are presented as though the spindle error was zero.

It is important to understand that the errors of the spindle and specimen combine vectorially, so that those of the spindle cannot be determined or allowed for by simple subtraction from the combined profile of spindle and test piece or spindle and specimen.

D.2 Methods of determining spindle errors

D.2.1 Two methods are used for separating the errors of the spindle from those of a test piece that is supposed to be truly round. They are known as the multi-step and reversal methods.

Both these techniques assume that the rotational errors of the spindle repeat every revolution and that random errors are reduced to a minimum by carrying out the tests in the suitable environment and by averaging over one or more revolution.

D.2.2 In the multi-step technique, which is applicable to radial and axial errors and combinations of these, the test piece is mounted on an indexing table. The accuracy of rotation of the indexing table is unimportant since it remains stationary whilst actual measurements are being made. Roundness data is recorded over, for example, 4 revolutions of the spindle to permit assessment of the random error to give some degree of averaging and the information is stored. The test piece is then indexed through 30° and roundness data is again recorded and stored before indexing the test piece to its next position.

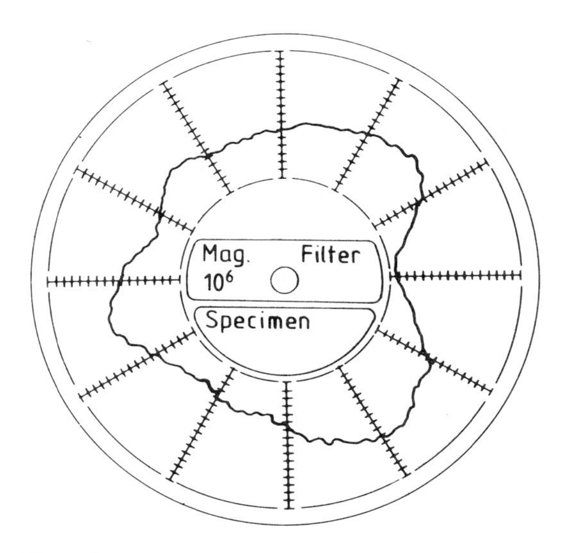

(a) Test piece rotation 0° relative to the instrument spindle

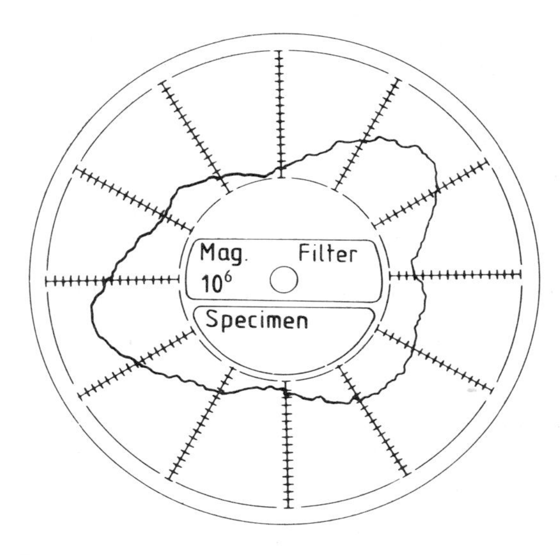

(b) Test piece rotation 60° relative to the instrument spindle

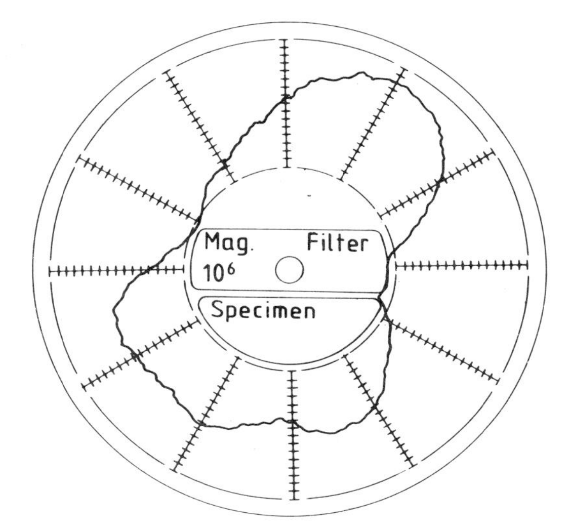

(c) Test piece rotation 120° relative to the instrument spindle

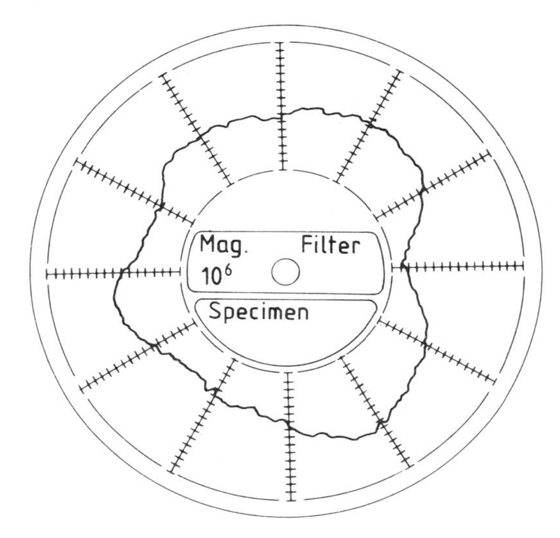

(d) Test piece rotation 180° relative to the instrument spindle

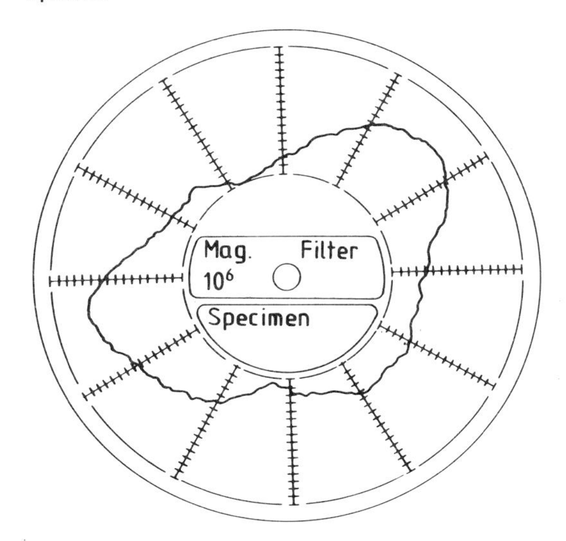

(e) Test piece rotation 240° relative to the instrument spindle

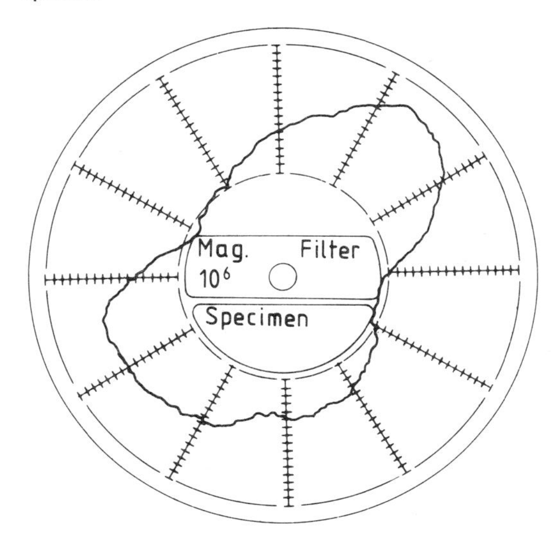

(f) Test piece rotation 300° relative to the instrument spindle

Figure 19. Illustrations of spindle errors and test piece out-of-roundness

The process is continued until the test piece has been indexed in 12 steps through 360° when additional data may be obtained which can be used to identify any system drift.

At each measuring position the data will be a combination of the rotational errors of the spindle and the out-of-roundness of the test piece. When the test piece is indexed through 30°, the errors will combine in different phases as illustrated in figure 19. A complete matrix of equations will therefore be obtained after indexing the test piece through 360° relative to the instrument spindle. This matrix can be solved in a computer and the individual errors of either the spindle or the test piece can be printed out or shown on the graph recorder of the instrument after reversion from digital to analogue form.

D.2.3 The reversal method, which is applicable only to radial errors, is one that has been commonly applied in many fields of metrology. The procedure is to record the profile of the test piece at the highest possible instrument magnification and then to record on the same chart the profile obtained after rotating the test piece through 180° with the instrument pick-up relocated 180° from its normal position relative to the spindle. The locus of the bisector of the two profiles represents out-of-roundness of the test piece. The rotational error of the spindle can be found in a similar manner but it is necessary to change the sign convention of the instrument for the second profile recording before plotting the bisector of the two profiles. The principle is similar to solving two equations of the form $x \times y$ and $x - y$.

The separated radial errors of the spindle and test piece are shown in figure 20.

D.2.4 Tilt, or coning error, can be taken into account by measuring the errors at two axial and/or two radial positions stated with reference to a feature or features of the spindle. The tilt error may then be expressed by giving these values and positions, or by giving the value at one of these positions and the rate of change therefrom, for example in micrometres per metre, assuming adequate linearity.

Appendix E

Determination of least squares centre and circle

From the centre of the chart (see figure 21), draw a sufficient even number of equally spaced radial ordinates.

In figure 21 they are shown numbered 1 to 12. Two of these at right angles are selected to provide a system of rectangular coordinates XX, YY.

The distances of the points of intersection of the polar graph with these radial ordinates, P_1 to P_{12}, are measured from the axes XX and YY, taking positive and negative signs into account.

The distances x_4 and y_4 of the least squares centre from the centre of the paper are calculated from the following approximate formula:

$$a = \frac{2\Sigma x}{n}$$

$$b = \frac{2\Sigma y}{n}$$

where

Σx is the sum of the x values;
Σy is the sum of the y values;
n is the number of ordinates.

(a) Spindle error

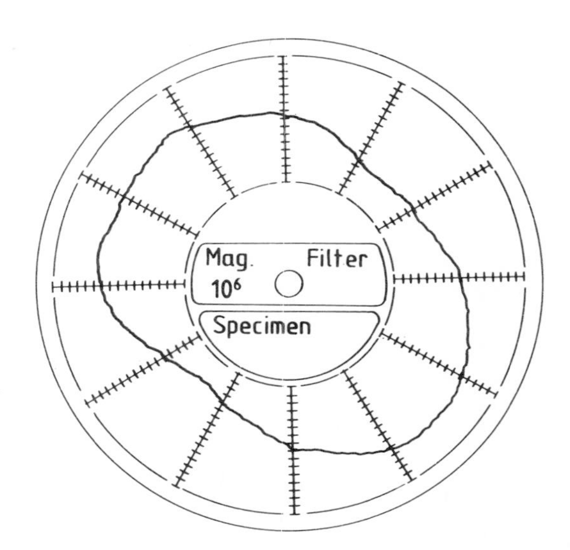

(b) Test piece out of roundness

Figure 20. Example of spindle error and test piece error

The radius R of the least squares circle, if wanted, is calculated as the average radial distance of the points P from the least squares centre, that is:

$$R = \frac{\Sigma r}{n}$$

where

Σr is the sum of radial values.

Appendix F

Rules for plotting and reading polar graphs

F.1 General. The rules given in **F.2** and **F.3** can generally be applied.

F.2 Plotting

F.2.1 To avoid excessive distortion due to polar plotting, the trace should generally be kept within a zone of which the radial width is not more than about a half of its means radius (see figure 22).

F.2.2 The eccentricity E should be kept within about 15 % of the mean radius of the graph for general testing, and to within 7 % for high precision testing.

F.3 Reading

F.3.1 Angular relationships are read from the centre of rotation of the chart.

For example, points 180° apart on the workpiece are represented by points 180° apart through the centre of rotation of the chart (see figure 23).

F.3.2 Diametral variations are assessed through the centre of rotation of the chart (see figure 24).

F.3.3 Radial variations are assessed from the centre of the profile graph but are subject to a small error that limits the permissible eccentricity.

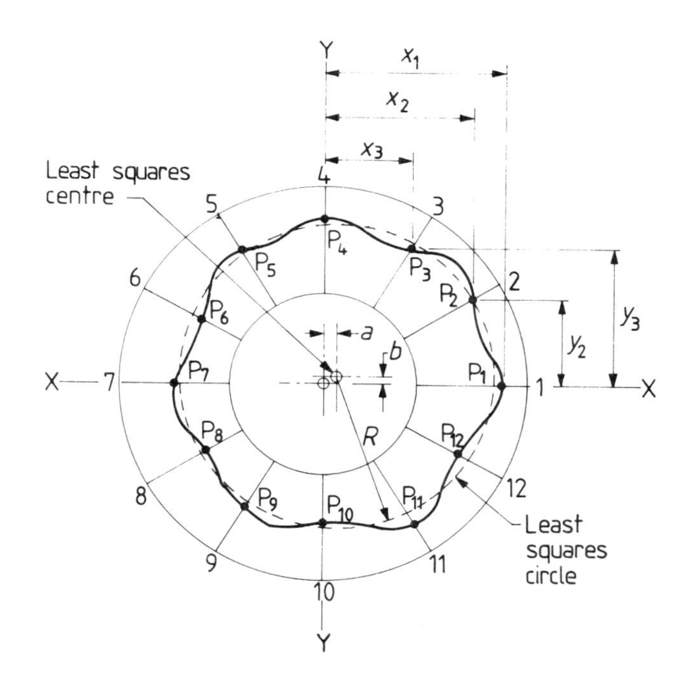

Figure 21. Determination of least squares centre and circle

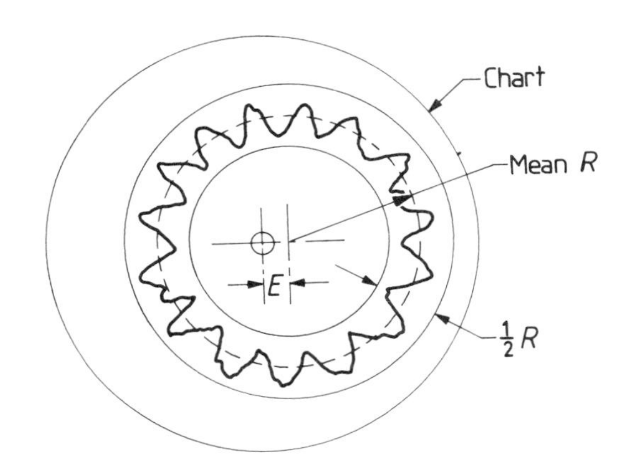

Figure 22. Distortion due to polar plotting

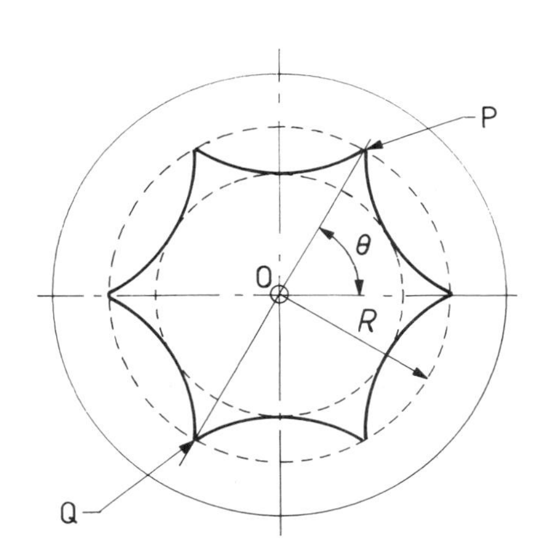

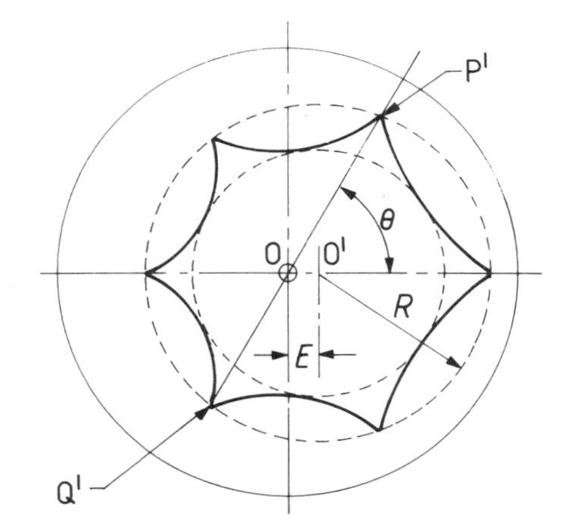

Figure 23. Diametral variations

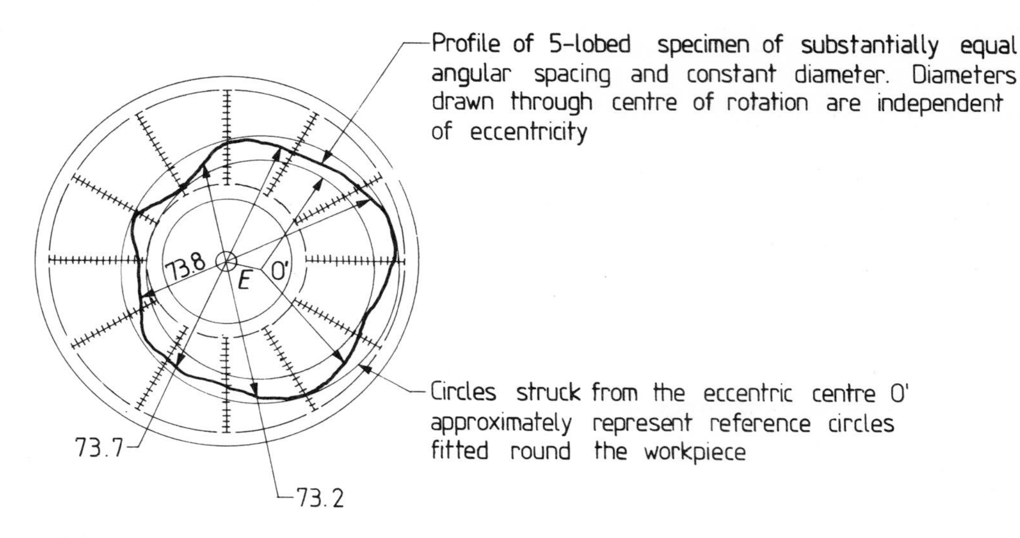

Figure 24. Radial variations

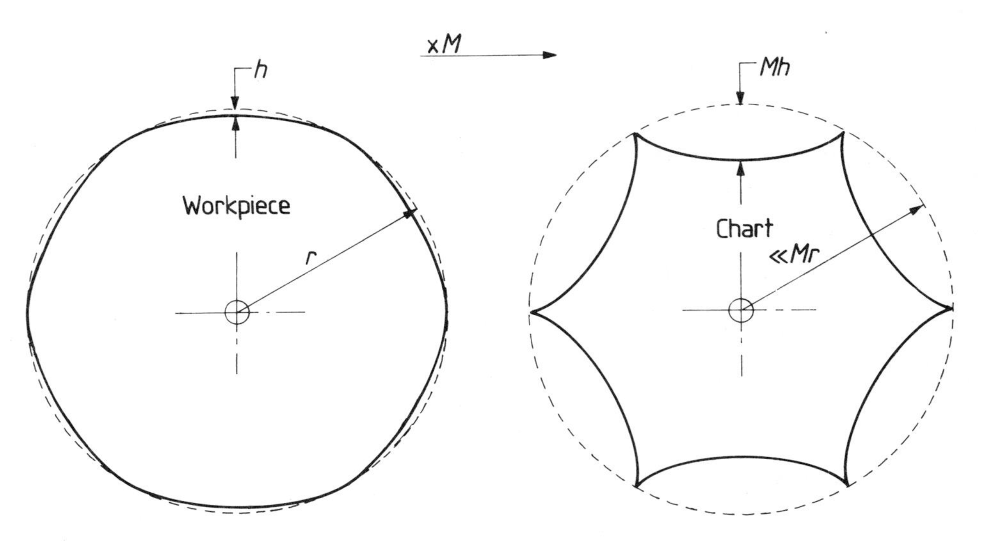

Figure 25. Error due to stylus displacement

F.3.4 It should be noted that, as a result of highly magnifying the radial variations without correspondingly magnifying the actual radius of the workpiece, portions of the surface that are convex may appear as being concave around the periphery of the displayed profile (see figures 18 and 25).

For routine or in-process inspection the procedure dealt with in BS 3730: Part 2 may either be needlessly accurate or the items under inspection may be too large to be accommodated. BS 3730: Part 3 proposes methods which may give faster and cheaper ways of assessing departures from roundness. It gives methods for the numerical assessment of departure from roundness by the combination of two-point and three-point measurement. The following methods are dealt with:

(a) Determination by means of two-point measurement (measurement of diameter).
(b) Determination by means of three-point measurement, summit (symmetrical or asymmetrical setting).
(c) Determination by means of three-point measurement, rider (symmetrical setting).

The assessed value will deviate from the true value. The difference between the measured value and the true value can be estimated with the help of Tables 2 to 8 in BS 3730: Part 3 (shown on pages 207–12), under the presumption that the undulation numbers are known and are of sinusoidal nature. For non-sinusoidal undulations, a theory for estimating such deviations is not yet available.

BS 3730

Assessment of departures from roundness

Part 3. Methods for determining departures from roundness using two- and three-point measurement

. . . .

2. Definitions

For the purposes of this Part of BS 3730 the definitions given in Part 1 apply, together with the following.

2.1 two-point measurement. Measurement between coaxial anvils, one fixed and one moving in the direction of measurement (see figures 1 and 5).

2.2 three-point measurement. Measurement between anvils, two fixed and one moving in the direction of measurement (see figures 2, 3, 4, 6, 7 and 8).

2.3 summit method. A three-point measurement in which the two fixed anvils are situated on one side and the measuring anvil is situated on the other side of the workpiece axis in the plane of measurement (see figures 2, 3, 6 and 7).

2.4 rider method. A three-point measurement in which the two fixed anvils are situated on the same side as the measuring anvil in the plane of measurement (see figures 4 and 8).

2.5 symmetrical (three-point) setting. A setting at which the direction of measurement coincides with the bisector angle between fixed anvils (see figures 2, 4, 6 and 8).

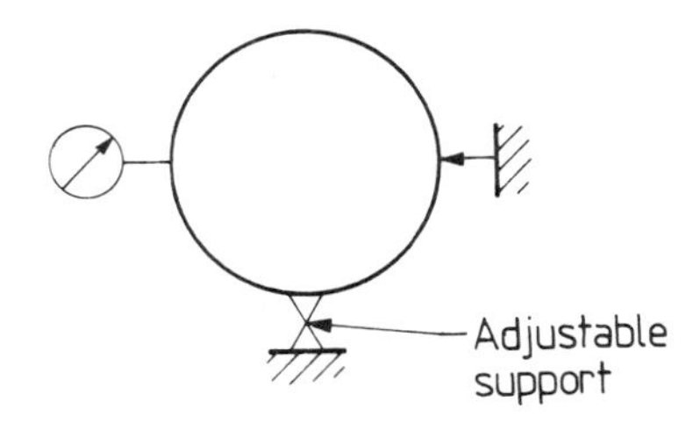

Figure 1. Two-point measurement

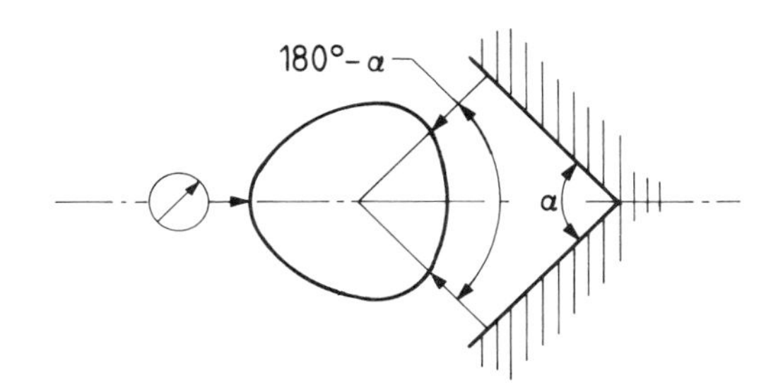

Figure 2. Three-point measurement, summit method, symmetrical setting

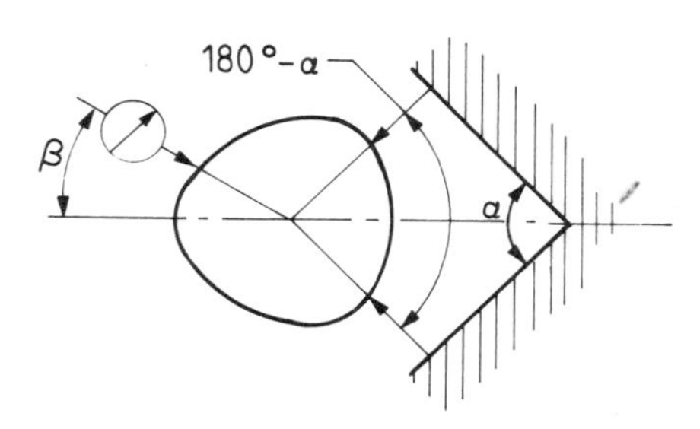

Figure 3. Three-point measurement, summit method, asymmetrical setting

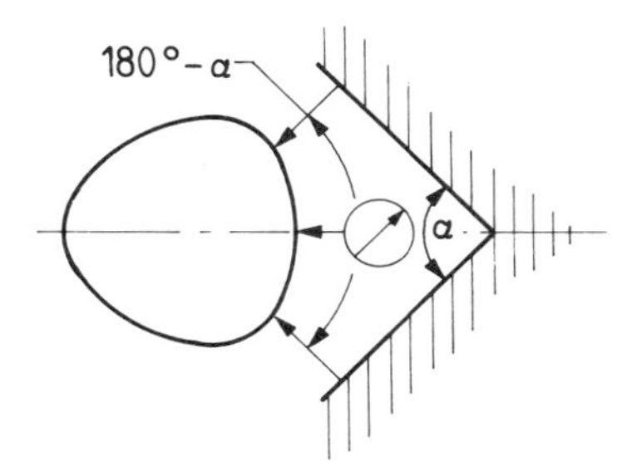

Figure 4. Three-point measurement, rider method, symmetrical setting

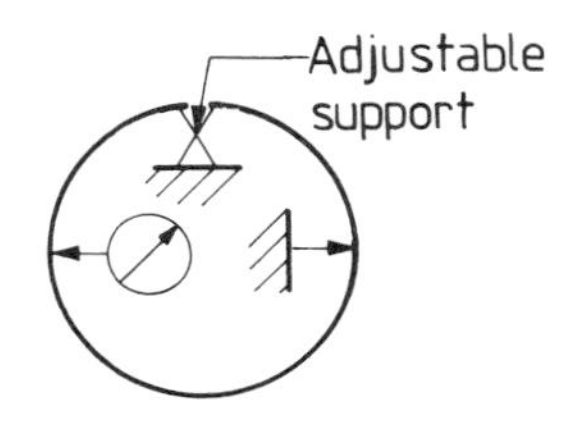

Figure 5. Two-point measurement

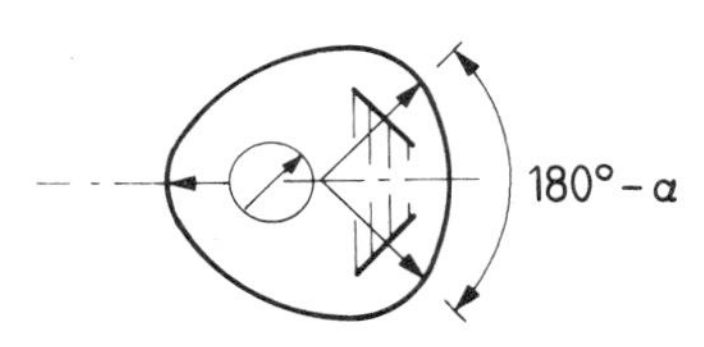

Figure 6. Three-point measurement, summit method, symmetrical setting

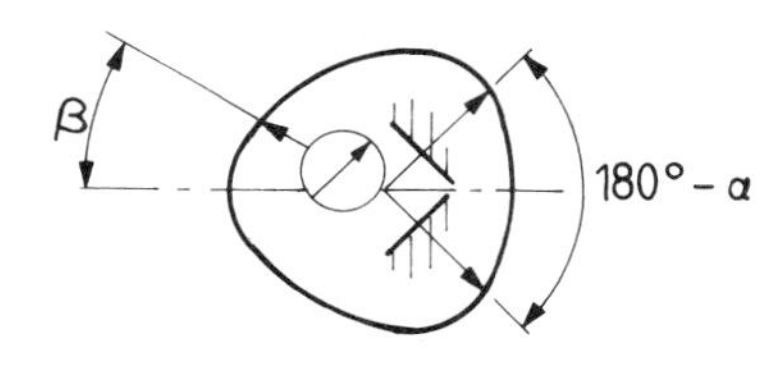

Figure 7. Three-point measurement, summit method, asymmetrical setting

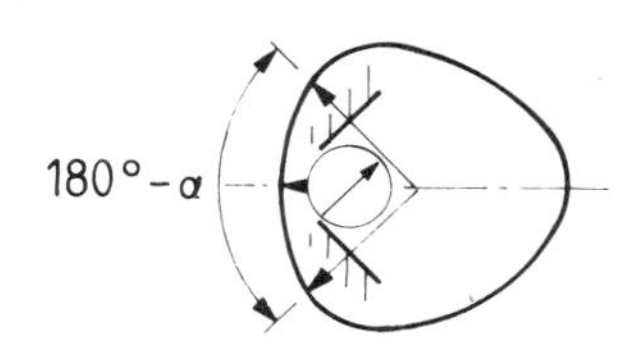

Figure 8. Three-point measurement, rider method, symmetrical setting

2.6 asymmetrical (three-point) setting. A setting at which the direction of measurement constitutes an angle with the bisector angle between fixed anvils (see figures 3 and 7).

3. Measuring conditions and instrument

3.1 Measuring anvil static force. The static measuring force shall not exceed 1 N.

NOTE 1. Preferably, the force should be adjustable and set at the lowest value that will ensure continuous contact between anvil and the surface being measured.
NOTE 2. For thin-walled workpieces, a high measuring force may affect the measuring result. Therefore it is necessary to reduce the force to the minimum value possible.

3.2 Measuring anvils. Depending on the form of the object, the measuring anvil shall be selected from table 1.

NOTE. Where the form of the object precludes the use of the anvils given in table 1, other anvils may be used.

3.3 Fixed anvils. Point or line contact shall always be used.

NOTE. The following is recommended:

(a) External measurement. V-support with a small radius. The median plane of the V-support should be in the same plane as the plane of measurement.
(b) Internal measurement. Sphere with a small radius. The median plane of the sphere should be in the same plane as the plane of measurement.

4. Procedure

In order to cover all possible form deviations and numbers of undulations always take one two-point measurement and two three-point measurements at different angles between fixed anvils.

NOTE. The measurement procedures may, under certain preconditions, be amplified (see tables 2, 3 and 4).

Select the angles between fixed anvils from the following:

symmetrical setting:
$\alpha = 90°$ and $120°$ or $\alpha = 72°$ and $108°$;

asymmetrical setting:
$\alpha = 120°$, $\beta = 60°$ or $\alpha = 60°$, $\beta = 30°$;

where
α is the angle between fixed anvils;
β is the angle between the direction of measurement and bisector of angle between fixed anvils.

5. Expression of results

Calculate the corrected value of the departure from roundness δ from the formula:

Table 1. Types of anvil

Surface form	Anvil radius	Surface radius
	mm	mm
Convex surface	Spherical: 2.5	All
Convex edge	Cylindrical: 2.5	All
Concave surface	Spherical: 2.5	≥10
Concave edge	Cylindrical: 2.5	≥10
Concave surface	Spherical: 0.5	<10
Concave edge	Cylindrical: 0.5	<10

$$\delta = \frac{\Delta}{F}$$

where
Δ is the measured departure from roundness, which is largest value obtained from the preceding two or three combinations of angles received in the required measurements;
F is the correction factor, with a value extracted from tables 2 to 8: as a first approximation F may be given a value of 2.

If the three-point measurement with symmetrical setting at 60° angles between fixed anvils is used when measuring workpieces with a known even or odd number of undulations, use the factor F given in table 8.

NOTE. This angle is useful as it gives measured values of higher correction factors than the other angles in this standard.

Tables 5, 6 and 7 give true factors F for any given number of sinusoidal undulations and measuring method and if the number of sinusoidal undulations is known, calculate the departure from roundness by using the F factors directly from tables 5, 6 or 7 as indicated in tables 2, 3 or 4.

It is not possible to calculate exactly the departure from roundness if the number of undulations is unknown and in these cases calculate a maximum, average and minimum value of δ from the formula above using the largest Δ value and the factors obtained from tables 2, 3 or 4.

When selecting F, determine the limit on the maximum number of undulations per revolution, according to whether the number of undulations is known and whether this number is an odd or even value, from tables 2, 3 and 4.

NOTE 1. For 90° and 120° settings the maximum number of undulations per revolution is 22, which assumes that a greater number of undulations than this will not have any appreciable effect on the F factor. For 72° and 108° settings the maximum number of undulations per revolution is determined by the fact that for 19 undulations the F factor cannot be determined.
NOTE 2. When using tables 5 to 8 other combinations of setting, besides those given in tables 2 to 4 can be made.

Table 2. Values of factor *F* for $\alpha = 90°$ and $\alpha = 120°$, symmetrical setting

	Combination of settings*				
	2, 3S 90° and 3S 120°	**2, 3R 90° and 3R 120°**	**2**	**3S 90° and 3S 120°**	**3R 90° and 3R 120°**
Number of undulations, n_s	**Factor, *F***				
n_s unknown, but assumed to be $2 \leq n_s \leq 22$	max. 2.41 av. 1.94 min. 1.00	max. 2.41 av. 1.96 min. 1.00	—	—	—
n_s even but unknown, but assumed to be $2 \leq n_s \leq 22$	—	—	2.00	max. 2.41 av. 2.08 min. 2.00	max. 2.41 av. 2.17 min. 1.58
n_s odd but unknown, but assumed to be $3 \leq n_s \leq 21$	—	—	—	max. 2.00 av. 1.80 min. 1.00	max. 2.00 av. 1.80 min. 1.00
n_s known and even	—	—	2.00	exact†	exact†
n_s known and odd	—	—	—	exact†	exact†

* The abbreviations in the table represent the following situations:
2 = two-point measurement;
3S 90° = three-point measurement, summit, $\alpha = 90°$;
3S 120° = three-point measurement, summit, $\alpha = 120°$;
3R 90° = three-point measurement, rider, $\alpha = 90°$;
3R 120° = three-point measurement, rider, $\alpha = 120°$.
† If multiplied by *F* factors given in table 5.

Table 3. Values of factor *F* for $\alpha = 72°$ and $\alpha = 108°$, symmetrical setting

	Combination of settings*				
	2, 3S 72° and 3S 108°	**2, 3R 72° and 3R 108°**	**2**	**3S 72° and 3S 108°**	**3R 72° and 3R 108°**
Number of undulations, n_s	**Factor, *F***				
n_s unknown, but assumed to be $2 \leq n_s \leq 18$	max. 2.70 av. 2.16 min. 2.00	max. 2.62 av. 2.11 min. 1.38	—	—	—
n_s even but unknown, but assumed to be $2 \leq n_s \leq 22$	—	—	2	max. 2.70 av. 2.16 min. 2	max. 2.70 av. 2.13 min. 2
n_s odd but unknown, but assumed to be $3 \leq n_s \leq 17$	—	—	—	max. 2.62 av. 2.06 min. 1.38	max. 2.62 av. 2.06 min. 1.38
n_s known and even	—	—	2	exact†	exact†
n_s known and odd	—	—	—	exact†	exact†

* The abbreviations in the table represent the following situations:
2 = two-point measurement;
3S 72° = three-point measurement, summit, $\alpha = 72°$;
3S 108° = three-point measurement, summit, $\alpha = 108°$;
3R 72° = three-point measurement, rider, $\alpha = 72°$;
3R 108° = three-point measurement, rider, $\alpha = 108°$.
† If multiplied by *F* factors given in table 6.

Table 4. Values of factor F for $\alpha = 60°/\beta = 30°$ and $\alpha = 120°/\beta = 60°$, asymmetrical setting and $\alpha = 90°$ symmetrical setting

Number of undulations, n_s	Combination of settings*						
	2 and 3S 60°/30°	2, 3S 60°/30° and 3S 90°	2 and 3S 120°/60°	2, 3S 120°/60° and 3S 90°	2	3S 60°/30°	3S 120°/60°
	Factor, F						
N_s unknown but assumed to be $2 \leq n_s \leq 10$	2	max. 2.41 av. 2.04 min. 2	max. 2.38 av. 2.08 min. 2	max. 2.41 av. 2.13 min. 2	—	max. 2 av. 1.6 min. 0.73	max. 2.38 av. 1.69 min. 0.42
n_s unknown but assumed to be $2 \leq n_s \leq 22$	—	max. 2.41 av. 2.04 min. 2	—	max. 2.41 av. 2.09 min. 2	—	—	—
n_s even but unknown and assumed to be $2 \leq n_s \leq 22$	—	—	—	—	2	max. 1.41 av. 1.22 min. 0.73	max. 2.38 av. 1.22 min. 0.42
n_s odd but unknown and assumed to be $3 \leq n_s \leq 9$	2	2	2	2	—	2	2
n_s odd but unknown and assumed to be $3 \leq n_s \leq 21$	—	2	—	2	—	—	—
n_s even and known	—	—	—	—	2	exact†	exact†
n_s odd and known	—	—	—	—	—	exact†	exact†

* The abbreviations in the table respresent the following situations:

2 = two-point measurement;

3S 60°/30° = three-point measurement, summit, $\alpha = 60°$, $\beta = 30°$;

3S 120°/60° = three-point measurement, summit, $\alpha = 120°$, $\beta = 60°$;

3S 90° = three-point measurement, summit, $\alpha = 90°$ (symmetrical setting).

† If multiplied by F factors given in table 7.

Table 5. Values for factor F for $\alpha = 90°$ and $\alpha = 120°$, symmetrical setting, for $n_s = 2$ to $n_s = 22$

Number of undulations, n_s	Combination of settings*				
	2	3S 90°	3S 120°	3R 90°	3R 120°
	Factor, F				
2	2	1	1.58	1	0.42
3	†	2	1	2	1
4	2	0.41	0.42	2.41	1.58
5	†	2	2	2	2
6	2	1	†	1	2
7	†	†	2	†	2
8	2	2.41	0.42	0.41	1.58
9	†	†	1	†	1
10	2	1	1.58	1	0.42
11	†	2	†	2	†
12	2	0.41	2	2.41	†
13	†	2	†	2	†
14	2	1	1.58	1	0.42
15	†	†	1	†	1
16	2	2.41	0.42	0.41	1.58
17	†	†	2	†	2
18	2	1	†	1	2
19	†	2	2	2	2
20	2	0.41	0.42	2.41	1.58
21	†	2	1	2	1
22	2	1	1.58	1	0.42

* See asterisk footnote to table 2.

† In this case the method gives no indication of deviation from roundness.

Table 6. Values of factor F for $\alpha = 72°$ and $\alpha = 108°$, symmetrical setting, for $n_s = 2$ to $n_s = 22$

Number of undulations, n_s	Combination of settings*				
	2	3S 72°	3S 108°	3R 72°	3R 108°
	Factor, F				
2	2	0.47	1.38	1.53	0.62
3	†	2.62	1.38	2.62	1.38
4	2	0.38	†	2.38	2
5	†	1	2.24	1	2.24
6	2	2.38	†	0.38	2
7	†	0.62	1.38	0.62	1.38
8	2	1.53	1.38	0.47	0.62
9	†	2	†	2	†
10	2	0.70	2.24	2.70	0.24
11	†	2	†	2	†
12	2	1.53	0.38	0.47	0.62
13	†	0.62	1.38	0.62	1.38
14	2	2.38	†	0.38	2
15	†	1	2.24	1	2.24
16	2	0.38	†	2.38	2
17	†	2.62	1.38	2.62	1.38
18	2	0.47	1.38	1.53	0.62
19	†	†	†	†	†
20	2	2.70	2.24	0.70	0.24
21	†	†	†	†	†
22	2	0.47	1.38	1.53	0.62

* See asterisk footnote to table 3.
† In this case the method gives no indication of deviation from roundness.

Table 7. Values of factor F for $\alpha = 60°/\beta = 30°$ and $\alpha = 120°/\beta = 60°$, asymmetrical setting, for $n_s = 2$ to $n_s = 22$

Number of undulations, n_s	Combination of settings*		
	2	3S 60°/30°	3S 120°/60°
	Factor, F		
2	2	1.41	2.38
3	†	2	2
4	2	1.41	1.01
5	†	2	2
6	2	0.73	0.42
7	†	2	2
8	2	1.41	1.01
9	†	2	2
10	2	1.41	2.38
11	†	†	†
12	2	0.73	1.01
13	†	†	†
14	2	1.41	0.42
15	†	2	2
16	2	1.41	1.01
17	†	2	2
18	2	0.73	2.38
19	†	2	2
20	2	1.41	1.01
21	†	2	2
22	2	1.41	0.42

* See asterisk footnote to table 4.
† In this case the method gives no indication of deviation from roundness.

Table 8. Values of factor F for $\alpha = 60°$, symmetrical setting

Number of undulations, n_s	Combination of settings*	
	3S 60°	3R 60°
	Factor, F	
2	†	2
3	3	3
4	†	2
5	†	†
6	3	1
7	†	†
8	†	2
9	3	3
10	†	2
11	†	†
12	3	1
13	†	†
14	†	2
15	3	3
16	†	2
17	†	†
18	3	1
19	†	†
20	†	2
21	3	3
22	†	4

* The abbreviations in the table represent the following situations:
3S 60° = three-point measurement, summit, $\alpha = 60°$;
3R 60° = three-point measurement, rider, $\alpha = 60°$.
† In this case the method gives no indication of deviation from roundness.

Appendix A

Worked examples

Example 1. A centreless ground workpiece is to be verified. It is known that a three-lobed shape is present.

Available measuring equipment measures two-point, 3S 72 ° and 3S 108 °.

Measured results obtained:

$\Delta_{2\text{-point}} = 1\ \mu m$;
$\Delta_{3S\ 72°} = 8\ \mu m$;
$\Delta_{3S\ 108°} = 3\ \mu m$.

Calculation of the departure from roundness: for 3 sinusoidal undulations, table 6 shows:

$F_{2\text{-point}}$ (not applicable)
$F_{3S\ 72°} = 2.62$
$F_{3S\ 108°} = 1.38$
$\delta_{2\text{-point}}$ (not applicable)

$$\delta_{3S\ 72°} = \frac{\Delta_{3S\ 72°}}{F_{3S\ 72°}} = \frac{8}{2.62} \approx 3\ \mu m$$

$$\delta_{3S\ 108°} = \frac{\Delta_{3S\ 108°}}{F_{3S\ 108°}} = \frac{3}{1.38} \approx 2\ \mu m;$$

$$\delta = \delta_{max} = \delta_{3S\ 72°} \approx 3\ \mu m$$

Approximation of the departure from roundness:

$F = 2$

$$\delta = \frac{\Delta_{max}}{2} = \frac{8}{2} = 4\ \mu m$$

Example 2. A turned cylindrical bore is to be verified. Available measuring equipment measures two-point, 3S 90 ° and 3S 120 °.

Measured results obtained:

$\Delta_{2\text{-point}} = 2\ \mu m$;
$\Delta_{3S\ 90°} = 30\ \mu m$;
$\Delta_{3S\ 120°} = 27\ \mu m$.

Calculation of the departure from roundness: by comparing the results obtained with the given *F*-factors (2-point near zero value, 90 ° and 120 ° values almost equal), table 5 indicates that there are 5 or 19 sinusoidal undulations.

For 5 and 19 lobes the following *F*-factors are given in table 5:

$F_{2\text{-point}}$ (not applicable)
$F_{3S\ 90°} = 2$
$F_{3S\ 120°} = 2$
$\delta_{2\text{-point}}$ (not applicable)

$$\delta_{3S\ 90°} = \frac{30}{2} = 15\ \mu m$$

$$\delta_{3S\ 120°} = \frac{27}{2} = 14\ \mu m$$

$$\delta = \delta_{max} = \delta_{3S\ 90°} = 15\ \mu m$$

Approximation of the departure fron roundness:

$F = 2$

$$\delta = \frac{\Delta_{max}}{2} = \frac{30}{2} = 15\ \mu m$$

Standards publications referred to in this chapter

BS 3730 Assessment of departures from roundness
- Part 1 Glossary of terms relating to roundness measurement
- Part 2 Specification for characteristics of stylus instruments for measuring variations in radius (including guidance on use and calibration)
- Part 3 Methods for determining departures from roundness using two- and three-point measurement

12 Air gauging

Introduction

Air gauging is a widely used inspection method which can be applied in a variety of ways. It is particularly suitable for providing feed-back signals in a closed loop system.

Basic principles

If a tube A which terminates in the jet J (Figures 12.1) is connected to a source of compressed air held at a constant pressure, air will flow through the jet to atmosphere at a constant rate. If the surface S is now moved towards the jet the escape of air will be impeded and the flow will begin to decrease. Continued advancement of the surface will reduce the flow until finally, with the surface in contact with the whole area of the jet, the flow ceases. This simple device permits detection of the movement of the surface normal to the jet by observing the change in air flow.

In Figure 12.2, a regulator R maintains the incoming air supply at a constant pressure and a variable-area flowmeter F measures the flow of air through the jet. The float, f, responds to the changes in flow, rising as the flow increases, and falling as the flow decreases. The flowmeter can be graduated in units of length to give a scale which can be used to measure the displacement of the surface S normal to the jet.

It is not, however, essential to measure directly the changes in air flow: these changes can be converted into changes in air pressure as shown in Figure 12.3.

A restriction 0, called the control orifice, is introduced between the regulator and the jet, and the air pressure between this restriction and the jet is measured by a suitable pressure indicator G. The indicator will register a lower pressure when the air flow through the jet increases, a higher pressure when it decreases; as before, the scale of the indicator can be graduated and used for gauging.

These are the basic principles of air gauging which can be used to build robust measuring instruments of extremely high accuracy and stability. The open jets of Figures 12.2 and 12.3, which never make contact with the work being inspected, may be described as non-contact gauges. These have the advantages of virtually eliminating wear of the gauge and the danger of damaging the surface finish of the part under inspection.

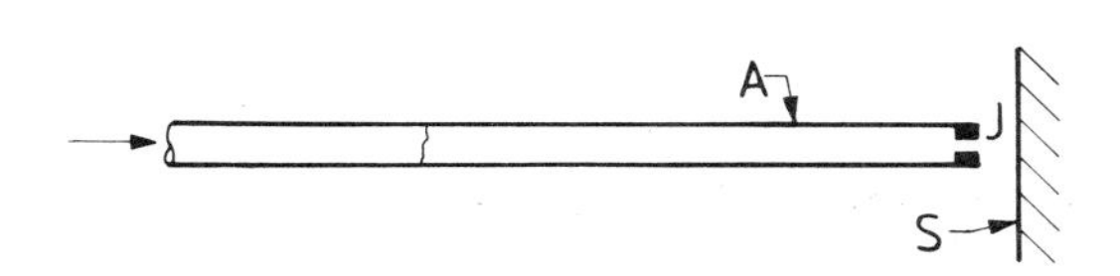

Figure 12.1 *Basic air gauge principle*

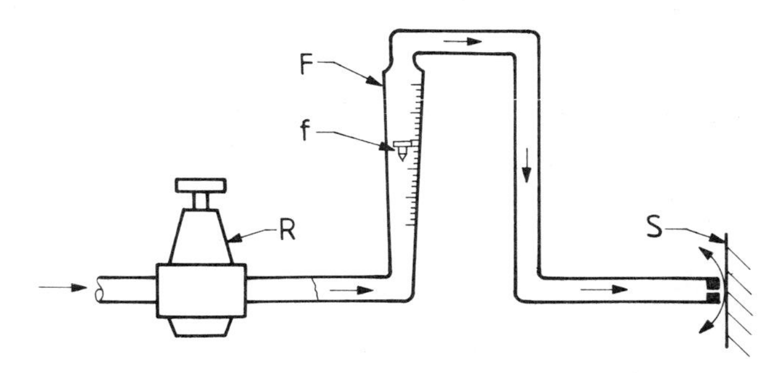

Figure 12.2 *Basic system with regulator and variable area flowmeter*

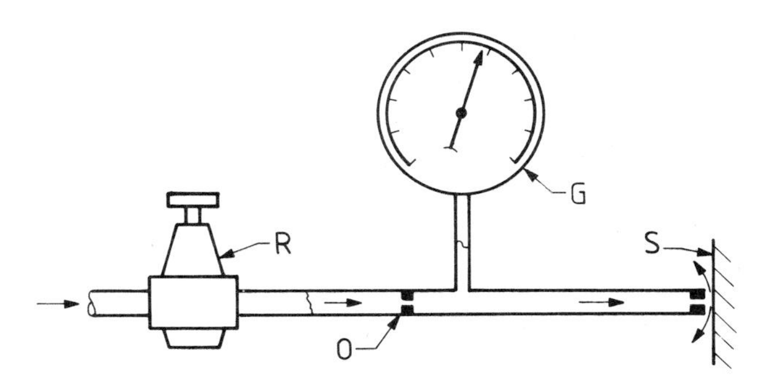

Figure 12.3 *Basic system with regulator, control orifice and pressure indicator*

Air gauging of rough or porous surfaces

For air gauging, as described above, to be successful the component under inspection needs to have a smooth, non-porous surface. For the inspection of rough and/or porous components it is necessary to use a contact gauging element. This may take one of several forms, as shown in Figure 12.4.

The contact gauging element provides a stylus which is in contact with the work being gauged and movement of the stylus changes the flow of air.

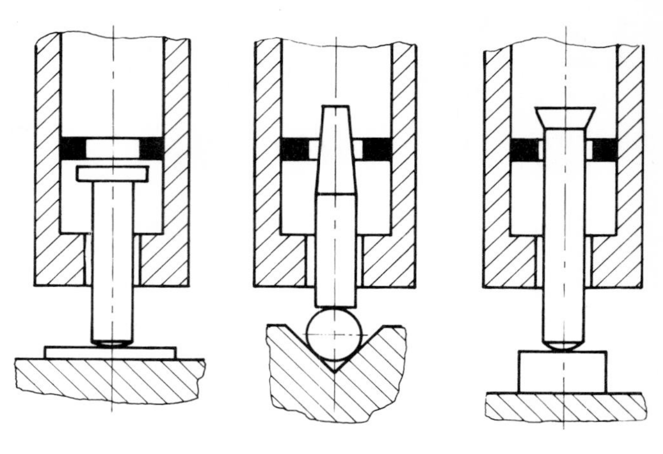

Figure 12.4 *Contact gauging elements*

Magnification

The magnification of an air gauging system, that is the ratio of the movement of the indicator index (the float in Figure 12.2 or the pointer in Figure 12.3) to the movement of the surface which produces it, can be relatively low (in the order of 1000) or very high (in the order of 100 000). Magnifications in the range of 5000 to 10 000 are common and permit accurate inspection of closely-toleranced components.

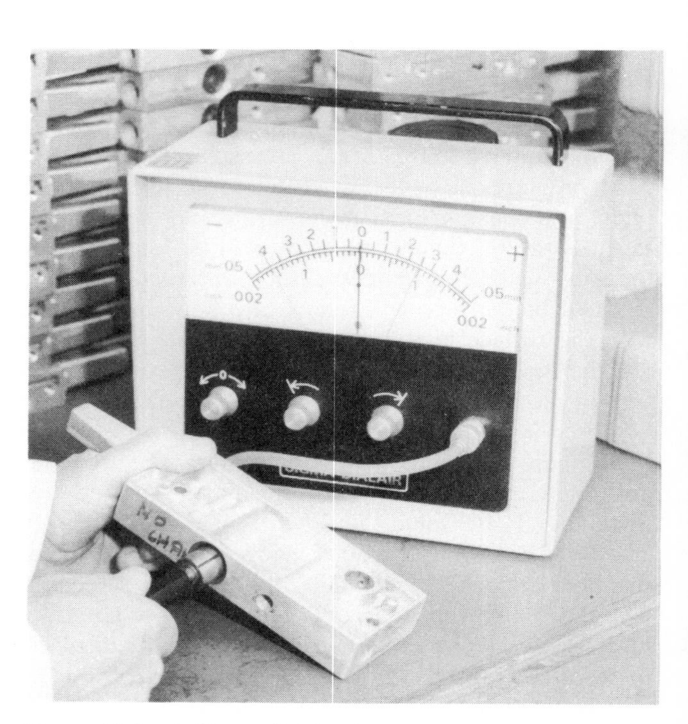

Figure 12.5 *A Sigma dialair air gauging indicating unit*
Reproduced by permission of Sigma Ltd, Letchworth

Air gauging system

An air gauging system comprises essentially an air gauge unit and one or more measuring heads. The air gauge unit contains the means to display the measured sizes, or to generate audible or visual signals based upon them (see Figure 12.5). It is also possible for conversion of the measurement into a signal to provide feed-back for automatic operations. Other elements are also contained in the air gauge unit depending upon the type; in particular, when the variable measured is pressure, it will contain the control orifice.

The measuring head may contain a single gauging element, either contact or non-contact, or two or more such elements. The form of the measuring head and the number of gauging elements used will depend upon the type of measurement to be made, for example, length or thickness, internal diameter, external diameter, etc.

Examples of application

Air gauging may be used in the inspection of dimensional and geometrical errors. The simplified examples in Figure 12.6 all employ a single gauging element.

In Figure 12.7, two or more gauging heads are used for measuring diameter (example (c) shows several gauging heads being used to measure average diameter).

Standardization

BS 4358, 'Glossary of terms used in air gauging with notes on the technique' was based upon a glossary produced by a group of British manufacturers of air gauging systems. During its preparation, the following objectives were kept in mind:

(a) to rationalize the terminology and eliminate the confusion arising from the use of different terms by different manufacturers;
(b) to provide users, and potential users, of air gauging equipment with an unambiguous means for communicating their requirements to manufacturers;
(c) to furnish a terminology which would find general acceptance and become the recognized terminology in the literature and in discussions of the air gauging method.

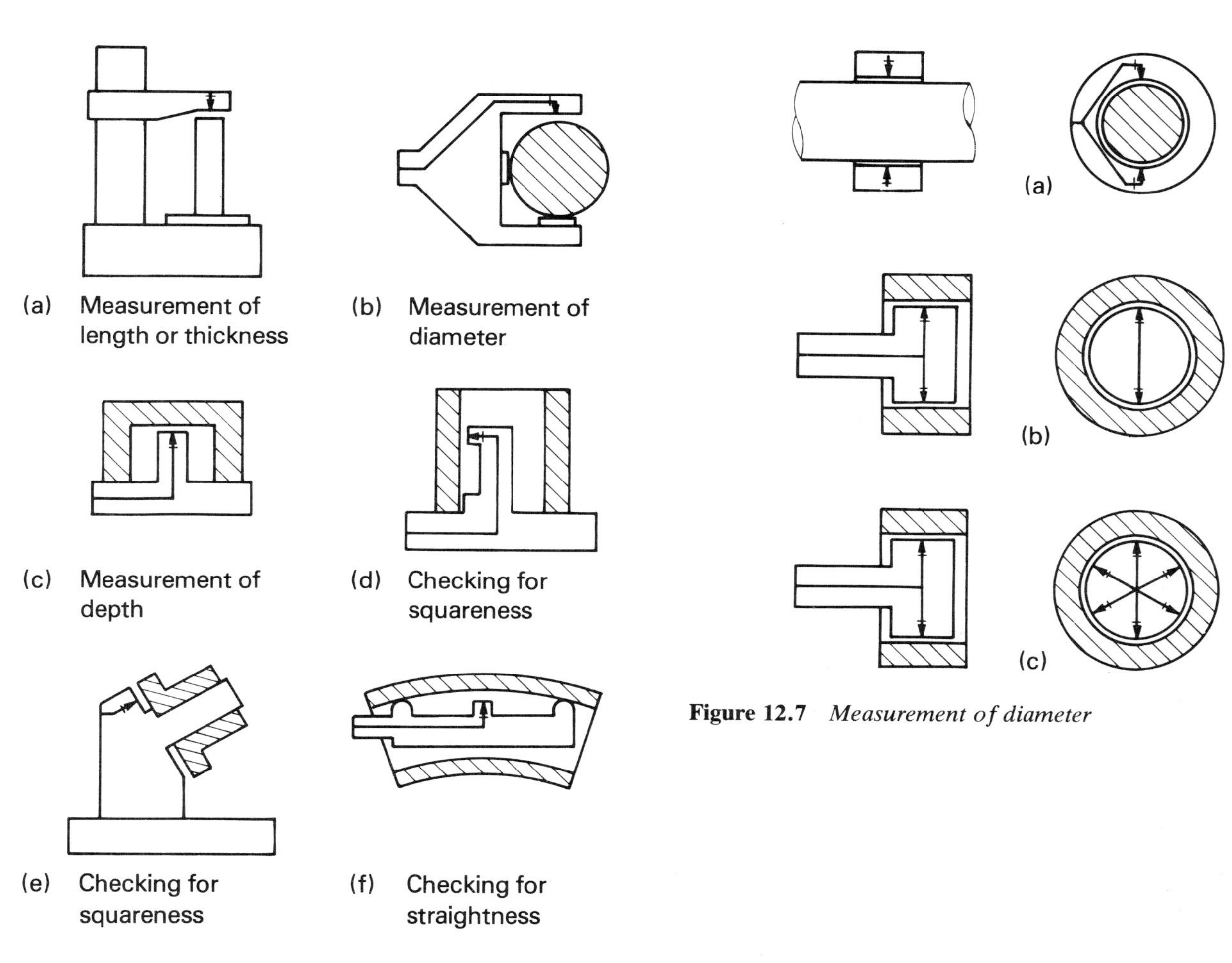

Figure 12.6 *Examples of application*

Figure 12.7 *Measurement of diameter*

BS 4358

Glossary of terms used in air gauging with notes on the technique

. . . .

1. General

air gauging (pneumatic gauging, *deprecated*). A precise measuring technique, using air drawn from a compressed air supply, which can be employed to measure size, or to check form or position.

. . . .

air gauging system. Measuring equipment employing air gauging based on the determination of flow or pressure and comprising an air gauge unit and one or more measuring heads as detailed in section **2** and sections **3**, **4** and **5** respectively. The air gauging system may be designed for measuring one variable only, or for the simultaneous measurement of two or more independent variables.

air gauging (flow) system. An air gauging system based upon the determination of flow.

air gauging (pressure) system. An air gauging system based upon the determination of pressure.

air gauge unit. That part of the air gauging system which displays the measured sizes or provides visual or audible signals or converts the measurements into signals for automatic operations. For details see section **2**.

measuring head. That part of the air gauging system which contains the gauging element(s) and which is presented to the work, or vice versa. For details see sections **3** and **4**.

gauging element. The element in the measuring head which detects dimensional change and provides the response which enables the air gauging system to measure that change. The gauging element may be an open jet or a contact gauging element, and the measuring head may have one, two or more elements depending on its type. For details see sections **3** and **4**.

open jet. A gauging element in the form of a jet of circular or rectangular section which does not contact the work being measured. The parameters of a jet are its internal and external diameters, when the section is circular; the external and internal length and breadth when the section is rectangular.

land width. The terminating wall thickness of an open jet.

jet recession. The amount by which an open jet is recessed behind the outer surface of the measuring head.

NOTE. Recession fulfils two objects:
(1) it protects the jet from mechanical damage:
(2) it maintains the essential minimum distance between the jet and the work being measured.

jet guard. A protective device to prevent mechanical damage to an open jet when the design of the measuring head does not permit the jet to be recessed.

contact gauging element. A gauging element fitted with a stylus which contacts the work being measured. See, for example, figure 4.

matched gauging elements. Gauging elements which have matched pneumatic characteristics so that they may be used in combination in a single measuring circuit.

asymmetry error. The error which may arise when the work being measured is asymmetrically disposed between gauging elements which are combined in one measuring circuit. This error is reduced to negligible proportions by correct design of the measuring head and the stipulation of its measuring range.

air escape duct. An air passage, e.g. a groove or hole, which provides an easy final escape to atmosphere of air which has emerged from a gauging element.

control orifice, *control jet*. The restriction used in air gauging (pressure) systems. In association with the gauging elements, the control orifice determines the magnification of the system.

supply pressure. The pressure of the supply from which the air gauging systems draws its air.

air filter. A device for cleaning the air supply to the air gauging system and which, ideally, removes all liquid and solid particles.

inlet restrictor. A fixed inlet orifice used in some air gauging (pressure) systems upstream of the pressure regulator to limit the quantity of air supplied to the system.

economiser orifice. The fixed orifice in the measuring head in some air gauging (flow) systems which limits the maximum air flow when the measuring head is disengaged from the work.

operating pressure. The constant pressure at which the air gauging system is operated.

operating pressure indicator. A pressure measuring instrument which registers the operating pressure and which is usually fitted to the air gauge unit.

pressure regulator. A device which keeps the operating pressure constant irrespective of pressure changes in the supply line and of changes of air flow in the air gauging system itself.

intermediate pressure, back pressure, *deprecated.* In air gauging (pressure) systems, the pressure between the control orifice and the measuring head.

simple pressure measurement. Measurement of the intermediate pressure in relation to the ambient atmospheric pressure or to the operating pressure as datum.

NOTE. This term and the next two terms serve to distinguish between the different methods currently used in air gauging (pressure) systems for measuring the intermediate pressure.

differential pressure measurement. Measurement of the intermediate pressure in relation to a pre-set fixed pressure obtained from a branch circuit having the same operating pressure as the measuring circuit. This fixed pressure is adjusted to be approximately equal to the intermediate pressure at the mid-point of the measuring range.

null-balance pressure measurement. Measurement of the intermediate pressure by a self-balancing system which brings the pressure in a secondary chamber to equality with the pressure being measured by means of a pressure regulating mechanism fitted with an appropriate indicating device.

datum setting control. The control which enables the indicator reading to be set to a datum, e.g. zero.

NOTE. This control will normally be used in association with a setting standard or a master setting jet.

magnification. The ratio of the movement of the indicator index (float, liquid surface, pointer) to the dimensional change which produces it, both being expressed in terms of the same unit of length. Thus, if a dimensional change of 0.001 inch causes the float in a variable-area flow-meter or the liquid surface in a pressure manometer to move 1.000 inch, the magnification is 1000. Magnification is therefore a measure of sensitivity.

ranging. The process of adjusting an air gauging system so that the movement of the indicator index (float, liquid surface, pointer) conforms to a pre-established scale at two or more points, i.e. so that the air gauging system has a specified magnification.

measuring range. The range of sizes (minimum to maximum) which the air gauging system is designed to measure and over which the indicating device is graduated.

magnification control. The control which enables the magnification of the air gauging system to be adjusted to a specified value.

NOTE. This control will normally be used in association with setting standards or master setting jets.

setting standard. A standard of appropriate form and of established size used for setting up the air gauging system so that its reading represents true size.

NOTE. Two setting standards, one to provide a reading near the lower limit of the measuring range and the other a reading near the upper limit, enable both magnification and datum to be set. A single setting standard allows the datum to be set when ranging has already been done.

master setting jet. A calibrated jet, which is substituted for the measuring head and used to set up the air gauging system so that it has the required magnification.

NOTE. When ranging, two master jets are used, one to provide a reading near the lower limit of the measuring range and the other a reading near the upper limit. A single master setting jet allows the datum to be set when ranging has already been done.

measuring signal. In measuring systems, the physical quantity which represents the measurement.

signal limiter. A device which may be used in air gauging (pressure) systems to limit the movement of the indicator index when the measuring head is disengaged from the work, i.e. when the measuring signal has been removed.

signal retainer. A device which may be used in air gauging (pressure) systems to retain the indicator display when the measuring signal has been removed.

signal inverter. A device which may be used in air gauging (pressure) systems for reversing the direction of the indicator movement relative to the measuring signal.

pressure amplifier. A device in which an input measuring signal is used to release from a local source an output signal of pressure greater than that of the input signal but bearing a definite relationship to it.

transducer. A device which converts a measuring signal expressed in terms of one physical quantity into an equivalent signal expressed in terms of another physical quantity.

differential air gauging. A technique whereby two individual measurements are combined so as to display their difference on a single indicator.

in-process gauging. Gauging carried out during processing, e.g. measurement of a workpiece whilst it is being machined.

post-process gauging. Gauging carried out immediately after processing, e.g. measurement of a workpiece immediately after it has been machined.

air gauging machine. An inspection machine employing air gauging in which all, or the greater part, of the complete process of loading, positioning, measuring, classifying and unloading the workpieces is carried out automatically.

NOTE. The machine may measure one or more dimensions of the workpiece and the acceptable workpieces may be classified by grading into size groups within the tolerance zone. Classification may be effected by marking the workpieces according to a symbol or colour code or by their physical separation. (See also **B.4**.)

air gauge size control. The control of size by a monitored automatic air gauge control system comprising a measuring head to measure the work and an air gauge control unit to generate the control signals which regulate the manufacturing process so that the work is brought to the desired size without the intervention of a human operator. See also **B.5.**

2. Air gauge units

Introduction. The air gauge unit contains all those parts of the measuring circuit which are used in conjunction with the measuring head to form the air gauging system. It provides the means to display the measured sizes or to generate signals based upon them for purposes of control or recording. The unit may contain one or more of the following items:

> Air filter, pressure regulator, operating pressure indicator, control orifice, magnification and datum setting controls, signal retainers, signal limiters, signal inverters.

Air gauge units may be subdivided into air gauge indicating units and air gauge control units.

2.1 Air gauge indicating unit

air gauge indicating unit. An air gauge unit which displays the measured sizes.

NOTE. In air gauging (flow) systems the unit contains a flow-responsive device which is usually a variable-area flowmeter, the measured size being then indicated by the position of the float in the tapered tube. In air gauging (pressure) systems the unit contains a pressure-responsive device, the measured size being displayed on a manometer type column indicator or a scale and pointer indicator.

Alternatively, in air gauging (pressure) systems, the results of measurement may be displayed by means of electric signal lights, announced by audible signals, or recorded.

single display air gauge indicating unit. An air gauge indicating unit for displaying the measurement of a single variable.

multiple display air gauge indicating unit. An air gauge indicating unit for simultaneously displaying the measurements of two or more variables.

special air gauge indicating units. Air gauge indicating units generally conforming to the basic definition but particularly designed for specific applications.

2.2 Air gauge control unit

air gauge control unit. An air gauge unit as described earlier which converts the measurements into signals for automatic control. In addition to this primary function, the unit may also contain means for displaying or recording the measured sizes.

single input air gauge control unit. An air gauge control unit which generates signals for the control of a single variable.

multiple input air gauge control unit. An air gauge control unit which generates signals for the simultaneous (or programmed) control of two or more variables.

special air gauge control units. Air gauge control units generally conforming to the basic definition but particularly designed for specific applications.

3. Measuring heads for dimensional measurement

Introduction. In this section will be found terms and definitions relating to measuring heads used for the measurement of length, internal diameter and external diameter. Terms relating to measuring heads for inspecting form and position are given in Section **4** and appendix A.

3.1 Length (including thickness and depth) and diameter

air gauge probe. A measuring head for general use fitted with a single gauging element which may be an open jet or a contact gauging element. When two or more are used in one measuring circuit, the gauging elements must be matched.

air gauge jet probe. An air gauge probe in which the gauging element is an open jet.

air gauge contact probe. An air gauge probe fitted with a contact gauging element.

air gauge comparator. A rigid structure fitted with a worktable and one or two gauging elements for measuring workpieces by comparison with standard gauges of closely similar size, the standard gauge and the workpiece being placed in turn on the worktable. The gauging elements may be open jets or contact gauging elements. When one gauging element is fitted, its distance from the worktable is adjustable to accommodate workpieces of different sizes and the worktable provides the measuring datum. When there are two gauging elements, the position of the upper one in relation to the worktable is adjustable but the lower one is fixed rigidly to the worktable which then serves only as a support for the workpiece.

The capacity of the comparator is the maximum size of workpiece it will accommodate.

external air gauge comparator. An air gauge comparator for measuring external dimensions.

internal air gauge comparator. An air gauge comparator for measuring internal dimensions.

3.2 Length (including height above a datum surface)

air gauge test indicator probe. A probe of small size, and therefore particularly suitable for measurements in positions of restricted access, which is fitted with a single contact gauging element. The probe is designed for clamping to a rigid stand and contact with the workpiece is made by means of a ball-ended stylus lever. This lever may be mounted on a friction pivot so that its angular relationship to the body of the probe can be adjusted.

3.3 Internal diameter

air plug gauge. A plug gauge which can be inserted into the bore to be measured and which is fitted with one or more pairs of gauging elements. The gauging elements may be open jets or contact gauging elements.

The air plug gauge is of fixed size and can be used to measure only diameters lying within its measuring range as specified by the manufacturer.

For the measurement of *single diameter* the air plug gauge is fitted with two gauging elements arranged diametrically.

By rotation and translation of the plug within the bore the position of the diameter measured can be varied both in azimuth and depth.

For the measurement of *average diameter* the air plug gauge is fitted with an even number of gauging elements, four or more, spaced equally around the circumference.

through-bore air plug gauge. An air plug gauge for measuring through bores. Since the gauge can be passed right through the bore the longitudinal position of the gauging elements is not of primary importance.

blind-bore air plug gauge. An air plug gauge for measuring blind bores. The gauging elements are placed so as to permit measurements to be made as close as possible to the bottom of the bore.

shouldered-bore air plug gauge. An air plug gauge for measuring shouldered holes. The gauging elements are placed so as to permit measurements to be made as close as possible to the shoulder.

multi-dimension air plug gauge. An air plug gauge for the simultaneous measurement of two or more spaced diameters which may be of equal or of different size.

special air plug gauges. Air plug gauges generally conforming to the basic definition but particularly designed for specific applications.

adjustable air bore gauge. A gauge of adjustable size which can be inserted into the bore to be measured and which is fitted with one or more gauging elements.

The adjustable air bore gauge is designed for the measurement of single diameter. It can be adjusted to cover a wide range of sizes and for each size will have a specified measuring range.

NOTE 1. Adjustable air bore gauges of three types are available, namely:

(a) gauges which are self-centring and self-aligning;
(b) gauges which are self-centring only;
(c) gauges which are self-aligning only.

NOTE 2. In present designs the gauging elements are of the contact type.

3.4 External diameter

air ring gauge. A ring gauge through which is passed the cylindrical work to be measured and which is fitted with one or more pairs of gauging elements. The gauging elements may be open jets or contact gauging elements.

The air ring gauge is of fixed size and can be used to measure only diameters lying within its measuring range as specified by the manufacturer.

For the measurement of *single diameter* the air ring gauge is fitted with two gauging elements arranged diametrically.

By relative rotation and translation of the shaft within the ring gauge the position of the diameter measured can be varied both in azimuth and along the axis.

For the measurement of *average diameter* the air ring gauge is fitted with an even number of gauging elements four or more, spaced equally around the circumference.

through-shaft air ring gauge. An air ring gauge for measuring plain cylindrical work. Since the work can be passed right through the ring gauge the longitudinal position of the gauging elements is not of primary importance.

close-to-shoulder air ring gauge An air ring gauge for measuring shouldered shafts. The gauging elements are placed so as to permit measurements to be made as close as possible to the shoulder.

multi-dimension air ring gauge. An air ring gauge for the simultaneous measurement of two or more spaced diameters which may be of equal or different size.

special air ring gauges. Air ring gauges generally conforming to the basic definition but particularly designed for specific applications.

air calliper gauge. A calliper gauge for the measurement of cylindrical work. It has one or two gauging elements, which may be open jets or contact gauging elements.

The air calliper gauge is designed to measure a single diameter and cannot conveniently be used to measure average diameter.

By relative rotation and translation of the gauge and the work, the position of the diameter measured can be varied both in azimuth and along the axis.

hand air calliper gauge of fixed or adjustable size. A hand air calliper gauge of fixed or adjustable size which can be applied by hand to the work to be measured.

The fixed size calliper gauge can be used to measure only diameters lying within its measuring range as specified by the manufacturer. The adjustable calliper gauge can be adjusted to cover a wide range of sizes, and for each size will have a specified measuring range.

multi-dimension hand air calliper gauge. A hand air calliper gauge, normally of fixed size, designed for simultaneous measurement of two or more diameters of equal or different size.

special hand air calliper gauge. Air calliper gauges for hand use conforming generally to the basic definition but particularly designed for specific applications.

fixture mounting air calliper gauge (non-floating) (fixed or adjustable size). An air calliper gauge of fixed or adjustable size designed for rigid mounting in a holding fixture, or to form part of a multi-gauging fixture, and which normally provides location for the work being measured.

The fixed size gauge can be used to measure only diameters lying within its measuring range as specified by the manufacturer. The adjustable gauge can be adjusted to cover a wide range of sizes, and for each size will have a specified measuring range.

fixture mounting calliper gauge (floating) (fixed or adjustable size). An air calliper gauge of fixed or adjustable size mounted in a holding fixture, or forming part of a multi-gauging fixture, so that it is free to move and is thus self-locating on the work being measured.

The fixed size gauge can be used to measure only diameters lying within its measuring range as specified by the manufacturer. The adjustable gauge can be adjusted to cover a wide range of sizes, and for each size will have a specified measuring range.

machine mounting air calliper gauge (fixed or adjustable size). An air calliper gauge of fixed or adjustable size designed for mounting on a machine tool to measure the workpiece during or immediately after machining. The calliper may for example be arranged for mounting on the wheelhead, the machine bed or the headstock, and may be of the floating or non-floating variety.

The fixed size gauge can be used to measure only sizes within its measuring range as specified by the manufacturer. The adjustable gauge can be adjusted to cover a wide range of sizes, and for each size will have a specified measuring range.

4. Measuring heads for the inspection of form and position

The workpieces made by the engineering industry are usually based upon a few fundamental geometrical forms such as the plane, sphere, cylinder and cone. Determinations of the sizes of these workpieces calls for measurements of length, diameter and angle, but if errors of form are present, i.e. the workpieces are not true to the geometrical forms they are meant to represent, the measurements may be wrong and so prove misleading when assessing functional qualities. Inspection therefore normally requires assessment of form as well as determination of size, a requirement to which the designer may have drawn attention in advance by specifying certain form tolerances in addition to size tolerances.

It is well to emphasize that the true measurement of form usually presents more difficulty than the true measurement of size, and this because two or more different errors of form may be present simultaneously. Thus, to take a simple example, a nominally circular, straight bore may be both out of round and tapered, and a test for taper which overlooks the out-of-roundness, and so ignores the latter's influence on the readings taken, will give misleading results. When checking form, irrespective of the technique used, it is therefore necessary to consider in advance what types of error may be present and to adopt methods of inspection which will not lead to incorrect conclusions. The inspection of the out-of-round, tapered bore can be satisfactorily made with a through-bore air plug gauge fitted with two gauging elements arranged diametrically provided (i) the diameter variation in each of several sections is first examined by rotation of the gauge without axial traverse and (ii) when axial traverse is made to check for taper, the gauge is kept in the same azimuth. By adopting this method, the out-of-roundness measurement is not invalidated by the taper and the taper test is not affected by the out-of-roundness. (In the rare case when the pattern of diameter variation rotates with depth, the taper can still be measured by rotating the gauge during the axial traverse so that it keeps in step with the pattern rotation, though the accuracy is likely to be impaired if the out-of-roundness is marked and the pattern rotates rapidly with depth.)

The first workpieces from a new production run must be critically examined in order to establish what different types of form error are present and to what degree. Experience of a given manufacturing process, or even with a particular machine tool, will suggest what should be expected; lobing in spherical or cylindrical parts made by centreless grinding, camber in plain holes finished by honing are examples in point. But when the nature of the form errors has been established and the production is under control, a relatively simple method of inspection can usually be devised which will be adequate to maintain control and be economically suitable.

Air gauging can usefully be introduced at this stage, and in appendix A of this glossary will be found information regarding measuring heads designed for the inspection and measurement of form. The use of these measuring heads offers an economic means of inspection, but it must be appreciated that like all other techniques with similar intention, it does not necessarily give the full information obtainable from a detailed, point-to-point examination. Where it has been considered desirable to do so, the limitations imposed by the presence of compound errors of form when using the measuring heads described in the appendix have been indicated.

Appendix A

Examples of air gauging techniques for checking common errors of form and position

A.1 Introduction. The aim of this appendix is to illustrate the application of air gauging techniques to the practical assessment of errors of form and position.

The various types of error of form and position dealt with in this appendix are separately enumerated in sections **A.2**, **A.3**, etc., below.

. . . .

In each section, immediately following the definition, illustrations are shown of recommended air gauging methods which may be used to make a practical assessment as to whether the component being examined is within the prescribed tolerance zone.

In the illustrations, the component being gauged is shown shaded. The gauging elements in the measuring heads are shown thus —+► and translation or rotation of the measuring head or of the component is shown thus——►. Those measuring heads for inspecting for taper which are marked with an asterisk (*) may also be used to determine variation of diameter along the axis. Those marked with a dagger (†) should be used

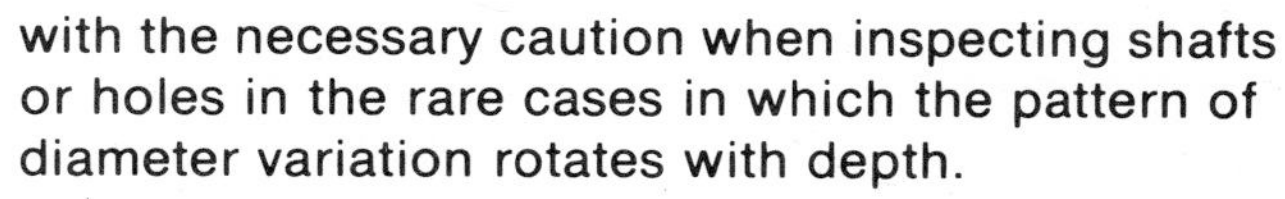

with the necessary caution when inspecting shafts or holes in the rare cases in which the pattern of diameter variation rotates with depth.

These illustrations, though including the majority of the measuring heads normally available, should not be regarded as exhaustive and they do not, of course, show measuring heads particularly designed for special applications. It should be noted that certain errors of form and position can be measured by means of external and internal air gauge comparators using conventional methods.

A.2 Straightness tolerance – of a line. The tolerance zone in any one plane is limited by two parallel straight lines in that plane a distance t apart.

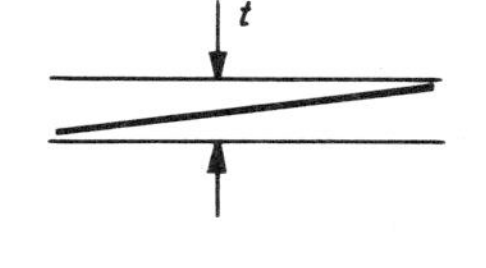

Air plug gauge
4 gauging elements in common axial plane

Figure 7. Straightness

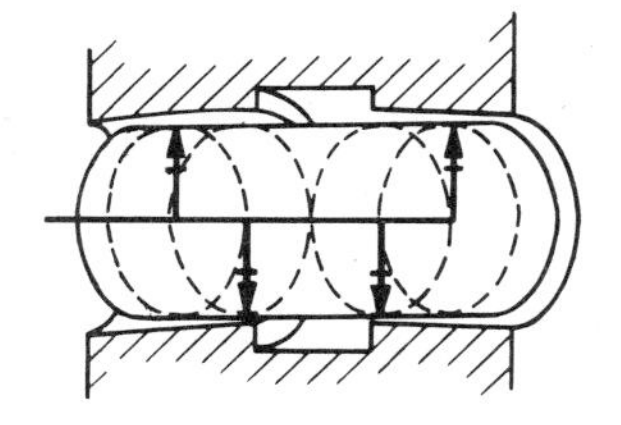
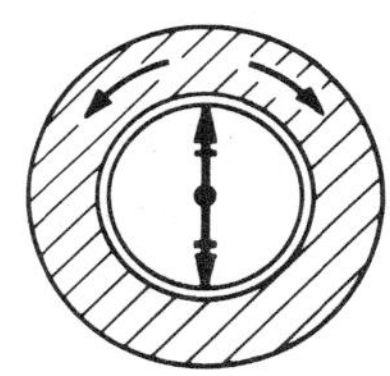

Air plug gauge
4 gauging elements in common axial plane, centre pair disposed to suit component conditions

Figure 8. Straightness

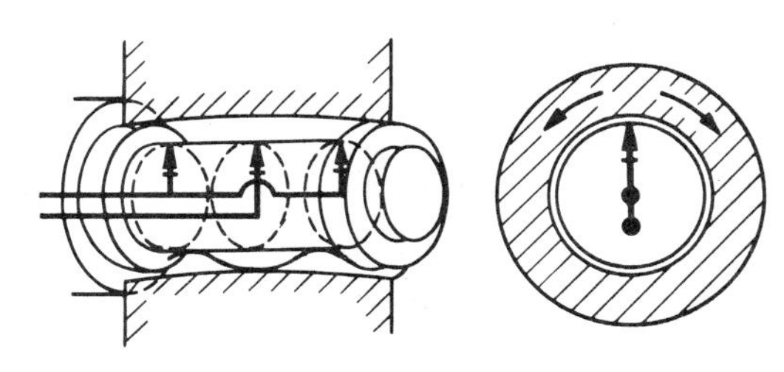

Air plug gauge
3 gauging elements in common axial plane

Figure 9. Straightness

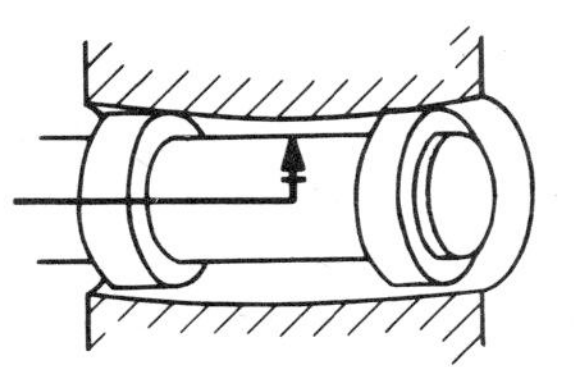
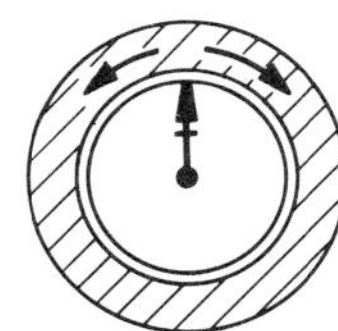

Air plug gauge
1 gauging element

Figure 10. Straightness

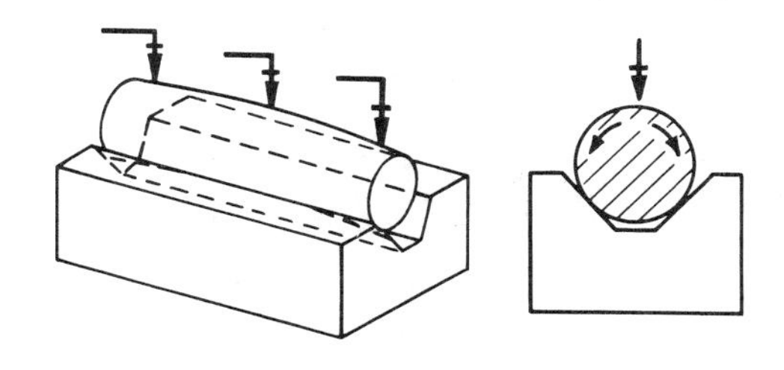

Vee block with gauging elements
n gauging elements

Figure 11. Straightness

A.3 Parallelism tolerance – of a line with reference to a datum line. The tolerance zone in any one plane is limited by two parallel straight lines in that plane a distance *t* apart and parallel to the datum line.

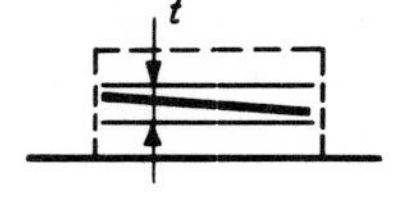

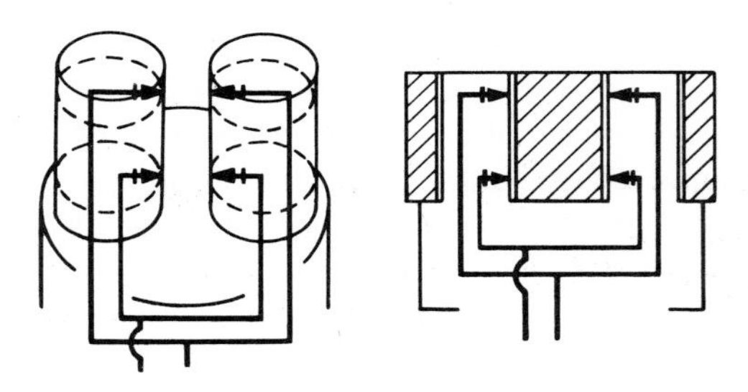

4 gauging elements in 2 pairs measuring thickness of material between holes

Figure 12. Parallelism in one plane of 2 holes of uniform diameters

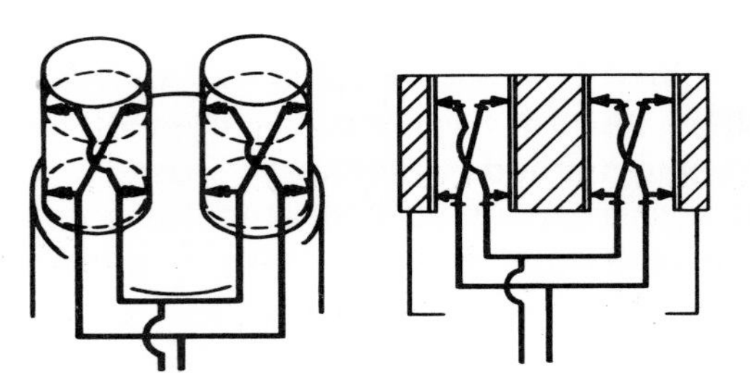

8 gauging elements in 2 sets of 4 giving compensation for diameter variation

Figure 13. Parallelism in one plane of 2 holes

NOTE. These illustrations indicate methods of measuring parallelism of two holes in one plane only. When it is required to measure parallelism in two planes, the measuring arrangements should be duplicated with the second system at right angles to the first.

A.4 Perpendicularity tolerance – a line with reference to a datum plane. The tolerance zone in any one plane is limited by two parallel straight lines a distance *t* apart and perpendicular to the datum plane.

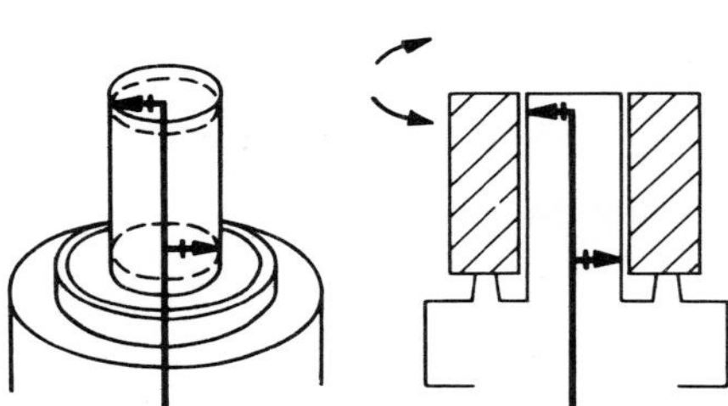

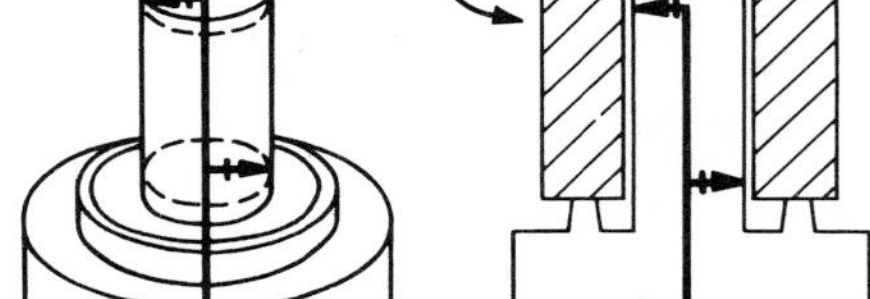

2 gauging elements

Figure 14. Squareness bore to face in any plane, diameter being uniform

4 gauging elements in 2 pairs giving compensation for diameter variation

Figure 15. Squareness bore to face in any plane

A.5 Flatness tolerance. The tolerance zone is limited by two parallel planes a distance *t* apart.

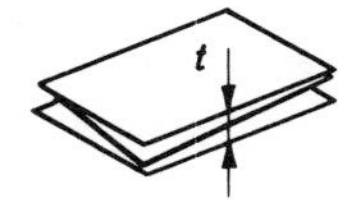

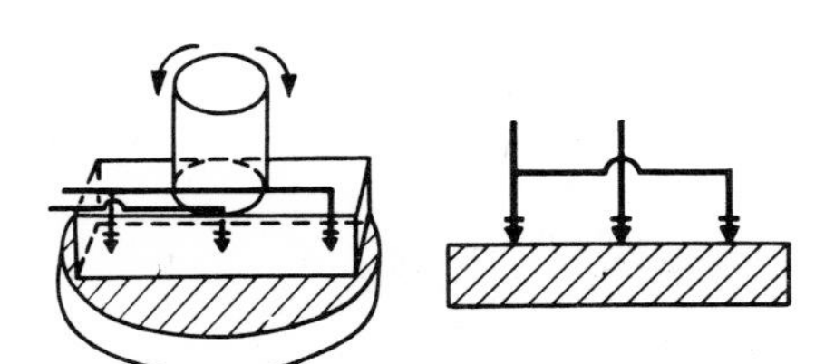

3 gauging elements in common plane

Figure 16. Flatness

A.6 Perpendicularity tolerance – of a surface with reference to a datum line. The tolerance zone is limited by two parallel planes a distance *t* apart and perpendicular to the datum line.

NOTE. This may be regarded as a corollary to **A.4**, q.v.

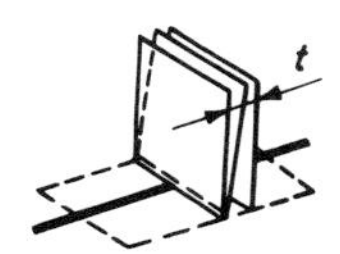

A.7 Perpendicularity tolerance – of a surface with reference to a datum plane. The tolerance zone is limited by two parallel planes a distance *t* apart and perpendicular to the datum plane.

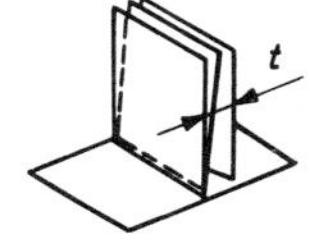

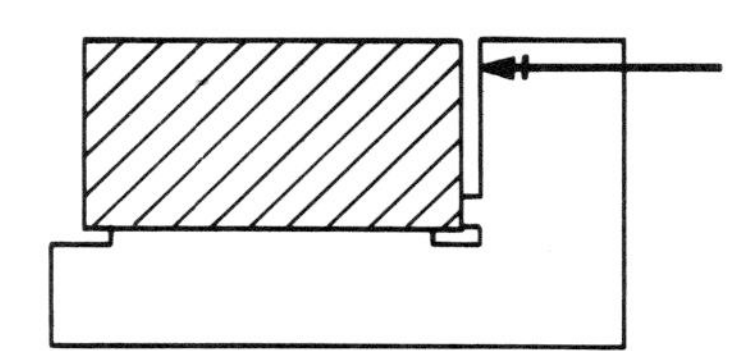

Reference square with 1 gauging element

Figure 17. Squareness when surfaces are flat

A.8 Angularity tolerance – of a line with reference to a datum line. The tolerance zone is limited by two parallel straight lines a distance *t* apart and inclined at the specified angle to the datum line.

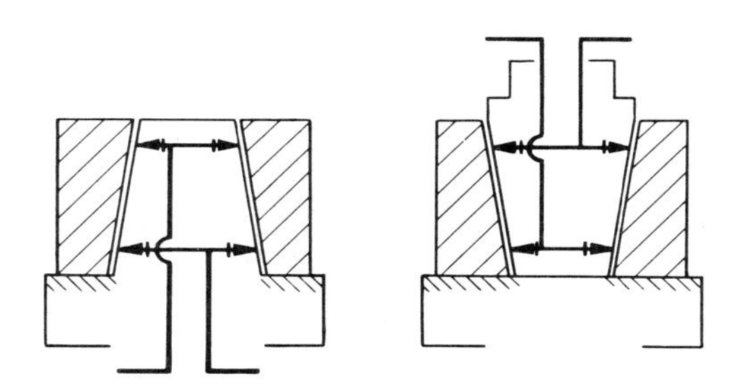

Tapered air plug gauge
4 gauging elements in 2 pairs

Figure 18. Error in design taper

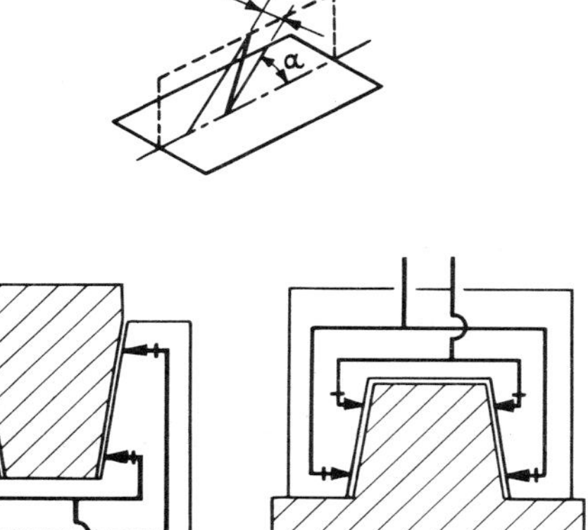

Tapered air ring gauge
4 gauging elements in 2 pairs

Figure 19. Error in design taper

A.9 Circularity tolerance. The tolerance zone is limited by two co-planar concentric circles a distance *t* apart.

For the purposes of this standard, two types of circularity error are distinguished, viz.:

(a) Ovality. The ovality of a nominally circular profile is the difference between the major and minor diameters of the profile.

(b) Lobing. The error of form when a nominally circular profile is of varying curvature but apparently constant diameter.

With this distinction in mind, the following methods are available, but see section **4**, paragraph **3**.

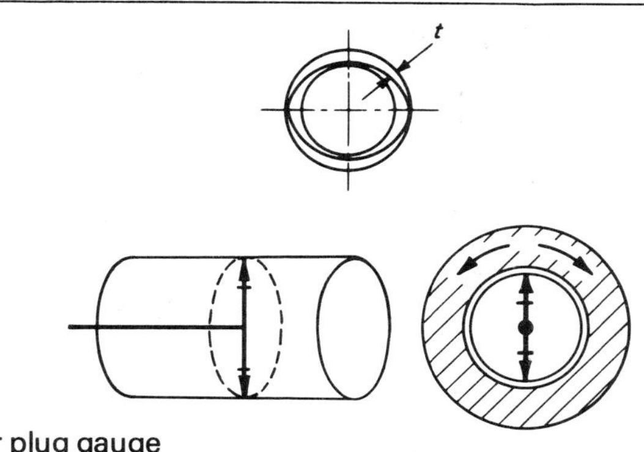

Air plug gauge
2 gauging elements

Figure 20. Ovality

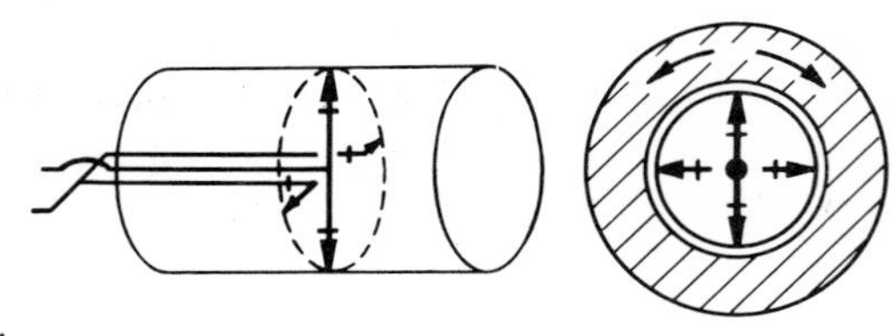

Air plug gauge
4 gauging elements in 2 pairs spaced at right angles

Figure 21. Ovality

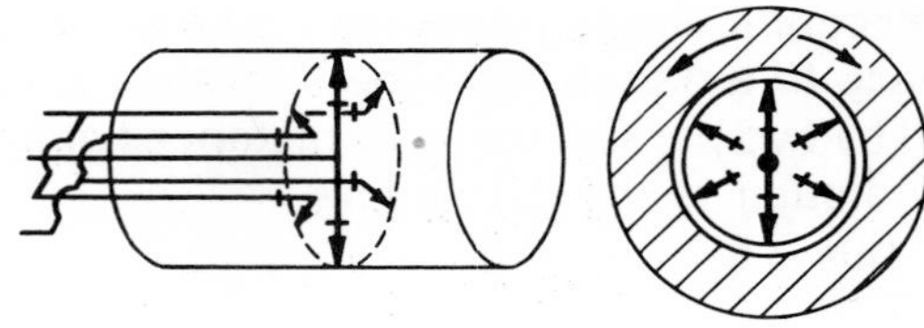

Air plug gauge
6 gauging elements in 3 pairs equally spaced in one plane

Figure 22. Ovality

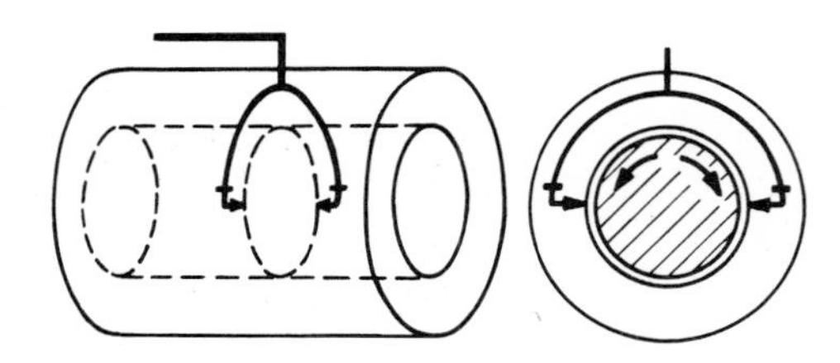

Air ring gauge 2 gauging elements

Figure 23. Ovality

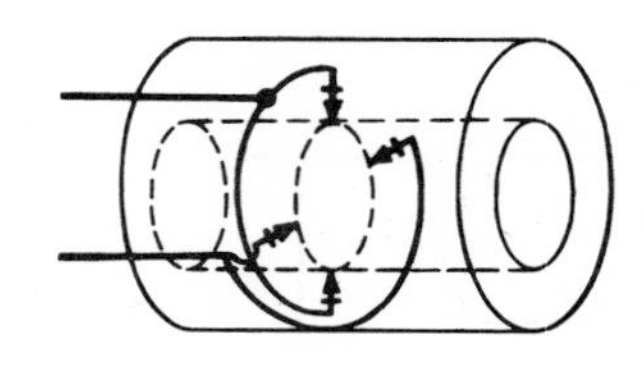

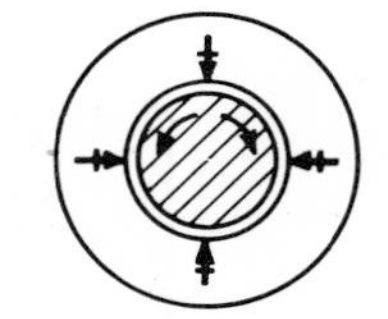

Air ring gauge
4 gauging elements in 2 pairs spaced at right angles

Figure 24. Ovality

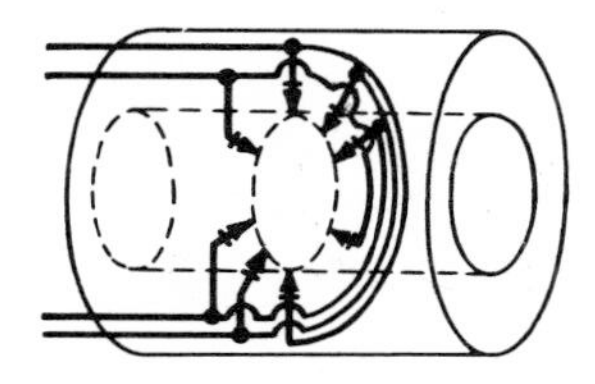

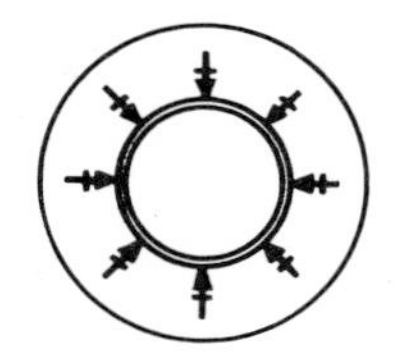

Air ring gauge
8 gauging elements in 4 pairs equally spaced in one plane

Figure 25. Ovality

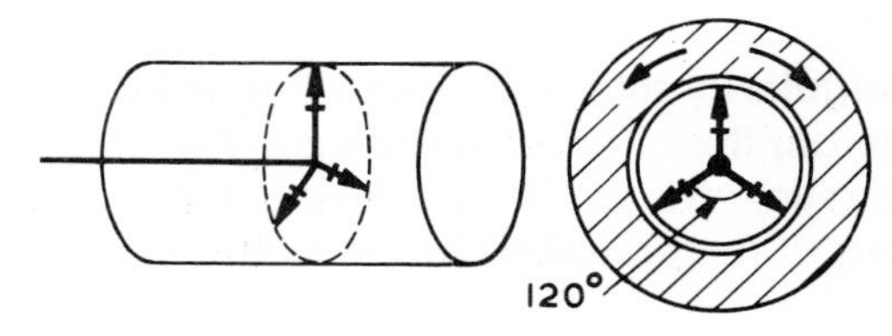

Air plug gauge
3 gauging elements

Figure 26. 3-point lobing

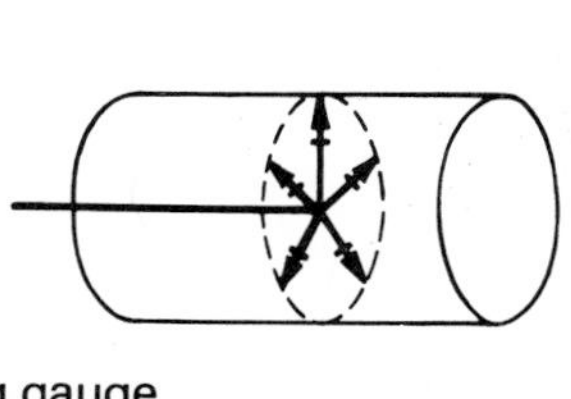

72° a

72° b

Air plug gauge
(a) 3 gauging elements
(b) 5 gauging elements

Figure 27. 5-point lobing

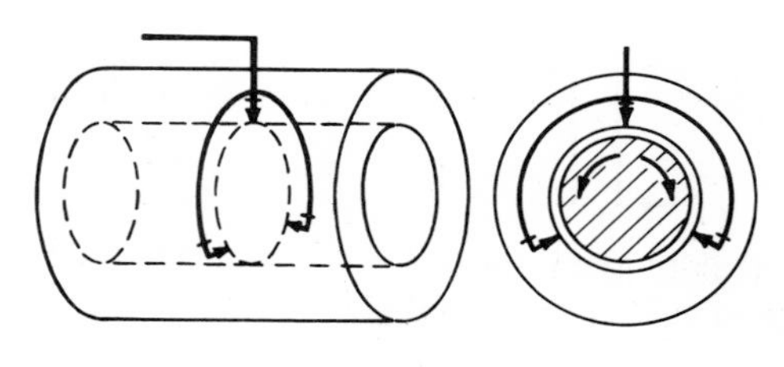

Air ring gauge
3 gauging elements

Figure 28. 3-point lobing

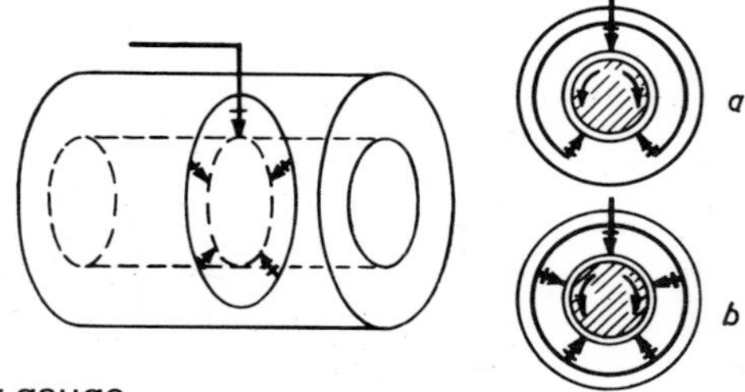

Air ring gauge
(a) 3 gauging elements
(b) 5 gauging elements

Figure 29. 5-point lobing

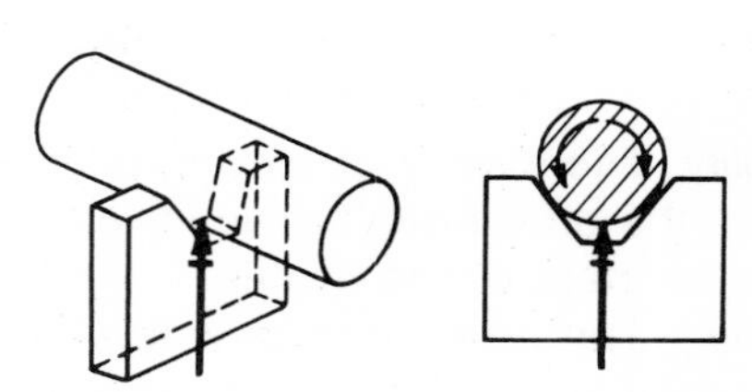

Vee gauge
1 gauging element

Figure 30. Ovality or lobing (comparative)

A.10 Cylindricity tolerance. The tolerance zone is limited by two coaxial cylinders a distance t apart.

Errors of cylindricity can arise as the result of one or more of the following errors:

(a) errors of straightness – see **A.2**.
(b) errors of circularity – see **A.9**.
(c) variation of diameter along axis.

For the purposes of this standard, inspection of errors of cylindricity is made by separate tests for these three types of error. Suitable methods are indicated in figures 7 to 11 for straightness (**A.2**), in figures 20 to 30 for circularity (**A.9**), and in the following figures 31 to 39 for variation of diameter along the axis.

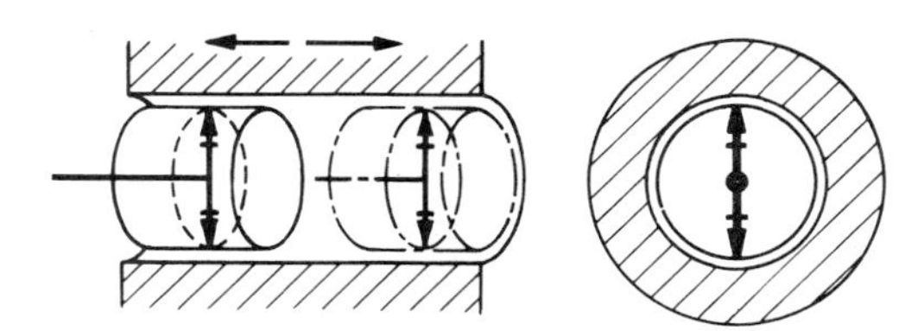

Air plug gauge
2 gauging elements

Figure 31. Taper*†

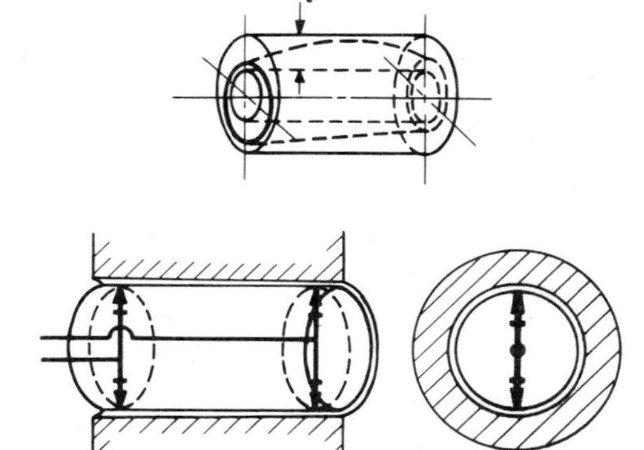

Air plug gauge
4 gauging elements in 2 pairs suitably spaced along axis

Figure 32. Taper†

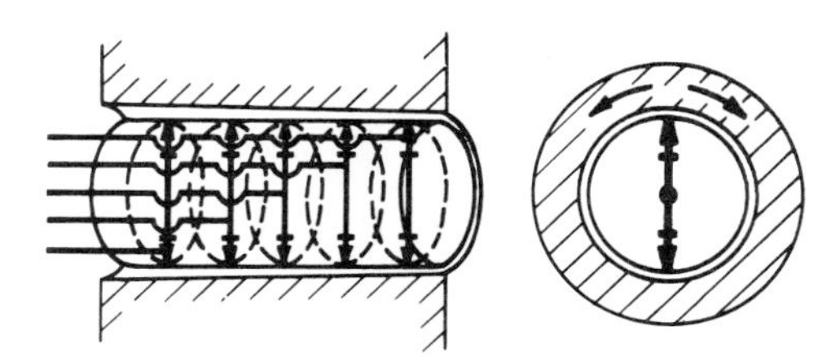

Air plug gauge
2*n* gauging elements in *n* pairs suitably spaced along axis

Figure 33. Variation of diameter along axis

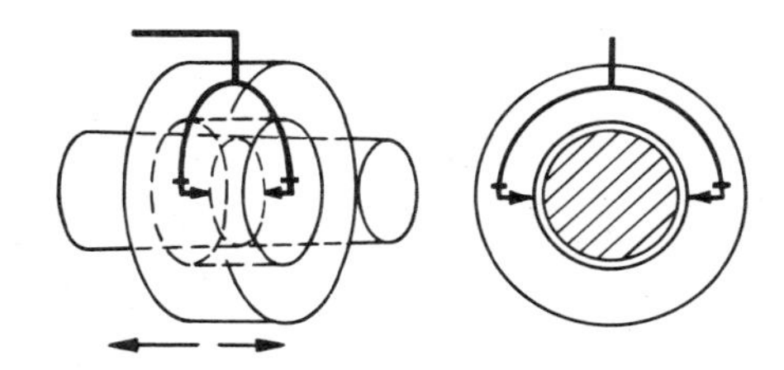

Air ring gauge
2 gauging elements

Figure 34. Taper*†

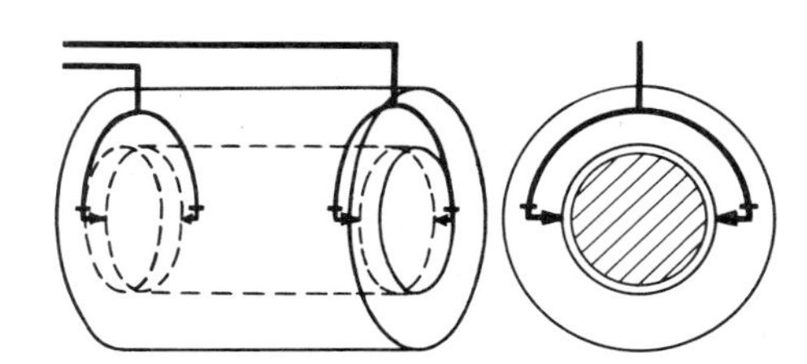

Air ring gauge
4 gauging elements in 2 pairs suitably spaced along axis

Figure 35. Taper†

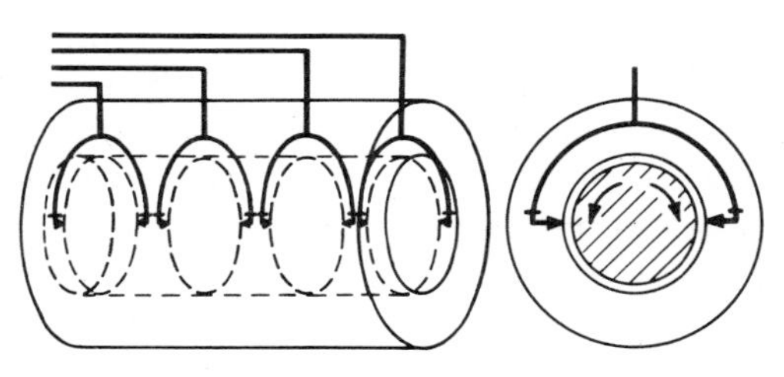

Air ring gauge
2*n* gauging elements in *n* pairs suitably spaced along axis

Figure 36. Variation of diameter along axis

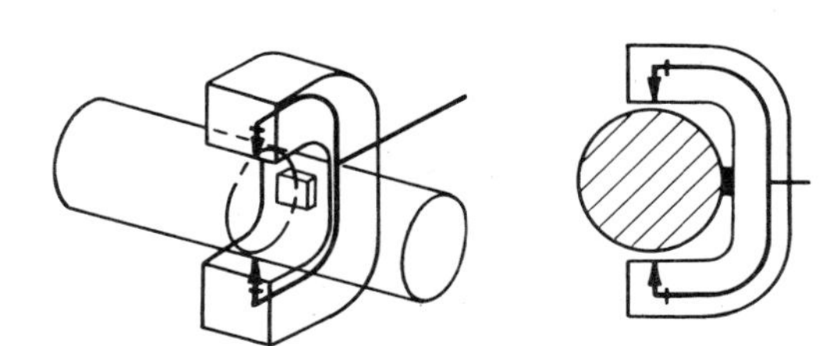

Hand air calliper gauge
1 or 2 gauging elements

Figure 37. Taper†

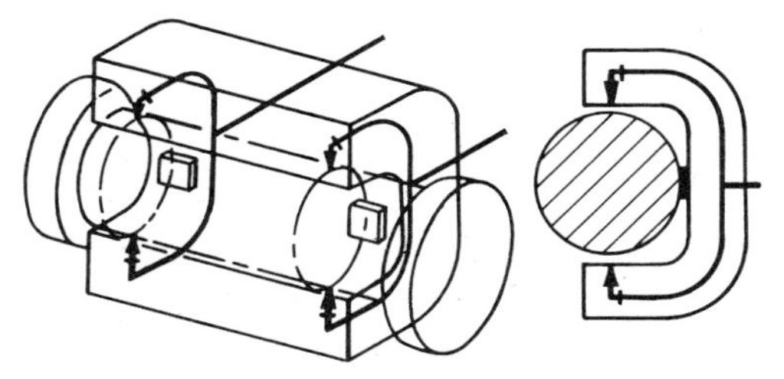

Multi-dimension hand air calliper gauge
2*n* gauging elements

Figure 38. Variation of diameter along axis

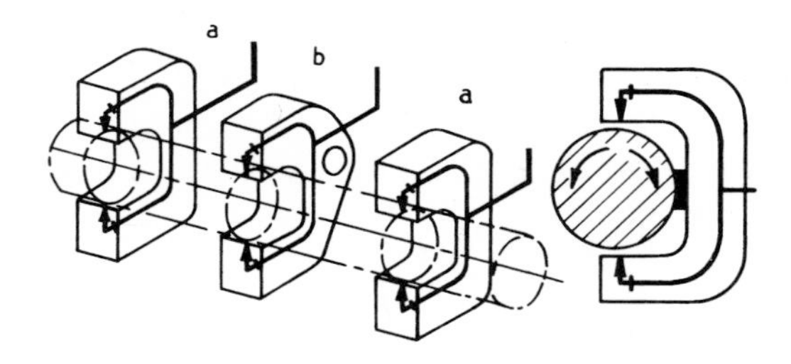

Fixture mounting air calliper gauge
(a) Non-floating
(b) Floating
2 gauging elements

Figure 39. Variation of diameter along axis

A.11 Concentricity tolerance. The tolerance zone for the concentricity of a circle B with reference to a coplanar datum circle A is the diameter *t* of the circle concentric and coplanar with A within which the centre of B lies.

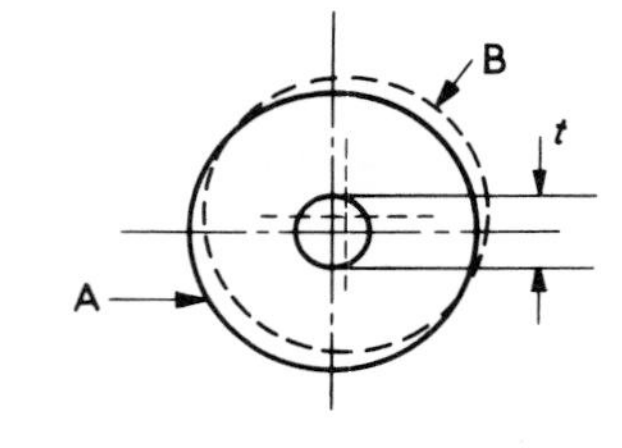

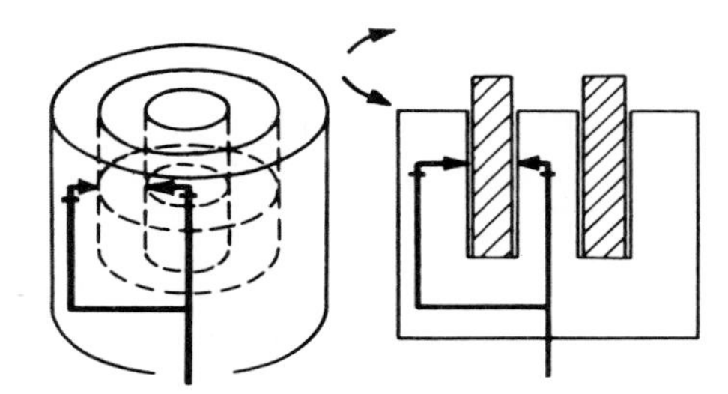

2 gauging elements measuring material thickness

Figure 40. Concentricity outside to inside diameter where there are no errors of circularity

---

A.12 Radial and axial run-out tolerance. The run-out tolerance represents the maximum permissible variation *t* of position of the considered feature with respect to a fixed point during one complete revolution about the datum axis (without axial movement in the cases shown in figures *b* and *c*). Except when otherwise stated this variation is measured in the direction indicated by the arrow at the end of the leader line which points to the toleranced feature.

The run-out tolerances specified may include defects of circularity, coaxiality or perpendicularity provided the sum of these defects does not exceed the specified run-out tolerance.

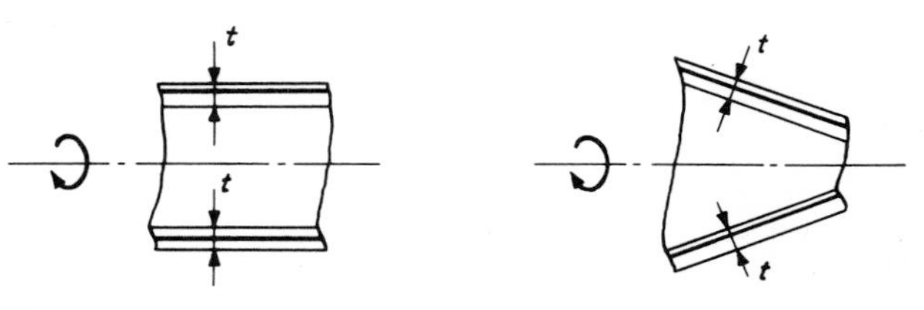

Figure *a*.

Figure *b*.

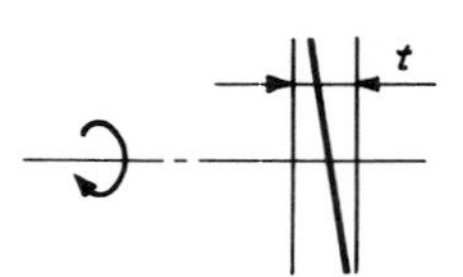

Figure c.

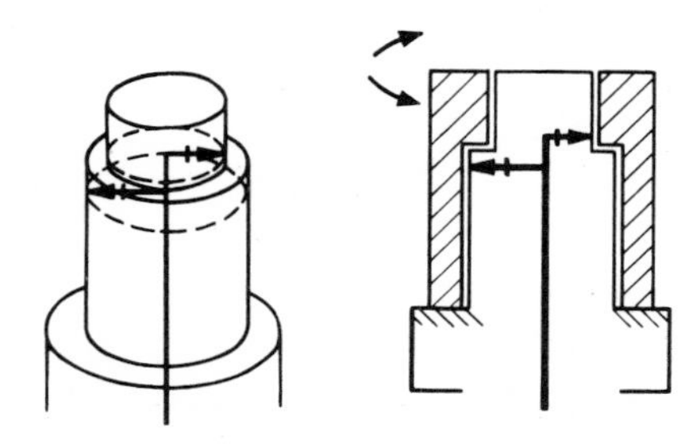

2 gauging elements
End face of component is datum and must be square to axis

Figure 41. Radial run-out (concentricity) of 2 inside diameters

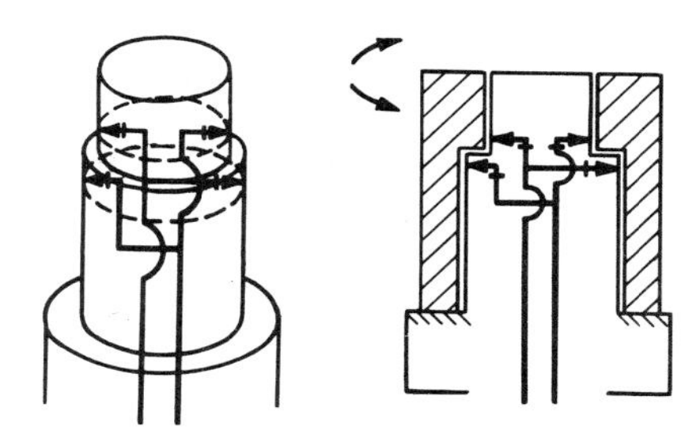

4 gauging elements in 2 pairs giving compensation for diameter variation
End face of component is datum and must be square to axis

Figure 42. Radial run-out (concentricity) of 2 inside diameters

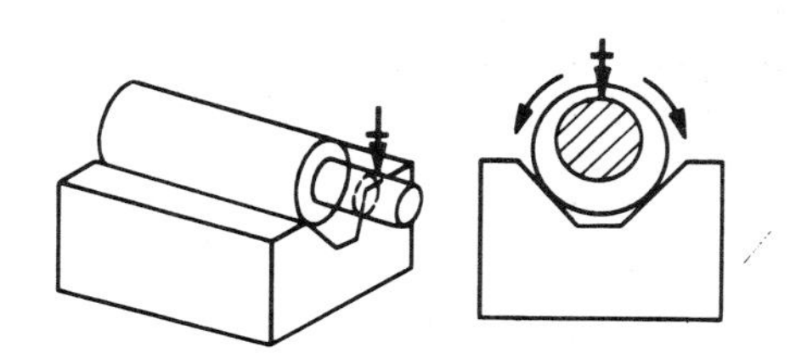

1 gauging element

Figure 43. Radial run-out (concentricity) of 2 shaft diameters

A.13 Positional tolerance of a line. The tolerance zone is limited by two parallel straight lines a distance *t* apart and disposed symmetrically with respect to the true specified position of the considered line if the tolerance is only specified in one plane.

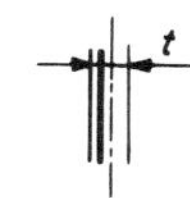

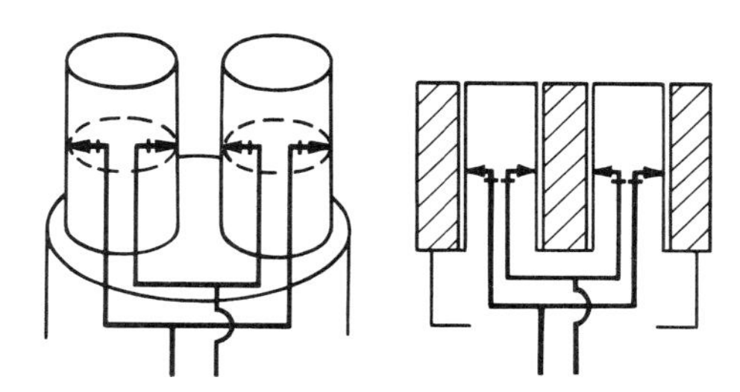

4 gauging elements in 2 pairs giving compensation for diameter variation

Figure 44. Centre distance of 2 holes

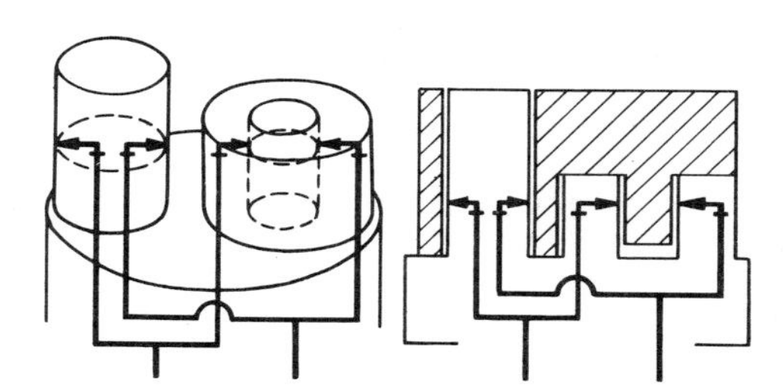

4 gauging elements in 2 pairs giving compensation for diameter variation

Figure 45. Centre distance shaft to hole

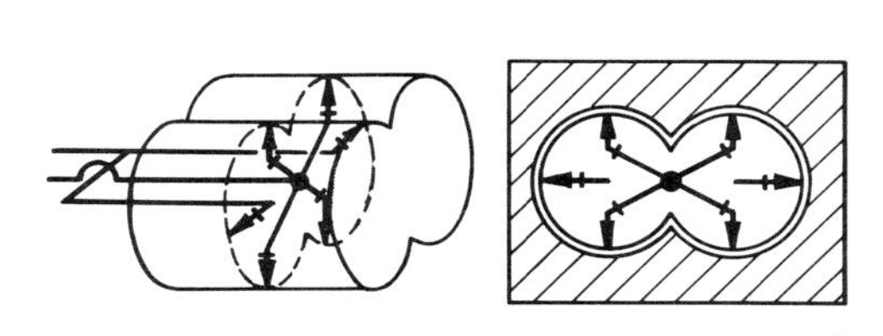

6 gauging elements giving compensation for diameter variation

Figure 46. Centre distance of intersecting holes

Appendix B

Other applications of air gauging

B.1 Introduction. In this appendix, examples are given of applications of air gauging which have not been considered in the earlier sections. The examples have been divided into six categories, viz. measurement, recording, automatic inspection, gauging for selective assembly, match machining and control of size. In the first category a few examples are given which, though not making strict use of the air gauging principle, employ related methods of measuring. It should also be noted that in some of the applications in this category the introduction of control, as a further stage, is readily achieved; thus, measurement of the travel of a moving member (see position detection) would be a first step in controlling the movement.

When considering the use of air gauging in the applications mentioned below it will be recalled that the method offers high sensitivity with stability of indication, that the measuring head can be very small and contact with fragile or easily damaged components avoided by the use of open jets, and that remote indication is readily arranged.

The examples given are not exhaustive; they are intended to illustrate the numerous additional ways in which air gauging can be useful in engineering, and they will probably suggest further applications to the reader.

B.2 Measurement. The following are some of the applications of air gauging methods to measurement.

B.2.1 *Position detection*

B.2.1.1 In machine tool applications, monitoring the position of the work or of slides, quills, clamping devices, the cutting faces of grinding wheels, etc.

B.2.1.2 Measuring the travel of a moving member.

B.2.2 *Fiducial indication.* To show when a given datum condition has been established, the air gauge being used as a sensitive null indicator.

B.2.3 *Continuous gauging*

B.2.3.1 Measurement during production of the thickness of materials (metals, plastics, paper, etc.) manufactured in strip or sheet form by a continuous process.

B.2.3.2 Measurement of the average diameter or mass per unit length of materials (textiles, plastics, metals, etc.) manufactured in thread or wire form by a continuous process.

B.2.4 *Measurement of fragile products.* Examples are the measurement of the diameter of cigarettes and cigarette filters during production. The use of non-contact measuring heads and low operating pressures are particularly valuable in this application.

B.2.5 *Contour and shape*

B.2.5.1 Complex shapes, using multiple measuring heads or a single measuring head following a path determined by a master profile.

B.2.5.2 Standard air gauge spherometer for radial measurement of spherical surfaces.

B.2.6 *Extensometry and dynamometry*

B.2.6.1 Standard designs of extensometers and dynamometers.

B.2.7 *Distortion*

B.2.7.1 Change of dimension or shape of structures.

B.2.7.2 Movement under load of a nominally rigid member of a machine, e.g. the bed of a machine tool.

B.2.7.3 Dimensional changes or distortion of moving members using non-contact measuring heads, e.g. the increase in diameter of turbine discs running at high rotational speeds.

B.2.8 *Jet calibration.* Calibration of fluid metering jets and determination of flow characteristics. Typical examples are carburetter jets and capillary tubing for refrigerators.

The dependence of flow characteristics on pressure and on the nature of the fluid must not be overlooked in this application of air gauging.

B.2.9 *Leakage and leakage rate*

B.2.9.1 Inspection, as an aid to production, of fluid control units to establish overall quality of fit by determining leakage rate, e.g. the leakage rate between spool and sleeve in hydraulic spool valves.

B.2.9.2 Fibre fineness tests and porosity tests of paper.

B.2.9.3 Checking porosity of castings and leakages of built-up assemblies, e.g. engine blocks.

B.2.9.4 Checking the clearance around the ball in ball-point pens.

B.2.9.5 Establishing the presence of an essential hole, e.g. in primer caps.

B.3 Recording. Recording of measurements made with an air gauging system based on the determination of pressure can be arranged by using a signal from the air gauge unit. Two schemes are in general use; the signal is fed to a pneumatic recorder, or a transducer is employed to convert the pressure signal into an analogue electrical signal which is then fed to an electrical recorder.

B.4 Automatic inspection. Air gauging machines for the automatic inspection of machined parts are available. They may be fully automatic, the complete process being carried out without the intervention of a human operator, or they may be semiautomatic, one or more operations, such as leading, positioning, being performed manually.

Although inspection is their primary function, air gauging machines may also be used to exercise some form of control of the production. In general, this control is limited to stopping the machining when one workpiece, or each of several consecutive workpieces, is outside tolerance. Examples of applications are to those machines, e.g. presses, broaching machines and multi-spindle automatic lathes, which do not admit any other control action.

B.5 Control of size. Air gauge size control may be used to provide control systems for manufacturing processes which have the necessary in-built facilities, e.g. certain machine tools and rolling mills.

Both in-process and post-process control are catered for, a typical example of the former being the application to plain cylindrical grinding, and of the latter to through-feed centreless grinding. On-off or multi-step controllers may be employed, control action being normally based on the measurement of one dimension only of the work, though in some applications information in regard to other dimensions is also provided.

Since air gauging can be used for position detection (**B.2.1**) a comprehensive system for machine tools, covering monitoring of the position of the work and of the various moving members, as well as the control of size, can be organized.

Standards publication referred to in this chapter

BS 4358 Glossary of terms used in air gauging with notes on the technique

13 Plain setting rings

Introduction

Numerous types of instruments are available for the measurement of internal diameters. They vary in design, application and accuracy of performance. Many of these instruments effect measurements by means of two-point contacts but some use three-point contacts and indicate a derived diameter, and others, notably air gauges, do not use contact at all (see Chapter 12). Plain rings intended for setting such instruments employing two-point contact, three-point contact or independent of physical contact are covered in BS 4064 'Plain setting rings for use with internal diameter measuring equipment'.

The first requirement for a setting ring is that its measuring surface is cylindrical to within close limits. The actual diameter of the ring is relatively unimportant providing its size is known, or known to be within specified limits.

When considering the diametral tolerances for setting rings it is important that the effect of possible departure from ideal roundness be clearly understood. Uniform diametral measurements of a cylinder, obtained by using two diametrally opposite measuring contacts do not necessarily mean that the cylinder is truly circular in section. It may suffer from departures from ideal roundness. Measurement of a cylinder by means of three-point contact, for example, may reveal form deviations undetected when only two opposite contacts are used (see Chapter 11).

BS 4064

BS 4064 relates to plain setting rings intended to be used, either singly or in combination, for checking the scales on internal diameter measuring machines. It provides for a range of metric rings from 2 mm to 150 mm in three grades of accuracy (viz grades AA, A and B) which is aimed at providing a series of rings that will serve for all types of internal measuring equipment, while at the same time avoids the high cost entailed in manufacturing unnecessarily close to a specified size.

BS 4064

Plain setting rings for use with internal diameter measuring equipment

Metric units

. . . .

5. Accuracy

5.1 *General.* All measurements shall be referred to the standard temperature of 20 °C.

NOTE. Care should be taken when using the rings to avoid excessive handling which might cause variations in size due to heating.

5.2 *Geometric form.* Within the middle half of the depth of the ring the cylindrical measuring surface, as measured with a two-point contact, shall be uniform in diameter, i.e. it shall be parallel, to within the values given in column 2 of the table, and when assessed in accordance with BS 3730 departures from ideal roundness shall not exceed the permissible values given in column 3.

The departure from roundness is defined as the difference in radii of two co-planar concentric circles, the annular space between which just contains the profile of the surface examined.

5.3 *Measured size.* The mean diameter of a setting ring shall be taken as the mean of at least four diameter measurements made at the mid-plane and it is recommended that this diameter does not depart from the nominal size of the ring by more than the amounts given in column 4 of the table.

If required, the measured size at a localized plane may be specified for grade AA rings.

5.4 *Accuracy of determination.* The measured size referred to in **5.3** above shall be determined with the accuracy specified for the grade in question in column 5 of the table.

Accuracy of setting rings

Unit = 1 micrometre (0.001 mm)

1			2			3			4			5		
			Geometric form — Uniformity of diameter as measured by two-point contact			Roundness			Recommended maximum departures of mean measured size from nominal size (measured in the mid-plane)			Accuracy of determination of measured size		
Grade			AA	A	B	AA	A	B	AA	A	B	AA	A	B
Nominal diameter in mm	Over	Up to and including												
	2	25	0.5	1.3	2.5	0.3	0.7	1.3	0.8	2.0	3.8	±0.3	±0.8	±1.3
	25	50	1.0	2.5	5.0	0.5	1.3	2.5	1.5	3.8	7.5	±0.5	±1.3	±2.5
	50	100	1.5	3.8	7.5	0.8	1.9	3.8	2.3	5.8	11.3	±0.8	±2.0	±3.8
	100	150	2.0	5.0	10	1.0	2.5	5.0	3.0	7.5	15	±1.0	±2.5	±5.0

6. Certificate and marking

6.1 *Certificate.* The manufacturer shall supply a certificate of measured size with each setting ring.

6.2 *Marking.* Each setting ring shall be legibly and permanently marked on the top face with the particulars given below.

The marking shall be applied in such a manner that it does not affect the accuracy of the setting ring.

(a) The manufacturer's name or trademark.
(b) The number of this British Standard, BS 4064.
(c) The grade, i.e. AA, A or B.
(d) A serial number.
(e) If required by the purchaser, the measured size of the ring.

. . . .

Appendix B

Recommended general dimensions of setting rings

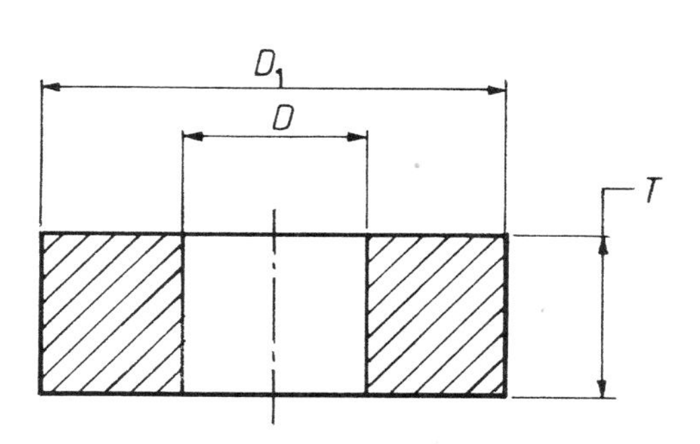

D			
Over	Up to and including	D_1	T
mm	mm	mm	mm
2	5	20	7
5	10	30	10
10	15	35	15
15	20	50	20
20	25	65	25
25	38	85	30
38	50	100	30
50	63	115	30
63	75	125	35
75	88	140	35
88	100	150	35
100	113	165	35
113	125	185	35
125	138	205	38
138	150	230	38

NOTE. Although the dimensions given in the above table differ in some respects from those specified for plain ring gauges in BS 1044 'Gauge blanks', the nearest corresponding standard blank will in most cases be suitable.

Standards publication referred to in this chapter

BS 4064 Plain setting rings for use with internal diameters measuring equipment

Index

Page numbers in italics refer to extracts from British Standards.

absolute error, *29*
accuracy, *39*, 42–3
accuracy class, *34*
air calliper gauge, *219–20*
air gauging, 213–15, *215–29*;
  miscellaneous applications, *228–9*
air plug gauge, *218*, *221*
Airy positions, 107–8, *108*
amplification, *33*
amplitude, *37*
analogue instrument, *27*
angle plate, 51, *51–4*
angularity testing, *223*
aperiodic instrument, *38*
arithmetical mean deviation, *185*,
  186–7
attenuation, *34*
autocollimator, *49*

backlash, gear, 160–1
base: gauge block, *106*; length bar, *111*
beam comparator, *47–8*
bench micrometer, 149
bevel protractor, 113
block squares: open form, *59–61*; solid
  form, *58–9*, *61*
box angle plates, *51–4*
British Calibration Service (BCS),
  17–20
British Standards, 14, 15; *see also at end of index for list of all abridged and/or quoted Standards*
British Standards Institution, 14, 15

calibration, *26*, *38*; systems, *40–2*
calliper gauges: adjustable, 134; air,
  219–20
centre point, gauge block, *106*
circularity testing, *223*
clamps, vee block, *74*
class index, *34*
comparator, *47–8*, 112, *112*
concentricity testing, *226*
constancy, *38–9*
constant of an instrument, *25*
conversion tables (metric/imperial), 11
crest, screw thread, *141*
cylindrical gauges, 132, 133
cylindrical squares, *57–8*, *61*
cylindricity testing, *225*

damping, *37–8*
datum error, *31*
dead band, *31*
decimal units, 16
design forms (screw threads), *141*, *142*,
  144
detector, *27*
deviation, *29*
dial gauges, 80, *80–4*
dial test indicators, 84, *84–7*
diameter measuring machine, 150
digital instrument, *27*
discrimination, *31*, *38*
dispersion, *30*
dividing heads, geometrical tests for,
  179
drift, *34*
dual flank testing, 161

end piece, length bar, *111*
engineers' parallels, *63–5*
engineers' squares, 54, *54–63*
errors: composite measurement, gears,
  164–5; in measurements, *29–30*; in
  measuring instruments, *29–33*; of
  form, 181
European Committee for
  Electrotechnical Standardization
  (CENELEC), 15
European Committee for
  Standardization (CEN), 15

feeler gauges, 79, *79–80*
fiducial value, *22*
fits, 120, 121–2, 125, 128
flank angles, *141*, 151–2
flatness testing, *222*; for drilling
  machines, *175–6*
full thread, *143*
fundamental deviations and tolerances,
  124–7
fundamental triangle, *141*

gain (of instrument), *33*
gap gauge, 132–3
gap setting gauge, *97*
Gauge and Tool Makers' Association
  (GTMA), 21
gauge blanks, 131
gauge blocks, 99–100, *100–4*;
  accessories, 104, *104–7*
gauge limits, *134–6*
gear measurements, 154–65
gear tooth size measurements, 160–1
gears: drawing information, 165;
  geometry, 154
General Conference of Weights and
  Measures (CGPM), 16, *23*
GO gauges, 121, 131–7, 145, *145–9*
GO limits, 121

hobbing, 154
holder, gauge block, *106*
hole-basis system of fits, 122, 128–9
hysteresis, *31*, *35*

index, *27*, *37*
indicating instrument, *26*, *37*
indication, *25*
influence quantity, *21*
inspection equipment, 44
instrument range, *37*
integrating instrument, *27*
International Electrotechnical
  Commission (IEC), 15
International Organization for
  Standardization (ISO), 15
International Organization of Legal
  Metrology (OIML), 21
international standard of measurement,
  28
International System of Units (SI), 16,
  17, *23*
ISO metric thread: gauges, *146–9*;
  profile, *145*
ISO system of limits and fits, 120–31

jaws, length bar, *111*
Joint Metrology Group, 21

lay (of surface pattern), 181, *185*, 189,
  190
lead (gear), 159
lead (screw thread), *142*
least squares centre, *204–5*
legal metrology, 21
length bars, 107–8, *108–11*
limit gauges, 131–3; limits for, *133–8*
limits and fits, 120–31

limits of error, *39*
linearity, *31*, *32*
lobed shaft, 192; testing, 212, *224*

machine tool tests, 167, *167–79*
major diameter (screw thread), *143*, 149
measurement standards, *28–9*
measurement systems, 16, *23*, *40–2*
measurements: angular, 113; errors in, 16, *29–30*; linear, 78
measuring instruments, *26–9*; accuracy, *38–9*; calibration, 38; design, *36–8*; errors in, *30–2*, 35; performance, *34–9*; sensitivity, *38*; uses and properties, *33–4*
measuring jaws; gauge blocks, *104–5*
measuring rules, engineers', 78, *78–9*
measuring transducer, *27*
meshing accuracy, *159–60*, 165
meter cut-offs, *187*, 189
metric vee blocks, *73–7*
metrology, 21, *21–34*
micrometer heads, 98, *98–9*
micrometer, 92, 95; external, *92–5*; internal, *95–8*
minor diameter (screw thread), *143*, 150
moving-index instrument, *27*

National Engineering Laboratory (NEL), 17
National Physical Laboratory (NPL), 14, 16, 17
national standard of measurement, *28*
National Testing Laboratory Accreditation Scheme (NATLAS), 17
non-linearity error, *31*
NOT GO gauges, 131–7, 145, *145–9*
NOT GO limits, 121

oval shaft, 192
ovality testing, *224*

parallelism testing, *222*; for grinding machine, *176–7*
perpendicularity testing, *223*
pitch (gear), 155–6
pitch (screw thread), *142–3*; errors, 151; measurement, 151
pitch diameter, *143*, 150–1, 152
pitch measuring machine, 151
plain setting rings, 230, *230–1*
plug gauges, 131, 132, *134–7*
polar graphs, rules for, *205–6*
precision, 42–3
primary standard, *28*
profile graph, 183
profile meter instrument, *185*

quality systems, 39–40
quantity, *21–2*
quantization error, *31*

radial run-out, *163*
random error, *29*
range (of instrument), *34*
recording instrument, *27*, *37*
reference material, *29*
reference standard, *28*
relative error, *29*
repeatability, *26*, *30*
response (of instrument), *25*
response time, *34*
ring gauge, 131, 133, *134–7*
rod gauge, 133
root (of thread), *141*
rotation (run-out), *179*
roughness, 181, 183, *184–5*, 188–9; comparison specimens, 189, 191
roundness, 192, *193–212*; assessment, *193–212*
roundness measurement: glossary, *193–6*; methods, *207–12*; stylus instrument, *197–206*
run-out tolerance, *226–7*

scale (of instrument), *27*, *37*
scatter, *30*
screw gauge, limits and tolerances, *146–9*
screw threads, 140, *140–5*, 145, *146–9*, 149–53; diameter, *143*, 149; form, 141, 151–2; inspection by measurement, 149–53; inspection gauging, *145–9*; internal, *145–9*, 152–3
scriber points, gauge block, *106*
secondary standard, *28*
sensitivity, *33*
sensor, *28*
setting gauges, *92*
settling time, *38*
shadow protractor, 152
shaft-basis system of fits, 122, 128, 130
SI units, 16, 17, *23*
sine bars and tables, 114, *114–16*
single flank testing, 164, *165*
span (of instrument), *34*
spindle errors, roundness testing, *202–4*
spirit level, 116–17, *117–19*; vials, 118–9
squareness testing, *222*
stability, *34*, *38*
standard of length, primary, 17
standard proportion gear, 154
standardization: aims, 14; development, 13–14
standards, measurement, *28–9*
steel rules, 78, *78–9*
stick micrometers, *95–8*
straightedges, 65, *65–73*, 67, 69, 71
straightness testing; by air gauging, *221*; for lathes, *172–4*
stylus measuring instruments, surface texture, 182; roundness measurement, *197–206*
surface plates, 44, *44–9*
surface texture, 128, 181–4, *184–7*, 188–91; assessment, 184–7; control, 165; designation, gears, 165; 189; measurement, 182–4; production time, 191
systematic error, *29*

taper parallels, 152
Taylor Principle, 131, 137
ten point height method, *185*, *187*
test methods, precision of, 42
thread form, *141*, 151–2
threshold sensitivity, *see* dead band
tolerances, 120–5; machine tool, *168–70*
Tomlinson surface recorder, 182
toolmaker's flats, *49–51*
tooth alignment measurement (lead error), 159, *159–60*
tooth profile: measurement, 156–7; tolerance, *157–8*
tooth thickness, 160–1, *161*
tooth-to-tooth composite tolerance, *161–2*
totalizing instrument, *27*
traced profile, *193*
transducers, *27*
transfer standard, *28*
travelling standard, *28*
true value, 42–3
truncation, basic, *142*
try squares, *54–7*, *61–3*

Unified thread, *142*
units of measurement, *22–3*

variability, 42
variation gauge, *44*, *46*
vernier callipers, precision, 87, *87–90*
vernier height gauge, 90, *90–1*

waviness, 181, 183–5
weighted mean, *26*
Weights and Measures Laboratory, 21
Whitworth thread, *142*
workpiece parameters evaluation, *196*
working standard, *28*

zero error, *32*

British Standards abridged and/or quoted in the text

BS 0 A standard for standards, 15
BS 308 Engineering drawing practice, 15, 131, 165, 189
BS 436 Spur and helical gears, 154, *155*
BS 817 Surface plates and tables, *44–9*, 87, 89, 106
BS 852 Toolmakers' straightedges, 106

BS 869 Toolmakers' flats and high precision surface plates, *49–51*, 106
BS 870 External micrometers, *92–5*
BS 887 Precision vernier callipers, 87, *87–90*
BS 906 Engineers' parallels, *63–5*
BS 907 Dial gauges, *80–4*
BS 919 Screw gauge limits and tolerances, 145, *146–9*
BS 939 Engineers' squares, 54, *54–63*
BS 957 Feeler gauges, 79, *79–80*
BS 958 Spirit levels, 116–7, *117–19*
BS 959 Internal micrometers, *95–7*
BS 969 Limits and tolerances on plain limits gauges, 131–2, 133–4, *134–8*
BS 1042 Methods of measurement of fluid flow in closed conduits, 28
BS 1044 Gauge blanks, 131, 132, 231
BS 1054 Engineers' comparators, 112, *112*
BS 1134 Assessment of surface texture, 55, 101, 165, *184–7*
BS 1643 Vernier height gauges, *90–1*
BS 1685 Bevel protractors, 113
BS 1734 Micrometer heads, 98, *98–9*
BS 1916 Limits and fits for engineering, 169
BS 2045 Preferred numbers, 46, 52
BS 2485 Tee slots, tee bolts, tee nuts and tenons, 52
BS 2517 Definitions for use in mechanical engineering, 140, *140–5*
BS 2634 Roughness comparison specimens, 189
BS 2643 Performance of measuring instruments, 16, *34–9*
BS 2795 Dial test indicators, 70, 84, *84–7*, 89
BS 2846 Guide to statistical interpretation of data, 30
BS 3064 Sine bars and tables, 114, *114–16*
BS 3509 Spirit level vials, 119
BS 3643 ISO metric screw threads, 140, 145, *148*, 150
BS 3696 Specification for master gears, 161, 165
BS 3730 Assessment of departure from roundness, 192, *193–212*
BS 3731 Metric vee blocks, 73–7
BS 3800 Methods for testing the accuracy of machine tools, 167, *167–72*, *174–8*
BS 4064 Plain setting rings, *230–1*
BS 4311 Metric gauge blocks, 99–100, *100–4*, 104; accessories, *104–7*
BS 4358 Glossary of terms used in air gauging, 214, *215–29*
BS 4372 Engineers' steel measuring rules, 78, *78–9*
BS 4500 Specification for ISO limits and fits, 120–5, 128, 133
BS 4500A Selected hole-basis (for fits), 128, *129*
BS 4500B Selected shaft basis (for fits), 128, *130*
BS 4656 Accuracy of machine tools and methods of tests, 167, 172, *172–9*
BS 4696 Method of determination of asphaltenes in petroleum products (precipitation with normal heptane), 29
BS 4727 Glossary of electrotechnical, power, telecommunications, electronics, lighting and colour terms, 27
BS 4778 Quality assurance terms, 42
BS 5204 Straightedges, *65–73*
BS 5233 Terms used in metrology, 21, *21–34*
BS 5317 Metric length bars and accessories, 90, 107–8, *108–11*.
BS 5497 Precision of test methods, 42
BS 5532 Statistics – Vocabulary and symbols, 42–3
BS 5535 Right angle and box angle plates, 51, *51–4*
BS 5590 Screw thread measuring cylinders, 150
BS 5750 Quality systems, 39
BS 5781 Measurement and calibration systems, *10*, 17, 39–40, *40–2*
PD 6461 Vocabulary of legal metrology, 21, *29*, *32*
PD 7306 Introduction to surface texture, 184
Handbook 22 Quality assurance, 43